E. Ciciotti · N. Alderman
A. Thwaites (Eds.)

Technological Change in a Spatial Context

Theory, Empirical Evidence and Policy

With 36 Figures

Springer-Verlag
Berlin Heidelberg New York
London Paris Tokyo
Hong Kong Barcelona

Prof. Dr. Enrico Ciciotti
Associate Professor of Urban
and Regional Economics
University of Venice
Ca'Tron - S. Croce 1957 - 30125 - Venice, Italy

Dr. Neil Alderman
Alfred Thwaites

Centre for Urban and Regional
Development Studies
University of Newcastle upon Tyne
Newcastle upon Tyne NE1 7RU, U.K.

ISBN-13: 978-3-642-75931-4 e-ISBN-13: 978-3-642-75929-1
DOI: 10.1007/978-3-642-75929-1

Softcover reprint of the hardcover 1st edition 1990

2142/7130-543210

PREFACE

This book represents a selection of contributions to the Regional Science Association, European Summer Institute, held at Arco, Lake of Garda, Italy, from 17th to 23rd July 1988 on the theme of 'Theories and policies of technological development at the local level'. As such it forms a complementary volume to the collection of invited papers entitled 'The Spatial Context of Technological Development', edited by Professors Riccardo Cappellin and Peter Nijkamp, and published by Avebury/Gower.

The Summer Institute covered a broad range of themes and contributions from over 100 researchers. It has therefore not been our intention to represent or replicate these themes directly, but to provide as wide an international coverage of current research in regional science in the field of technological change as possible. The papers to be found in this book are drawn from the European Community, Eastern Europe, Scandinavia, North America and elsewhere.

Inevitably, as the book was to be published in English, most of the editorial work has been undertaken at the Centre for Urban and Regional Development Studies at Newcastle University. In so doing, we have attempted to standardize the English as far as possible, whilst trying not to lose the characteristics of the styles of the individual authors.

We should like to thank Riccardo Cappellin and Peter Nijkamp for inviting us to edit this particular volume and also Professor Carlo Borzaga and his team for their work in organizing the Summer Institute. Special thanks must go to Sue Robson and Denise Weites for their exceptional hard work retyping the various manuscripts, correcting proofs and laying out the tables and figures etc. This project has involved them in long periods of 'after-hours' work and required great patience, both with us and our recalcitrant laser printer.

We should also like to acknowledge the assistance of Anne Rooke of the Newcastle University Geography Department with some of the figures and Ash Amin, whose knowledge of the Italian economy helped us to express a number of ideas in better English. The financial support of the UK Economic and Social Research Council and the University of Newcastle upon Tyne is also gratefully acknowledged. Last, but not least, we should like to thank the authors themselves for their patience with what has been a long project and for their prompt responses to our repeated requests for clarification or modifications. We hope that the result does justice to their work.

Enrico Ciciotti (Milan)
Neil Alderman (Newcastle)
Alfred Thwaites (Newcastle)

CONTENTS

LIST OF TABLES

LIST OF FIGURES

1. INTRODUCTION

Enrico Ciciotti, Neil Alderman and Alfred Thwaites

History demonstrates that from the earliest times there have been spatial variations in political, social and economic advances and that the differences between regions in the level and pace of economic growth and development have accelerated since the industrial revolution. It is only in the twentieth century, however, that attention has focused on trying to establish the extent of, and reasons for, such economic variations and concerted action been taken at a variety of spatial scales to try to overcome, or at least reduce, the disparities. At the same time, leading regions have striven to maintain the momentum which has led to their success.

Twenty to thirty years ago the alleviation of regional distress was seen in terms of transfer or aid payments to less favoured regions from core economies and in terms of redistributing the manufacturing base of global and national economies, focusing in particular on industrial movement. This involved both attracting inward investment to the less favoured regions, and constraining developments within core regions. This was frequently accompanied by the construction or reconstruction of the physical infrastructure in less favoured regions. It was hoped that these processes would help to create new linkage patterns inducing further expansion and growth, particularly in terms of employment and income. In many senses such approaches to the regional problem could be viewed as attempts to produce clones of successful regions.

The view that the regional problem could be overcome by such methods was dispelled by the recession which hit most nations in the late seventies and eighties. The reliance on a narrow base, mainly of manufacturing enterprises, was further questioned when the subsequent recovery of output growth was not accompanied by employment growth to the same degree as that experienced previously. These revelations prompted a search for a more adequate conceptualization of the economic development process and the identification of key factors promoting economic development above and beyond the supply of low cost land, capital and labour.

The result has been a shift away from a reliance purely on the peripheral or branch plants of manufacturing enterprises located in core nations or regions towards attempts to identify and reconstruct local comparative advantage in a variety of sectors. Such processes cannot be seen simply in local terms, but must be viewed within wider national and international economic contexts in which satisfying demand, institutional, networking, and supply conditions is essential. Moreover, a more holistic view of development has emerged in which the synergy between various economic activities (manufacturing, services, institutions etc.) at a variety of spatial and functional scales is fully acknowledged. Such a concept of local economic development thrusts greater responsibility upon the actors in local areas to play a full part in their future development through the generation, absorption and use of new knowledge, to produce new economic activities that can be exploited in the market place. Key components of such a strategy are technology and technological change. It is posited that a major feature of a dynamic local economy is likely to be the creation and sustenance of a local innovative milieu.

The importance of technological change to economic growth and development has long been recognized; the search for new knowledge and its incorporation into the economic development process has become a professional activity (Freeman, 1982). Research intensive nations, industries and enterprises appear to lead in economic advance (OECD 1986; Freeman, 1982), measured in terms of sales growth (Odagiri, 1985) export success (Rothwell, 1979, 1980), productivity growth (Mansfield, 1968) or profitability (Schott, 1981). At the same time, technological change can provide the basis for new industries and the rejuvenation or modernization of older sectors. New products and services tend to substitute for old ones and also tend to erode markets held by those producing or supplying under old technological regimes. New processes can give cost, quality, delivery, and reliability advantages over the non-adopters.

These possibilities for enhancement or threat to further economic progress are relevant at all spatial levels, including the regional or local area level, where the generation, adoption and use of technology by manufacturing or service enterprises can be seen as a competitive weapon. However, the capability of a region to create a dynamic local innovative milieu may be constrained by a lack of other technological changes in public sector industries, institutions and infrastructures (eg. telecommunications). The creation of dynamic local milieus requires, therefore, a consideration of the generation, adoption and use of technology as part of a set of synergetic and inter-linked layers of relationships between different elements of the local and extra-local economy.

Evidence of the spatial dimension to technological change suggests that it is unevenly distributed between and within national and local economies, and that lower levels of technological advances are frequently associated with depressed levels of economic performance. At the international level the member states of the European the Community are concerned over the leads in technology in certain sectors held by the United States and Japan (CEC, 1986). Similarly, Boretsky (1975) claimed that the United States could only maintain its military and economic lead if it retained its technological lead over other nations. Numerous studies carried out in Europe and elsewhere have also found that less favoured regions often lag in the generation and absorption of technology in manufacturing activities (Thwaites et al, 1981, 1982; Ciciotti, 1983; Rees et al, 1984; Brugger and Stuckey, 1987; CEC, 1987 a & b). Others, such as Foord & Gillespie (1985), have shown that in the offices of manufacturing units and in service enterprises located in less favoured regions of the UK the take up and use of new information technologies (e.g. microcomputers, word processors) is lower than in similar units located in core regions. Similarly Gillespie et al (1984) discovered that numerous less favoured regions across Europe tended to have inadequate levels of telecommunications infrastructure and as a result were unable to maintain a high level of modern information and communication services. Such poor technological performance and supporting infrastructure may inhibit further economic development in currently less favoured regions, particularly in the light of the increasingly competitive conditions which are emerging at the global scale and work out at the local level in terms of output, employment, income, occupational structure etc.

At a very general level, Dicken (1986) observes that during the past twenty years the world seems to have entered a period of economic turbulence which is neither easy to interpret, nor to identify the emergence of predictable outcomes, except to suggest that more change is on the way. The less favoured areas must therefore be seen within an ongoing process of change where few of the forces have been played out. One of the processes of change is that, at a world scale, in both manufacturing and services there are an increasing number of national contributors to output and trade. For example, between 1963-82, while world output grew, the share of world manufacturing output held by the Developed Market Economies fell by 13 percentage points, with the shares of the Centrally Planned Economies and the Developing Countries increasing (Chapman & Walker, 1987). The new entrants added not only a contribution towards the globalization of production, but also acted to increase competition in unprotected home and overseas markets.

A notable feature of the post war era has also been the rapid increase in world trade in manufactured goods. Thus any area, industry or firm participating in such markets is likely to have to fight harder to retain or improve its present position. One method of improving competitive strength is through the use of more advanced technology.

A further feature of the changing world economy is rapidly changing customer tastes and demands (Mensch, 1979), which means that conventional methods of operation, economies of scale and price competition are no longer necessarily adequate for competitive success. Increasingly, firms are turning to product and market differentiation as a strategy for securing or improving market share and competitive position. Factors such as quality and reliability are becoming as important as price (Leonard and Sasser, 1982). Firms, therefore, have to be increasingly sensitive to the needs of their customers and to display the flexibility to respond to these changing circumstances of demand. This frequently requires them to adopt newer technologies and new approaches to the organization of production and distribution, and to develop new methods of linking with markets.

Concurrent with these market changes is the rapid rate of technological change itself, which demands increasingly rapid research and innovative responses from firms, as well as opening up new opportunities and creating new competitive pressures. In many product groups the pace of technological change is increasing, resulting in shorter product cycles and greater pressure for new product development to keep up with the state of the art in technology. In addition, product change frequently needs to be accompanied by process change which increases the demands on research and development staff and engineers. Closer links between research, design and production teams are becoming increasingly important and complex (Thwaites and Alderman, 1988). Technological changes in the form of new products or production processes are also taking many firms into what for them are uncharted areas of operation (e.g. manufacturers producing and selling software).

Technological change is essentially based upon the generation and transfer of knowledge. If the pace of technological advance needs to quicken, these enterprises require ready and swift access to knowledge and its providers both from within and without the firm.

In addition, firms striving to ensure their place in present and future productive systems are adjusting their functional structures to meet the new challenges. Such adjustments tend to push them towards increasing the proportion of their total activities in service production (taken in its broadest sense), as opposed to those directly engaged in physical production. Some of these services are provided in-house, but increasingly firms seem to be looking outside the organization for their service needs and transferring some of these activities into the market place (U.N.C.T.A.D., 1988). Thus, as firms themselves change their own demands for external services, they are becoming more specialist and more sophisticated. This in turn is changing the nature of relationships between different types of enterprise as such services play a greater role in competitiveness. What is emerging is that technology, manufacturing and service activities are increasingly seen as parts of a symbiotic relationship in local economic growth (Dicken, 1986).

The development of the services industry has been facilitated by (and in some cases is dependent on), developments in technology, particularly information and communication technologies. The emergence of the information economy (Hepworth, 1989) potentially supplies enterprises with many new economic goods (i.e. products, processes and services) and new ways of organizing their intra- and extra-corporate functions through the use of information technology in the workplace. These can take the form of management information systems, local area networks (LAN) and wide area networks (WAN) etc. As the infrastructure of WANs, and LANs is developing to form networks of great technological and economic potential, it is important that less favoured regions are not bypassed or are only able to join on unfavourable terms.

From the foregoing, the conclusion is reached that technological change is an important component of economic growth and development at a variety of spatial scales. It also seems evident that trends in the international economy will place greater pressures on industry and enterprise, largely irrespective of location, to improve its performance, notably in technological terms as part of competitive strategy. Evidence suggests that the currently economically depressed or less favoured regions of the world are failing to establish and/or maintain their position in the technological rankings table and as a result there is a danger that they will continue to lose out in the economic growth stakes, thereby further dividing nation from nation, region from region.

If we wish to intervene to reduce disparities in technological and economic progress, or simply to analyse the causes of such disparities, it is necessary to explain why such differences in development have come about, which are the key factors in local technological development and where are the leverage points which can be used to develop a new and more positive trajectory for less favoured areas. The contents of this book are offered as contributions to various aspects of this very important debate. In European terms this is a very timely contribution given the prospect of considerable challenges and change to enterprise, organizations and institutions in the run up to implementation of the proposals covered by the Single European Act (CEC, 1987c). Implementation of the Act is intended to increase competition between firms to the benefit of the Community as a whole, but which will go some way towards creating a Community of competing regions rather than the present international competitive forces and in which some regions could make gains while others lose in relative terms.

The papers in the book were selected from over 100 contributions presented at the Summer Institute of the European Regional Science Association held 17-22 July 1988 in Arco, Italy, with the theme of 'Theories and Policies of Technological Change at the Local Level[1]. This collection of papers is not intended to present a comprehensive review of material in this field, but represents a cross-section of issues from theory to policy, and evidence from a wide variety of national contexts. The contributors to this book range from very experienced researchers to emerging scholars with different disciplinary and institutional backgrounds. We think that individually and collectively the papers in this volume will contribute to the continuing debate about technological change and local economic development and that all readers should find something new and interesting to attract their attention.

Theoretical and Conceptual Issues

The first theme covered in the book is associated with the conceptualization of economic development at the regional or local level with a primary focus on its interface with technology. Rather surprisingly, and despite the long term interest in local economic development, few fundamental contributions to economic growth theory have emerged over the past twenty years and even fewer incorporating a technological dimension.

Crevoisier, in the opening chapter of the book, addresses these issues, indicating that the traditional models available to us in regional economies '... have displayed certain limits with regard to explaining changes in the economic map of Europe and the United States.' Such models emerged from particular forms of production based on functional logic. Such a system of economic activity relies on some element of technological stability and predictability of markets which permits economies of scale. This leads to a concentration of functional activities within companies which in turn concentrate in core regions. The 'centre' of the company acts as the driving force in the system, allocating economic activities to regions in response to corporate benefit. Crevoisier questions the ability of this functional model to cope with rapid change and uncertainty and to explain current trends in the regional and wider economy and argues that, under these new conditions, technological change requires networks of individuals and organizations interacting with a locally driven dynamic or territorial logic. These networks are able to react to new economic environments speedily and in innovative ways; however, Crevoisier suggests that we cannot totally reject the concept of spatial economic differentiation while production based on functional logic continues as a major force in the world and there is a need to integrate functional and territorial logic.

The difficulty of producing all encompassing models of local economic development is illustrated by Asheim. In the case of Norway he claims that the basic component of the functional logic model (i.e. mass production) has been largely absent except in energy intensive export sectors and in Norway there is high dependence on the small and medium sized firm (SME). Asheim's aim is to present a theoretical framework within which the diffusion of innovations among SMEs operating in industrial districts can be analysed.

Asheim presnts an interesting critique of the 'flexible specialization' thesis, arguing that the 'flexible specialization' of today cannot be equated with the craftsmen of yesteryear but is a new form of manufacture based on new technology. Thus flexible specialization consists of both intra- and extra-firm characteristics, including entrepreneurship and a local culture supportive of such activity. Asheim argues that in an economy dominated by SMEs, if historical cultural and environmental conditions are favourable, then 'system areas' can emerge where there is a balance between competition and co-operation together with rapid innovation and technological diffusion.

Asheim goes on to discuss the threats to such systems and the necessary conditions for achieving or maintaining stability. Hence, although industrial district developments may be highly desirable in overcoming the problems of underdevelopment or industrial decline, they will not be established in many regions because of the regions' inability to satisfy the necessary local intra- and inter-firm conditions.

Any study of comparative economic and technological performance at the regional level quickly reveals the wide diversity of experience and capabilities upon which to build. In terms of technological advances, some regions have the capability to generate and adopt the most advanced techniques while others are attempting to overcome inertia and to reorient existing activities towards more modern sectors. Still others have yet to establish themselves in terms of modern technological generation and rely on the transfer and use of technology from more advanced industrialized nations and regions. Many researchers are questioning the adequacy of current models of economic growth for less favoured regions or countries and this subject provides the focus for the papers by Kim and Rabellotti.

Kim suggests that technological progress in underdeveloped countries (UDCs) is to a large degree dependent upon local technological capability (i.e. technological knowledge embodied in workers and organizations) and to a lesser extent upon capital investment or technology embodied in capital. The technological capability of a country places limits on the range of technology/capital combinations that a firm can profitably exploit in a UDC and goes some way towards explaining the current nature of investment undertaken in UDCs by international corporations. Kim identifies the dilemma that, while for long term development and greater economic independence it is necessary to raise technological capability over time, the public good nature of this technological capability makes this difficult.

Rabellotti argues that any strategy for economic and technological development of currently underdeveloped countries should consider those organizational forms of production that are most likely to enhance the development process. The most appropriate organizational forms for UDC development lie between the traditional make or buy dichotomy and include many possibilities for co-operative agreements between local firms, but perhaps more importantly, between local and foreign firms. Recognizing the supply and market weaknesses of the UDC economy, appropriate organizational forms could be selected which maximize local content and technological capability and provide a mechanism for learning and enhancement in the future.

In many UDCs the branches of a foreign firm or a single enterprise often dominate local economies, as a single buyer or monopsonist. However this situation is also experienced in the more advanced industrial nations, notably in more rural or isolated areas peripheral to the core of the national economy. In the paper by Bradfield it is argued that the existence of a monopsonist in an area will inherently lead to a slower rate of technological advance in production processes than in competitive markets with identical characteristics, because the monopsonist is able to achieve lower marginal and variable costs than his competitive counterpart and this will influence his tendency to innovate. The impact of capital intensive technological change (production processes) upon competitive (high wage) and monopsonistic (low wage) regions will also, Bradfield argues, be differentiated. In the former, technological change may have the effect of changing the industrial mix, while in the latter the effect may be to lower incomes, employment and increase dependence on core regions of the economy.

In the concluding paper of this section, Pompili focuses attention on the city and urban system as appropriate spatial units for analysis. Pompili reviews a wide range of theories and models of development from those based on mainstream economic theory, biological-evolutionary theories and game theory through to the more recent family of SOUDY models which highlight the role of innovation in urban development. Pompili is critical of those who would abandon earlier approaches and goes on to suggest theories of imperfect competition which provide a path to revitalization of some of the older theories without discarding the insights gained from them. These explorations and synthesis allow him to conclude that urban development will be uneven with the bursts of activity resulting from innovation attracted by supply side conditions in the city.

Methodological Issues

Theoretical and conceptual advances often require parallel developments in methods of enquiry and data analysis techniques if they are to be satisfactorily tested and interpretated. There is also a continual requirement for methodological and analytical advance within existing fields of research. The second part of this book contains three papers which make substantial contributions in this respect.

The paper by Sanchez-Garcia outlines the problems associated with the satisfactory incorporation of technological change within an input-output framework. The focus is upon accommodating diversification into new product areas within an existing economy. In particular, the author addresses the problems of taking account of the demands new products place on input sectors of the economy before the product contributes to output. The suggested solution to this problem is to adopt a 'hibernating' anology in which new products/sectors lie dormant until brought to life by expenditure on R & D, capital investment etc. Through such a mechanism diversification into new products can be accommodated as can the retirement and exit of obsolete products from the system.

The diffusion of new technology has been of interest to academics and policy makers for many years. Empirical evidence demonstrates that the pattern of diffusion through a population of firms or enterprises is typically 'S-shaped'. The problems in translating such evidence into predictive models of diffusion at the regional or local level using a probit regression approach provides the focus of Alderman's paper. The first set of problems addressed are associated with empirical measurement and in particular the use of survey data in the construction of such models. The dynamic context of technological change leads to a second set of problems, because the general rate of technological change appears to be increasing and the technique under study may be changing its functional capability, cost and operating profile as it diffuses over time. Despite the apparent weaknesses, tests of the model's forecasts, based on survey evidence, suggested they were superior to naive forecasts or linear extrapolations of the same data.

The third paper in this section by van der Knaap and van Geenhuizen discusses a research methodology at the micro-level of the firm based on a longitudinal analysis of firm histories. The research is concerned with the temporal relations between product-market strategies and the locational behaviour of firms within the context of a dynamic market. The authors acknowledge that neither longitudinal studies nor business histories are new methods of enquiry, but indicate that few such studies have included a spatial dimension, or evaluation, in the context of innovation. This method reveals the broad development path of the firm and the principal factors influencing its course and direction. The authors identify some of the limitations to this approach on the basis of a pilot study, but also show how the method points up weaknesses in conclusions reached via other methodological approaches.

Regional Issues: Some Empirical Evidence

It would be wrong to conclude from any book which is divided into sections that each section is mutually exclusive. This book is no exception in this respect and while the third set of papers are distinguished by their relationship to particular countries, or regions of countries, they contain their fair share of theoretical, conceptual and methodological rigour.

The value of technological advance to a national or local economy is not simply its contribution to the establishment of new industries and products, but also for its potential contribution to productivity growth (Schott, 1981). Christiansen, in a detailed examination of regional industrial performance in Denmark explores productivity differences between industry groups, classified by their level of technology, and between regions within industry groups in an attempt to establish whether regional or technological factors dominate local labour productivity performance. Christiansen's results demonstrate that the greatest productivity gains have occurred in the technologically advanced industries while, within each industry grouping, inter-regional variations in productivity have been slight.

In the next chapter we turn from the highly productive economy of Denmark to the problems of national and regional growth in Hungary; an economy which has tended to address its problems through the extension of economic activity and only comparatively recently has turned to intensification based on technological change as a source of economic advance. Klekner presents national and regional figures for R & D activity which shows a high level of concentration around the capital Budapest and to a lesser extent in other University towns around the country. Past attempts to decentralize and disperse economic activities seem to have had little impact on research activities. The paper goes on to allocate industrial income and its growth to regions and sub-regions. By bringing together the technological and local area performance data some possible associations between income per capita and research efforts are suggested. However, these relationships remain tenuous; instead capital investment per employee is found to have the most positive effects on local economic change.

The problems of industrial development in rural areas in Israel provides the focus of Felsenstein and Bar-El's paper, but have their parallels in many other countries. The authors claim that traditional models of rural industrialization based upon agriculture do not integrate rural town or village activity into the national pattern of more technologically advanced industrial growth, but rather integrate them into the local economy. Similarly, trickle down effects tend to benefit second tier cities before the villages and the establishment of branch plants of larger manufacturing units in rural areas does not necessarily lead to full integration into the national economy and further economic progress either. The authors present an analysis of the characteristics of nearly 200 firms in Israel that reveals locational rigidities that constrain rural participation in the national growth trajectory and which might be overcome through the design of suitable policy.

A more traditional approach to the investigation and analysis of regional variations in technological change is adopted by Todtling in the Austrian case. Based on interviews at 149 plants manufacturing in the metal and steel machinery, electrical and electronics industries in five Austrian regions the author's analysis shows that there is a tendency for the pre-conditions for innovation and process adoption to vary spatially. The lowest levels of technological capability and change are found in the older industrial and peripheral areas of the country. These findings are in accordance with those from a number of other European countries, but arguably are particularly worrying for a country which on the whole is not in the forefront of technological advance and finds itself peripheral to the main markets in Europe.

A wide ranging literature review, reflected in an extensive bibliography, provides the basis for the early section of Belussi's paper on the diffusion of innovations in traditional sectors in the Veneto region of Italy. Her critique of past studies leads to the conclusion that for many diffusion is seen as a series of events rather than as a process. Such a view does not accommodate easily the changing nature of the techniques under study nor their proper role in long term enterprise development. It also does not capture the complexity of the actual search, learning and decision-making process which accompanies or even causes innovation and adoption in firm-specific situations. The empirical research reported in the paper attempts to overcome some of these shortcomings through a detailed examination of the interdependency between the adaptation of exogenously derived technological change and the internal innovative capability of firms.

Policy Issues

The recognition that technological advance could lead to economic advance obviously interests those concerned with policy design. The fourth and final part of this book is devoted to a discussion of policy issues at four spatial scales: the supranational, national, regional and metropolitan scales.

The primary concern of the European Community in technological terms has been focused upon competition with the United States and Japan. Such concerns have led to the establishment of Community-wide programmes supporting pre-competitive research such as RACE, ESPRIT and BRITE, the participants of which are selected upon excellence in performance. However, the spatial distribution of scientific and technological activities is currently uneven and it is feared that competition and the establishment of the single European market may lead to further concentration of activity and widening regional disparities in both technological and economic development.

Using Belgium as an example, Quevit shows that there are currently a variety of technological trajectories at a sub-national level. The regional technological capability is influenced by local environments, including the degree of sectoral specialization in technology intensive sectors, the size of enterprises, and centres engaged in the demand for and supply of appropriate technological skills (e.g. Universities, research centres). A review of the regional impact of aspatial Community technology policies suggests that they tend to reinforce existing technological advantages and rarely create new centres of strength in currently less favoured areas. A critique of current policy leads to suggestions for improvements in policy design and implementation.

Rather different pressures for revitalization of the national and regional economy based upon technological advance seem to have been at work in Poland. Ciechocinska's fascinating description of the evolution of the Polish economy shows deep contrasts with experience in the West. She charts the change from an economy growing by increasing the scale of operations to one based upon the intensive use of resources through the adoption of new technology and methods of operation and the conditions that have led to a poor technological profile of Polish industry in comparison with western industry. The description of recent efforts to create more favourable conditions for economic initiatives and the explicit role of new technology in this is highly pertinent given the current climate of change in Eastern Europe and the shift away from centralized control and ownership.

Tolomelli's paper reflects on ten years of experience in Emilia-Romagna of local government intervention to promote technological and economic advances in local industry. What is remarkable about the Emilia-Romagna story is the way in which the local government support mechanisms have been sensitive to the changing nature of economic activity over time and have adjusted their strategy accordingly, as the relationships between large and small firms, large firm reorganization and the demand for the development of entrepreneurial cultures have proceeded. This adjustment is perhaps made possible through their belief in public/private concord, their vision of their role as coordinators rather than dictators and beliefs in institutional co-operation accompanying their endeavours to fit the strategies of Emilia-Romagna within wider regional and national goals.

In the final chapter of the book Correia also emphasises the need for policy intervention in his look at the case of the Lisbon metropolitan region. Even within the comparatively successful Lisbon region there are wide variations in technological capability. Correia discusses the scope for municipal intervention to overcome such local problems and concludes that the municipality has the opportunity to act as a catalyst in future developments by creating networks of actors in the local economy and harmonizing the needs of physical and economic development.

Given the wide variety of issues and approaches documented in this volume it is perhaps of no surprise that no single message emerges. Rather it is the diversity of ways in which technological change has implications for the development of local economies and the variety of policy mechanisms that are therefore needed in response to this that this book illustrates. Hopefully it provides a flavour of the breadth of research within regional science that is being directed at the whole question of technological change in a spatial context.

References

Boretsky, M. (1975) Trends in US technology: a political economist's view, American Scientist, 63, 70-82.

Brugger, E.A. and Stuckey, B. (1987) Regional economic structure and innovative behaviour in Switzerland, Regional Studies, 21, 241-254.

CEC (1986) Proposal for a Council regulation concerning the Framework Programme of Community activities in the field of research and technological development (1987 to 1991), COM(86) 430 final, Commission of the European Communities, Brussels.

CEC (1987a) Research and technological development in the less favoured regions of the Community (STRIDE), Commission of the European Communities, Brussels.

CEC (1987b) Science and Technology for Regional Innovation and Development in Europe, Commission of the European Communities, Brussels.

CEC (1987c) Single European Act, Commission of the European Communities, Brussels.

Chapman, K. and Walker, D. (1987) Industrial Location, Blackwell, Oxford.

Ciciotti, E. (1983) La diffusione regionale delle innovazioni in Italia: il caso delle macchine utensilli a controllo numerico, paper presented at the IV AISRE conference, Fireize, November.

Dicken, P. (1986) Global Shift: industrial change in a turbulent world, Harper and Row, London.

Foord, J. and Gillespie, A.E. (1985) Reorganization, new technology and office jobs, Discussion Paper No.75, CURDS, University of Newcastle upon Tyne.

Freeman, C. (1982) The Economics of Industrial Innovation (2nd ed.), Pinter, London.

Gillespie, A.E., Goddard, J.B., Robinson, J.F., Smith, I.J. and Thwaites, A.T. (1984) The effects of new information technology on the less-favoured regions of the Community, Regional Policy Series No.23, Commission of the European Communities, Brussels.

Hepworth, M.E. (1989) Geography of the Information Economy, Belhaven, London.

Leonard, F.S. and Sasser, W.E. (1982) The incline of quality, Harvard Business Review, 60(5), 163-171.

Mansfield, E. (1968) The Economics of Technical Change, Longman, London.

Mensch, G. (1979) Stalemate in Technology: innovations overcome depression, Ballinger, New York.

Odagiri, H. (1985) Research activity, output growth and productivity increase in Japanese manufacturing industries, Research Policy, 14, 117-130.

OECD (1986) R & D, Invention and Competitiveness, Science and Technology Indicators No.2, OECD, Paris.

Rees, J., Briggs, R. and Oakey, R. (1984) The adoption of new technology in the American machinery industry, Regional Studies, 18, 489-504.

Rothwell, R. (1979) The relationship between technical change and economic performance in mechanical engineering, in Baker, M.J. (ed.) Industrial Innovation: technology, policy, diffusion, Methuen, London.

Rothwell, R. (1980) Innovation in textile machinery, in Pavitt, K. (ed.) Technical Innovation and British Performance, Methuen, London.

Schott, K. (1981) Industrial innovation in the UK, Canada and the United States, British-North American Committee, London.

Thwaites, A.T. and Alderman, N. (1988) The location of industrial R & D: retrospect and prospect, paper presented at the European Summer Institute of the Regional Science Association, Arco, Italy.

Thwaites, A.T., Oakey, R.P. and Nash, P.A. (1981) Industrial innovation and regional development, final report to the Department of the Environment, CURDS, University of Newcastle upon Tyne.

Thwaites, A.T., Edwards, A. and Gibbs, D.C. (1982) Interregional diffusion of production innovations in Great Britain, final report to the Department of Trade and Industry and the Commission of the European Communities, CURDS, University of Newcastle upon Tyne.

U.N.C.T.A.D. (1988) Trade and Development Report, 1988, United Nations, New York.

PART I - THEORETICAL AND CONCEPTUAL ISSUES

2. FUNCTIONAL LOGIC AND TERRITORIAL LOGIC AND HOW THEY INTER-RELATE IN THE REGION

Olivier Crevoisier

Introduction

This contribution aims to examine what writings in the field call functional logic and territorial logic. Functional logic roughly corresponds to the post-war mode of production. Much has been written about this mode, which is generally called 'Fordism' (Aglietta, 1987; Lipietz, 1977). However, since the late 1960s, the theoretical models used in regional economics have displayed certain limits with regard to explaining changes in the economic map of Europe and the United States. Territorial logic endeavours precisely to take into consideration the spatial turnabout (Aydalot, 1986b) which has been developing since then. Territorial logic will mainly be studied in connection with the idea of the milieu. In this context, we shall present a first attempt to systematize the problematic of the field.

Functional Logic

As the aim of this chapter is not to present functional logic, we shall merely describe briefly the mechanisms thereof.

Functional logic is based on a set of components made up of large markets and a foreseeable development of technology. The coherence of this set is maintained through very strong organization within companies and the establishment of relatively rigid ties between the company and the outside. Associated with this set is a specific way of managing and using the capital and wage ratio. Along with major companies and trade unions, the managers are major players; thanks to their position as intermediaries between capitalholders and salaried workers, they hold the power, and they impose their plan, which may be summed up as the growth of the company.

In general, the emphasis in functional logic is on taking advantage of economies of scale in production, stepping up the capital stock per worker and incorporating technological progress into the production process. To master this expansion, production structures must be set up internally whose various actors have well-defined roles to play that may be called functions and which are only coherent in the light of the functioning of the set as a whole.

The conditions which make it possible to use these procedures are basically as follows:

- there must be a certain amount of production standardization, which allows for market expansion;

- technology must either be stable or its development foreseeable if investments are to be recovered over a known period of time;

- in the production sphere, it is necessary to reconcile increased production, the adaptation of production techniques and return on capital and the wage ratio by strictly defining jobs (including the division between design-related jobs and execution-related jobs), required skills and procedures;

- lastly, the company must have sufficient financial strength to allow it to grow; this strength is both the means and the condition of the functioning of the system.

Effects of functional logic

i) Polarization effect

Owing to the factors on which it is based (particularly growth and innovation in large companies) and the resultant means, functional logic leads to several different types of concentration which have spatial ramifications:

- technical concentration: the size of the company is increased so as to simplify production processes and benefit from economies of scale and in order to integrate several stages of the production process at the same place;

- concentration of power: the process of functional integration always implies that structures are set up and rules defined with a view to ensuring the coherence of the organization as a whole. This goal always leads to a loss of autonomy by the various sub-sets to the benefit of a central body;

- economic concentration: the economic clout (in terms of revenue paid out, consumption of intermediate goods, etc.) of the largest companies is tending to increase;

- market concentration: this schema implies that there is a gradual shift from a market of free competition to an oligopolistic market.

ii) Specialization effect and spatial division of labour

Functional logic operates by making increasingly fine role distinctions with regard to both companies and employees. Separating design-related work from execution-related work and hierarchizing roles leads to corporate behaviour which seeks to take advantage of its environment by paying the lowest possible price for the various types of work it needs (for example, by hiring or purchasing goods manufactured in lower-cost areas) (Aydalot, 1976).

The paradigm of the spatial division of labour

Functional logic therefore leads to a type of spatial configuration called the spatial division of labour. The spatial economy and the way in which it finds expression in terms of economic policy, namely, the regional economy, have led to a discourse on space and practices. Practices give rise to a justificatory discourse which itself affects practices. This dual dialectic has formed an ideology of economic space which functions like a paradigm, i.e. as a unique means of representation accepted by all. Debates on economic space obtain information in the context of the paradigm, which is itself never called into question. By denying the specific nature of territories, the paradigm reflects a growth model which favours expansive accumulation (integration of agents and constant expansion of the space subjected to one's rules) (Pecquent, 1987a).

In all cases, this paradigm supposes that the centre is the driving force into which everything else must fit; it is the centre which creates the technologies and lifestyles and the only way the periphery can develop is by tapping and assimilating as rapidly as possible the elements from the centre (Aydalot, 1986b).

The Crisis of the Spatial Division of Labour and the Appearance of the Concept of the Network

The paradigm of the spatial division of labour has proved incapable of explaining the new spatial dynamics which have been noted since the early 1970s; new concepts have therefore been developed to take such factors into account.

We shall examine here the concept of the network, because much material has been published on this subject (Pecqueur, 1987; Planque and Py, 1986) and because it marks a moment of transition in reflection with regard to regional economies. In an initial phase, the concept of the network made it possible to reformulate to a certain degree the former paradigm of polarization and the functional division of labour; it has developed since then and is now at the heart of what we now call territorial logic.

As this transition has been taking place, the overall economic context has been undergoing far-reaching changes. The appearance of new competitors with lower production costs is pushing firms to find new products or procedures which enable them to opt out of the productivity race.

At the same time, as a result of the development of new technologies, particularly in the fields of electronics, EDP, microtechnology and new materials, it is now possible to make considerable improvements to a great many of the products we know and the procedures we use (Perrin, 1986). Thus, there exists an **innovative potential**. Furthermore, this innovative potential is very loosely defined: owing to the flexible nature of these technologies, many different uses can be envisaged for extremely diversified applications, and the production process in a great many old and new sectors may be restructured.

Owing to the above, technological innovation has become paramount. Moreover, it can no longer develop within the operating framework of a well-structured body. To take the case of the automotive sector: technological progress moves in a well-defined direction; the aim is to manufacture automobiles which are more and more comfortable and economical, pollute less, go faster, cost less, etc. However, in the capital goods industry, clothing, EDP, etc., there is little or no sense of direction and the dynamics depend on mechanisms whereby the size of the company or its organizational structure may become real hindrances.

Consequently, the specific strategies of the various individual or joint actors once again become key factors as far as innovation is concerned. An individual who acts dynamically and completes a project is viewed as being situated in a **network** of persons or bodies with whom he interacts; this network enables him to put together the various pieces which are needed for innovation (Perrin, 1986; Soulage, 1987; Pecqueur, 1987).

The individual manipulates his network. The logical consequence is that all one has to do to create additional innovative potential is to enrich the network. These are all policies of technopoles which aim to set up at the same site companies, a research centre and even commercial or administrative departments. It was thought that it was enough to bring all these elements together for each network to enrich itself from the capabilities of the others. Yet this vision is functional insofar as the various companies or bodies under consideration are seen as pieces of a puzzle which merely have to be brought together in one place for the puzzle to form. However, this vision, like policies aimed at attracting high-tech companies, overlooks the fact that the links underlying a network are cultural, not structural. **Most of the net-**

works studied in regions which have become dynamic already existed before these regions experienced strong growth: links between entrepreneurs or technicians in former industrial regions, links between former students in high-tech regions, and family, trade union or professional links in Italy. Thus, with territorial logic, innovation is the product of the emphasis placed on historically constituted know-how and technological culture thanks to internal dynamics. It does not spring from a planned upstream-downstream ordering of companies and bodies.

Territorial Logic and the Milieu

We now examine the components and functioning of territorial logic.

Although there is a wealth of material on this subject, it is quite clear that even if all the authors situate themselves at the same level of reality, many different terminologies and concepts are used. In this chapter, we shall attempt to make an initial summary of the material among these contributions which we feel refers exclusively to territorial logic. Among the notions used are the local industrial fabric (Thomas, 1987), the localized ecosystem (Planque, 1983; Pecqueur 1987), the industrial district (Garofoli, 1986), the local industrial system (Raveyre and Saglio, 1984) and the milieu (Aydalot, 1986a and b; Maillat, 1988).

In the light of recent developments concerning the problematic, it seems appropriate to list these notions, particularly that of the milieu, which this paper develops (for more developments and details, see Maillat and Perrin, 1989).

The milieu - a 'black box'

Judging from the wealth of new terms, it would appear that there is a level of analysis which remains ill-defined and where interesting economic phenomena may be observed. This level hardly fits traditional delimitations: it is trans-sectoral, does not cover a well-defined category of agents and does not have specific geographical limits. Notwithstanding this, vocabulary provides some indications. The word milieu suggests a framework of analysis with both an inside and an outside. Up until now, efforts to understand the milieu have focused on its effects (innovations, production of technology, etc.) rather than its composition and internal functioning. Yet it would appear that the time has come to define this framework better and to pinpoint its contents.

The milieu as a constituted whole

The aim is thus to describe the components of the milieu. To simplify matters, these components may be divided into three groups: the territorial production system, the actors and the cognitive dimension of the milieu.

i) The territorial production system

The territorial production system is structured round three fields between which flows circulate (Maillat and Vasserot, 1988):

(a) The territorial production apparatus which is made up of companies; in this connection, it is appropriate to identify the characteristics which are relevant to the problematic studied. For example, a key factor from the standpoint of innovation is knowing whether the functions of corporate R & D are localized in the region in question. In this connection, another fundamental element is company structure (groups, independent establishments, secondary firms, etc.) and capital ownership (from inside or outside the region). A milieu must not be subjected to the logic of filtering so that companies may be in a favourable position as far as innovation is concerned. In this respect, in view of the fact that their roots are generally stronger in the region, small and medium-sized companies have an important role to play.

In many regions, companies maintain ties with each other. A certain amount of inter-company coherence therefore emerges which may take the form of sub-contracting relations of dependency, sub-contracting of specialty, complementarity, competition, interdependence, partnership, etc. This may lead to certain corporate configurations, such as a network, a subsidiary with a major company - either upstream or downstream -, a host of very small firms, etc. Another element which is important today is the presence of company services. It would appear that industrial activities can no longer develop without incorporating new service functions at all stages (research, management, marketing).

(b) The labour market, whose structuring effect on the milieu is worthy of note. If one considers that individuals are mobile and that employment is the main factor governing their insertion in the milieu, it may be said that only attractive jobs are likely to attract and hold the labour force. Moreover, the important thing from an individual's point of view is not access to a good job, but rather to a set of jobs which are likely to allow him to fulfil his professional aspirations. With regard to unskilled labour, it appears that the key factor is not so much the number of jobs, but rather the possibility of achieving stability.

Thus, if the jobs which companies offer do not have certain attractive characteristics, they will not manage to keep employees, and hence not train them, either in training institutions or in-house.

(c) The scientific apparatus, which covers all of the milieu's structures and institutions of training and research. This set is important to the reproduction of skills and know-how and above all to their updating based on technological development and market needs. Today, the focus is on technological policies. A milieu does not need high technology to be dynamic: it is much more important for it to know how to use technology by adapting it to suit its own needs. Thus, the scientific apparatus has become a key component of the milieu insofar as it is capable of providing the milieu with appropriate technology.

ii) The actors

The various actors of the milieu are companies, socio-professional groups, public authorities and individuals. It is important to list these actors to underscore that many interests come into play here and that many strategies come into confrontation in a milieu (see below). One might think that the joint aim is regional development; this results in reality from a host of economic and political projects, which are aggregated according to a complex process. Thus, the milieu comprises an original combination of planned and quasi-planned elements (Brun, 1985a).

iii) The cognitive dimension

All of the above elements are, a priori, very heterogeneous and show no signs of coherence. Coherence is obtained through the cognitive dimension of this body: the various actors do not act as robots within the various elements described. If one attempts to analyse the process, one might say that these actors start by perceiving the various elements of the milieu. This perception is more than mere eye contact, and inevitably requires understanding. Consequently, a company is not a building which contains machinery and people with a view to manufacturing a given product. It is rather an entity which has specific characteristics and a past and which plays a role viewed as such in the productive system, in the balance of power, and in the technological, economic and social context of the milieu, whether it be as a company or through the people who work there or run it. This points to an entire set of relations. Yet this perceived set of relations must be systematized so as to avoid logical contradiction and to become understandable. Let us suppose, for example, that a given company which the milieu perceives as very competent, technologically speaking, endeavours to introduce a new technology and that this attempt is a total failure. Given this company's representation in the milieu, the milieu will attribute this failure a priori to the new technology. It may be noted that, if no other factor comes into play, this assessment will become final without the technology in question having been examined as such.

It is clear that the various elements of the milieu are reordered or restructured within an entire network of coherent mental relations which endow them with meaning. This is how a **coherent system of representations is formed.** Thus, when the environment changes, new technologies emerge, or cultural changes occur, the actors who belong to a given milieu make a more or less selective reading of these trends which is designed to make it possible to re-position changes within systems of representation.

These systems of representation have a collective dimension owing to the fact that they are constructed in a context of permanent interaction between the individuals of a community and their environment (Farr, 1984; Bassand and Hainard 1985). Once the elements of the milieu have been perceived, then understood thanks to their integration in the system of representations, the next step in the process is the elaboration of strategies. This stage, like the previous ones, is reached by each actor, according to his position and his own interests. As a result, the various strategies are inevitably in a more or less harmonious relationship with each other. However, even when conflicts do occur, the coherence of the milieu as a whole is not affected, as this coherence is not defined in relation to the interests present but rather in relation to a space of reference and a procedure of rationalization which is common to all actors. If we take the example of an employee in a workshop who makes a production improvement, the distribution of the benefits of this innovation creates a conflict of interest between employee and company. Nevertheless, this innovation occurs in a common place, in a clearly defined and shared technological context, and it is only possible to understand the importance of this innovation through a shared knowledge of the concrete situation in question.

Finally, the last stage is putting these various strategies into action. At this level, the process may or may not lead to the reorganization of the constituent elements of the milieu.

An example: know-how

Among these representations, the different types of know-how are worthy of note. These different types, as they appear in the milieu, are rarely totally objective entities. Thus, the different types of know-how are representations, products of the perception of the various elements of the milieu, technology and markets in general. For example, know-how, and more specifically its technological aspect, is constituted by all the capabilities which allow the application of skills or reinforce their applicability; it is therefore not limited to the possession of skills, but includes their arrangement based on the concrete conditions of their application as well as the elaboration and implementation of strategies. It may be noted that the conditions of application follow the same process of arrangement. We thus see that know-how can, inter alia, be characterized by objects which enter the field of perception. For example, the productive act gives rise to the elaboration of know-how which is directly related to the means, work objects and skills of an individual; individual know-how tends towards the knack which one has. The types of know-how which are developed through contact with training or research institutions have a less physical dimension and are closer to a reformulation or a reappropriation of knowledge with a view to adapting it to production. Lastly, know-how transmitted in professional circles comes from contacts with one's peers, through observation, communication and emulation. In general, know-how is the body generated by these different sources and it is the characteristic of being a body which brings out the primary feature of know-how, namely, its **integrative ability:** the ability to integrate science, the market, equipment and raw materials with a view to adapting them to production, all this based on the opportunities and constraints dictated by the specific characteristics of the production apparatus, the labour market and the technological culture of the milieu.

It should be noted in this connection that systems of representation tend to be stable. In general, individuals seek to preserve their system of representation. Of course, representations are also transformed by the pressure of events, but this transformation is not simultaneous and always interacts with existing representations. In this way, reformulating the old representations makes possible the process of breaking off/filiation (Aydalot, 1986a and b) and ensures that there is continuity between old and new situations.

As Planque (1983) points out, actors' behavioural changes are subjected to changes in their system of representation. By modifying the systems of representation which these actors create from their milieu and environment, cultural and/or technological development modifies the behaviour of these actors and their actions and, by extension, the organization and dynamic of the milieu.

A first definition of the milieu

When conceived in this way, the milieu is a **process** of continual perception, understanding and acts. This process constitutes the milieu as such. **The milieu groups together in a coherent whole a production system, a culture and actors. The coherence between the various actors lies in their common approach to situations, problems and opportunities** (Crevoisier et al, 1989). The corporate spirit, organizational practices, corporate behaviour and the way in which technology is used (know-how) are all integral components of the milieu. It may be noted that this definition does not a priori refer to a geographical entity. However, the milieu is deduced from several elements which themselves have regional ties. It is in this connection that the milieu must not be confused with the region. Even though the milieu spills over its borders in places, in particular when it maintains regular relations with some agents located on the outside, there are some elements (e.g. a branch of a major firm) which, although physically established in the region, maintain too few ties with it to form a part of the milieu.

Milieu and innovation

i) The perception of the milieu and of the environment

The systems of representation which create the regional milieu do not, however, develop solely in relation to the various elements of the milieu. It should be added in this connection that there is a second set which determines representations: the exterior, that is, everything that is outside the region. Indeed, one's self-perception is always derived from comparisons with others. This brings us back to the first concept of the milieu which is described above: the milieu as a framework for anaysis which defines simultaneously the interior and the exterior.

Two processes may thus be distinguished: the integration of activities is promoted within the regional milieu, while the process of differentiation is begun vis-à-vis the exterior (Crevoisier et al, 1989).

(a) **In the regional milieu,** the similarity of systems of representation makes for better communication between actors since problems are tackled in a like manner. In addition, these systems make it possible to integrate and inter-relate the elements of the milieu so as to form a more or less inter-related and finalized whole. This integration provides a means of increasing the intensity of relations and complementarities between the economic functions performed by the regional population (Planque, 1983). In this way, the effective heterogeneity of the milieu is reduced and differences or incompatibilities are inter-related. This integration makes it possible to relate the various elements of the milieu in a way which can lead to innovative processes. When this happens, one speaks of a network effect; this network is composed of all persons one either knows or can establish contact with and it can play a key role in the dynamic of a region, particularly with regard to corporate spirit, industrial diffusion and partnerships.

(b) **Outside the regional milieu,** this integration and network effect leads to a differentiation between the milieu and the other entities around it. The enhancement of local potential and the highlighting of the role of local entrepreneurial ability lead to an original model for regional development. This model implies that the fate of a region is not necessarily dictated by its natural resources or ability to attract branches of major firms, but that it depends on its innovative abilities, creativity and will to act (Maillat and Vasserot, 1988). Today, when comparative advantages are a product of a specific make-up of labour and the milieu, one can say that this differentiation can lead either to innovative processes which highlight the milieu's resources, or to the preservation of obsolete skills.

ii) Representations and innovation

It is generally recognized today that innovation stems from the association of information and resources (capital, skills, corporate spirit, creativity). As we have seen, the milieu is centred round an integrated body of resources in the broad sense of the word. Nevertheless, some elements are missing from the milieu and must be found elsewhere. Innovation is thus the integration, into the milieu, of key resources. These resources may be generated by the development of the milieu itself or come from the exterior. The social and technological context is changing: new behaviour patterns and new paradigms are emerging. These changes in the environment are perceived within the milieu through systems of representation. This enables the actors to detect the tensions and convergences which will, over time and according to their point of view, lead to either problems or opportunities.

Representations provide a means of putting changes in the environment into immediate perspective. This leads to a direct relationship between the movements of the environment and the resources of the milieu, given that each actor integrates external developments and re-orders them in order to render them compatible and usable with regard to the elements of the milieu. Consequently, one can say that representations make possible, or prevent, the appropriation of external technological and social changes.

iii) Characteristics of the innovative milieu

Innovation is formulated as a process of integrating elements which are decisive to the milieu's resources. It therefore implies a recasting of the milieu's representations. The innovative milieu is characterized by the integration of internal dynamics and external changes. The problem is therefore to be posed in the following manner: the milieu, to be innovative, must fulfil two non-antagonistic characteristics, namely, opening up to the outside, and achieving a high degree of integration on the inside.

(a) The milieu is innovative because it is capable of opening up, of gathering information, or even resources, from the exterior. The milieu must therefore open up with regard to the diversity of the general context, remain aware of and receptive to change so as to enrich itself. Yet this opening up implies risks, for how, in the mass of available information, can one pinpoint what information can be useful? How can one choose, among the infinite number of productive combinations possible, the most advantageous one? There is a danger of dispersing the energy and will of actors and using them up on methods which are already widespread.

(b) The milieu is innovative because its resources are organized, ordered and associated by means of economic, cultural and technological ties and the related representations. It is this process of providing order which makes resources usable in a new productive combination. It is also through the representations which the milieu has generated that one can apprehend and filter information and link them to the milieu's resources. The danger of an overly specific and overly integrated milieu is that too strong an emphasis on internal relations leads to closing off and, over time, to a loss of competitive ability. Innovation is thus strongly conditioned by the characteristics of the milieu. Among these characteristics is the degree of opening onto the exterior versus integration on the interior. The milieu is constantly torn between two antagonistic tendencies: opening and loss of its specificity on the one hand and closing off and depletion on the other hand.

Creativity is the process whereby the actors in a given milieu work out a new or original product which is suited to the constraints and purpose of the situation. The milieu can only achieve creativity when it meets these conditions of opening up/closing off, for it is at this moment that it supplies resources which can be mobilized and manages to link well-defined purposes to them.

The Inter-Relationship of the Two Types of Logic

Merely describing functional and territorial logic is not enough: today, there is agreement that each of these two paradigms describes a strong economic dynamic, that each has its strengths and weaknesses, and must accordingly be viewed within the same schema to arrive at a satisfactory explanation for regional dynamics (Pecqueur, 1987a; Soulage, 1986; Aydalot, 1986a; Gallianio and Gilly, 1985; Perrin, 1986).

We may now see how material on the subject has dealt with the way in which these two types of logic inter-relate. At the same time, a recapitulative table attempts to systematize these various contributions.

Using an input-output table (Figure 2.1) which shows the four possible ways in which the two types of logic can inter-relate (territorial>>territorial, territorial>>-functional, functional>>functional, functional>>territorial), it is possible to present a typology of industrial regions and regional trajectories. Given the present state of formulation of the functional and territorial dynamics, there is no unified theoretical framework. Consequently, this typology does not aim to be explanatory, but rather to show that, by going in this direction, it is possible to construct a theory explaining current spatial developments.

Before proceeding to a more detailed description of this table, some concepts need to be clarified, above all those developed by Ratti and D'Ambrogio (1989), which are based on a distinction between three different kinds of space for the company:

- **market space:** this is determined by the company's market relations, which are represented spatially;

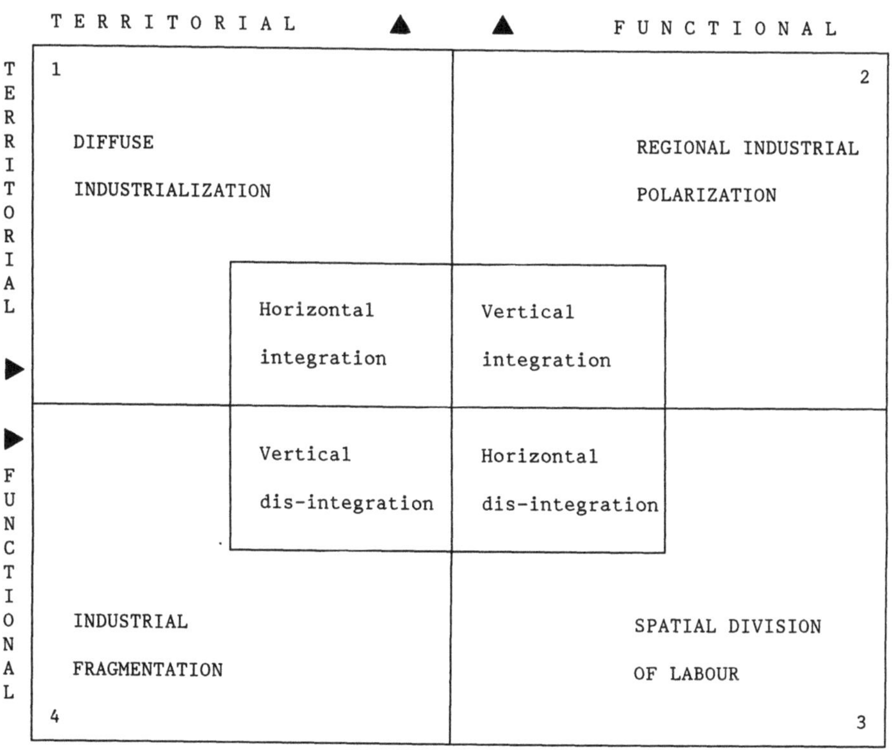

Source: IRER/OC

Figure 2.1: Recapitulative table

- **the company's production space:** here, the aim is to describe the production segments and to localize them in the (highly probable) event that spatial division of labour occurs;

- **support space:** this describes the relations determined by the organizational characteristics of the production factor and by the elements of the company's strategy which concern its relations with its partners, as well as those which relate to ties with private and public actors.

This space is comparable to the milieu. (Ratti and D'ambrogio, 1989).

Second of these concepts, the system of Swyngedouw & Anderson (1987), is based on the intensity and nature of links in a region. A distinction is made between the four following types of regions and corporate strategies:

- **horizontal integration:** links exist between the various actors in the region;

- **vertical integration:** the actors in a region have organized themselselves around a sector, branch or major firm;

- **horizontal dis-integration:** the companies and institutions in the region maintain links separately with other establishments in other regions;

- **vertical dis-integration:** the companies in the region do not have links with each other, and are each organized autonomously.

Box No.1: Diffuse industrialization

By its nature and functioning, the territorial/territorial box corresponds to the regional milieu as described above. The system is based on representations and know-how which are constituted historically through the sharing of the same values, experiences and technological culture, and is subject to various competition/cooperation types of rules. This system favours entrepreneurship and innovation which take advantage of and enhance the resources and opportunities of the milieu.

The archetype of this category is Silicon Valley in the Fifties and Sixties. Today, the regions in northeastern and central Italy (the NEC model, Brusco, 1986; Fua and Zacchia 1983; Garofoli, 1986) are the best example. In Ratti's typology, this stage is the one where the support space is very developed and covers the production space. The market space is broad because most of the firms have direct contacts with the end market.

For Swyngedouw & Anderson, this is the case where there are many diversified links between the various actors in the region: this is **horizontal integration.**

Box No.2: The regional industrial pole

When an industry matures and mass markets develop, pressures to cut costs, combined with economies of scale, act as incentives for concentration and integration. The delocalization of assembly lines begins, taking advantage of the existence of low-wage zones, while R & D units are kept in central regions (Swyngedouw and Anderson, 1987).

When expressed in the terms used up to this point, this consists of a functionalization of the territorial dynamic. As firms undergo mutual structuring, **vertical integration** occurs. Although the support space remains very strong, we have left the area of an economy where the driving forces are primarily local and other economic spaces develop, especially the production space; the market space remains limited for most firms, since their main outlets are located in the region, and expands for those which are located the furthest upstream in the region.

This category covers regional industrial poles which usually have a long tradition of crafts or semi-crafts before they set themselves up as industries. Some examples are Peugeot in Franche-Comté, the watch-making industry in the Jura Arc in Switzerland (Maillat and Vasserot, 1987), Fiat in the Piedmont, Silicon Valley today, wool manufacturing in Prato, Italy, (Ritaine, 1987) and the iron and steel industry in northeastern England.

Box No.3: The phase of maturity and spatial division of labour

This phase is that of pure functional logic. It characterizes firms which have sufficient mastery over their technology to be able to delocalize production or research units in regions which offer localization advantages relative to these various functions - in particular, low labour costs.

This category covers regions marked by the spatial division of labour; central regions, such as Paris and London, where strategic functions are concentrated (management, research and marketing); and peripheral regions which have not mastered their regional development, such as Brittany, Ireland or the Mezzogiorno in Italy.

There are no internal ties in these regions; the production space has been broken up to take advantage of wage differences, while the market space is vast and is controlled by major firms located upstream. The support space breaks up into partial segments (Ratti and D'Ambrogio, 1989). This corresponds to **horizontal dis-integration.**

Box No.4: The industrial redeployment phase

Today, major firms, rather than delocalizing a single function - manufacturing, research, etc. - as in the spatial division of labour, are tending to create companies which completely dominate production from conception to sales. These are all **fragmentation** strategies (Shutt and Whittington, 1987; Maillat, 1988) or **vertical dis-integration** strategies used by major firms.

There are two ways to territorialize a company:

First, major firms localize subsidiaries in the same place, as is the case with Sophia-Antipolis in the south of France. Relations between these enterprises and the parent company may take several forms:

- '**Decentralization** of production: large plants are broken up, but retained under the same ownership, by hiving off into smaller plants or by creating new subsidiary companies;

- **Devolvement:** large firms cease to own units directly, but retain revenue links with them, i.e. licensing or franchising;

- **Dis-integration** of production and innovation: large firms cease to own units of production or innovation, but retain control through market power (...) or, latently, through the power to repurchase the units'. (Shutt & Whittington, 1987, p17).

In this case, the support space is something which is planned by large firms, i.e. the idea is to take advantage of the presence of several enterprises in the same place, for example, to set up training structures, develop infrastructures to make it easier for technicians and their families to settle there, etc.

The second way of re-territorializing capital consists of taking over small, innovative, fast-growing firms. In France, large industrial groups have set up complex strategies to track dynamic small and medium-sized companies in order to encourage them to cooperate to the point of letting themselves be taken over. This is what Perrat (1987) calls capital regionalization. Moreover, this phenomenon is facilitated by the fact that this type of firm has difficulty securing funding to finance its growth (Planque and Py, 1986).

Autonomous regional development

An in-depth study has been done on the topic of autonomous regional development (Perrin, 1974). This theory is divided into three phases: start-up, polarization and deconcentration. The first phase takes place in areas with a crafts base which develops and where new firms are set up; this phase corresponds to industrial diffusion. However, the birth of production centres is not enough; diffusion must develop and spread. The second phase is a strengthening and structuring of the industrial system, which becomes capable of ensuring its own expansion in an autonomous fashion and over time. Lastly, the third phase is deconcentration which occurs when activities are allocated according to regional specialization.

We can thus see that this series of stages corresponds in the table to a shift from Box 1 to Box 2, and from Box 2 to Box 3. Some critics of this schema (Brun, 1985b) say that this system deals with the development of an infant industry, and that, in the case of developed countries, the first phase does not exist. Yet one may wonder whether cases such as northeastern and central Italy are not precisely an example of a new type of territorial dynamism in a developed country. Finally, it is true that this type of schema does not take into consideration the growing complexity of the productive system and of the development of new activities from existing industries. In our opinion, the incorporation of a fourth phase, i.e. vertical dis-integration would make it possible to take such factors into account. When major firms try to establish themselves in fields which they do not master, they avoid relying on the functional division of labour. They buy up high-growth small or medium-sized companies in this field, or set up a firm ex nihilo. The best illustration of this is what has been happening in recent years with robotics. Automobile manufacturers, which were and remain by far those which have the greatest use for robotics, either moved into production - as was the case with Renault and its failed experiment with Renault Machine Tools - or bought up autonomous firms.

As we can see, some regional trajectories may be plotted on the table by moving from one box to another. However, care must be taken not to read any kind of determinism into theories such as those of autonomous regional development or the product cycle applied to regions. A milieu can transform itself and move from one box to another, but this is rarely the case.

Conclusions

Today, the construction of a globalizing theory on spatial dynamics must take into consideration the new mechanisms which have been presented and grouped together in this paper under the idea of milieu'. Moreover, these mechanisms are giving rise to so much theoretical and empirical research that it would be unjustified to consider them as negligable. However, one must not think that the forces which have by and large determined spatial economic dynamics over the past forty years are no longer at work. Furthermore, visions which hold that the local or regional level is the sole economic driving force have been denounced, and rightly so, on many occasions. Today, the best solution seems to be to look for a formulation which can cover the two paradigms in a coherent way, without considering them as **a priori** antagonistic or incompatible. One way would be to start with an analysis of the regional production system, to conceive of it as an object with a certain amount of autonomy - at the level of inter-industry relations, know-how, technology, etc. - and to determine how this system is mixed into the international dynamic while remaining a full actor.

In any event, thoughts on this subject should also be extended to cover the search for more appropriate policies. The schema depicted shows that the spatial behaviour of firms varies according to their situation and the elements on which they rely in the region vary accordingly: a firm which applies functional logic will, above all, be interested in a policy of spatial redistribution of activities, in particular subsidies and tax advantages designed to attract firms; in the case of diffuse industrialization, a compány will be favoured by measures designed to promote revitalization, such as innovation aid, the creation of company think-tanks and training efforts; a region which is in the integration phase needs institutions to provide guidance, such as technological supervisory bodies; as for regions which welcome establishments which are relatively independent of large firms, it would be interesting to study how ties and locomotive effects could be created between these establishments, for example, through regional activities.

References

Aglietta, M. (1987) A Theory of Capitalist Regulation: the US Experience, Verso, London.

Aydalot, P. (1976) Dynamique spatiale et développement inégal, Economica, Paris.

Aydalot, P. (ed.) (1986a) Milieux innovateurs en Europe, GREMI, Paris.

Aydalot, P. (1986b) L'aptitude des milieux locaux à promouvoir l'innovation, in: Federwisch J. and Zoller, H. (eds.) Technologie nouvelle et ruptures régionales, 41-58, Economica, Paris.

Bassand, M. and Hainard, F. (1985) Dynamique socio-culturelle régionale, Presses Polytechniques Romandes, Luasanne.

Brun, R. (1985a) Approche systémique, industrie et région, Revue d'économie régionale et urbaine 1, 119-126.

Brun, R. 1985b Système productif régional et développement économique, Cahiers de l'institut d'économie régionale de Bourgogne et Franche-Comté, Dijon.

Brusco, S. (1982) The Emilian model: productive decentralization and social integration, Cambridge Journal of Economics 6, 167-184.

Crevoisier, O., Maillat, D. and Vasserot J. Y. (1989) L'apport du milieu dans le processus innovateur, in Maillat, D. and Perrin, J. C. (eds.), Actes du colloques GREMI II, Ascona.

Farr, R. (1984) Les représentations sociales, in Moscovici, S. (ed.), Psychologie sociale, 379-389, Presses universitaires de France, Paris.

Fua, G. and Zacchia, C. (eds.) (1983) Industrializzazione senza fratture, Il Mulino, Bologna.

Galliano, D. and Gilly, J. P. (1985) L'impact des groupes sur les systèmes productifs régionaux: une approche méthodologique, Laboratoire d'Etudes et de recherches en Economie de la production, Toulouse.

Garofoli, G. (1986) Le développement périphérique en Italie, Economie et humanisme 289, 30-36.

Lipietz, A. (1977) Le capital et son espace, Maspero, Paris.

Maillat, D. and Vasserot, J. Y. (1986) Les milieux innovateurs: le cas de l'arc jurassien, in Aydalot P. (ed.), Milieux innovateurs en Europe, GREMI, Paris, 217-246.

Maillat, D. (1988) SMEs, Innovation and Regional Development, paper presented at the European Summer Institute of the Regional Science Association, Arco, July.

Maillat, D. and Perrin, J. C. (eds.) (1989) Actes du colloque GREMI II, Ascona,

Maillat, D. and Vasserot, J. Y. (1988) Economic and Territorial Conditions for Indogenous Revival in Europe's Industrial Regions, in Aydalot, P. and Keeble, D. (eds.), High Technology and Innovative Environments, 163-183, Routledge, London.

Pecqueur, B. (1987a) Evolution de l'analyse de l'espace économique, Problèmes économiques 2009, 16-20.

Pecqueur, B. (1987b) Tissu économique local et systèmes industriels résiliaires, Revue d'économie régionale et urbaine 3, 369-378.

Perrat, J. (1987) Groupes multinationaux/PMI: Des liens nouveaux pour quel développement régional?, Revue d'économie régionale et urbaine 3, 467-474.

Perrin, J. C. (1974) Le développement régional, Economica, Paris.

Perrin J. C. (1986) Un bilan théorique et méthodologique, in Federwisch, J. and Zoller, H. (eds.), Technologie nouvelle et ruptures régionales, Economica, Paris, 205-210.

Planque, B. (1983) Innovation et développement régional, Economica, Paris.

Planque, B. and Py, B. (1986) La dynamique de l'insertion des PME innovatrices dans leur environnement: Problématique et propositions méthodologiques, Centre d'économie régionale, Aix-en-Provence.

Ratti, R. and D'Ambrogio, F. (1989) Processus d'innovation et intégration au milieu local: le cas du Tessin, in Maillat, D. and Perrin, J. C. (eds.) (1989) Actes du colloque GREMI II, Ascona.

Raveyre, M.F. and Saglio, J. (1984) Les systèmes industriels localisés: éléments pour une analyse sociologique des ensembles de PME industriels, Sociologie du travail 2, 157-176.

Ritaine, E. (1987) Prato ou l'exaspération de la diffusion industrielle, Sociologie du travail 2, 139-156.

Shutt, J. and Whittington, R. (1986) Fragmentation strategies and the rise of small units: Cases from the north West, Regional Studies 21, 13-23.

Soulage, B. (1987) Le développement local endogène: possibilités et limites, Revue d'économie régionale et urbaine 3, 361-368.

Swyngedouw, E. and Anderson, S. (1987) Le schéma spatial de la haute technologie en France, Revue d'économie regionale et urbaine 2, 321-349.

Thomas, J.N. (1987) Innovation et territoire, Revue d'économie régionale et urbaine 3.

3. INNOVATION DIFFUSION AND SMALL FIRMS: BETWEEN THE AGENCY OF LIFEWORLD AND THE STRUCTURE OF SYSTEMS

Bjφrn T. Asheim

Introduction

The aim of this chapter is to present a theoretical **framework** for analysis of some aspects of the process of innovation diffusion both among small and medium sized firms in industrial districts and to such firms from the surrounding businesses and social environment. In this theoretical chapter, special attention will be paid to the role of socio-cultural and political-institutional factors in constraining or enabling innovation diffusion in different regions. The framework will be used in the next step of a continuing research project at the University of Oslo to provide a comparative analysis of small firm enterprise in two industrial districts, one in Italy (Reggio-Emilia in Emilia Romagna), the other in Norway (Jaeren, south of Stavanger), both dominated by the same industrial structure (farm machinery industry). The project will try to identify the regional conditions for (a) flexible specialization and (b) regional systems of production. This will serve as a background for an analysis of the general national and specific regional conditions in Norway for (c) the development and diffusion of flexible specialization, and (d) the establishment and building up of regional systems of production based on flexible specialization as creative and innovative milieus for industrial restructuring and renewal (Asheim, 1988a).

To be able to evaluate the size and diffusion of flexible specialization as a form of production, it is necessary to carry out concrete research, i.e. theoretically informed case studies. A number of such studies have already been made in Italy and other West-European countries (Brusco, 1986; Fua, 1983; Garofoli, 1982). However, such studies have not been carried out in Norway, and this project will, in addition to its comparative aspect, represent a contribution to the assessment of flexible specialization as an alternative model for future industrial development in Norway (Asheim, 1988a).

Epistemologically, the research project is based on a **realist approach**, i.e. a combination of abstract, theoretical and concrete (intensive) research (Sayer, 1984; Asheim, 1987). This chapter focuses on the abstract part, i.e. on the theoretical work of conceptualization and abstraction, which will later be used in the comparative analysis of regional conditions for small firm enterprise based on flexible specialization in Italy and Norway (i.e. the concrete, intensive part of the research). In abstract research events are only considered as possible outcomes, while concrete research studies the actual outcome (Sayer, 1984).

The Research Environment

The international and national context of the project can briefly be summarized as follows.

Internationally the situation in the late 1970s and the beginning of the 1980s showed certain interesting tendencies with regard to the performance of the manufacturing industries:

(a) a market saturation in the demand for durable consumer goods combined with an increased competition from the Newly Industrializing Countries (NICs);

(b) a technological maturity of mass production as the dominant form of production resulting in a de-industrialization of the western (OECD) world (with the exception of Japan);

(c) a technological development (especially process innovations using information technology) which has lead to an increased profitability of small firms and production units due to a partial elimination of the technical rationale behind the internal economies of scale (Asheim, 1985);

(d) a growing demand for customized/semi-customized products. This has resulted in small firms being able to compete successfully with larger companies in national and international markets leading to a growth of the small firm sector, which in some countries (e.g. Italy) to a certain degree has offset the stagnation in areas dominated by large firms during this period of industrial restructuring (Sabel, 1982).

The domestic, Norwegian context presents a quite different picture due to the fact that mass production never became the dominant method of production in Norway. The Norwegian industrial structure can more accurately be characterized as a 'dual' one, i.e. divided into two sectors: the export-oriented large-scale industries based on cheap hydro-electric energy (primarily electro-chemical and electro-metallurgical industries), and the small-firm sector producing consumption and capital goods for the domestic market (Asheim, 1988b). Even if this picture has been modified through a gradually increasing internationalization of the Norwegian economy over the last thirty years, the industrial structure is still to a certain degree 'dualistic' with rather fundamental structural problems, and not properly adapted to international market conditions. On the one hand Norwegian export is still heavily dependent on energy-intensive products and raw materials (especially oil products during the last 15 years) amounting to between 80 and 85 per cent of the total value of exports. And on the other hand the small-sector manufacturing industry shows both low production of value added and low productivity, and is lagging behind in technological development. Both the transfer of technology from abroad and the development of new technology in Norway is too slow, especially in the field of production technology (process innovations). The number of industrial robots per 1,000 employees in Norwegian manufacturing industry is one, while in Sweden the number is 2.5. These problems result in 'mainland-Norway' having a deficit of between 30-40 billions N.kr. in the balance of payment, which highlights the great importance of the oil-sector in the Norwegian economy.

Thus, Norway is in great need of a **reindustrialization** based on expansion in branches other than the traditional ones, primarily in various mechanical industries, so that the dependence on the oil sector can be reduced and the balance of payment problems solved. To reach these goals a working group with representatives from the Norwegian Association of Manufacturing Industries and the national trade union has estimated that value added in the manufacturing industries has to increase by at least 50 per cent towards the year 2000, which emphasizes the need for higher value added and productivity in all branches. This means that bigger efforts have to be made to stimulate technological development both by individual firms or companies and by government. Industry has to invest more to increase its competence and technological capacity in order to be able to restructure and become more productive, and the government must allocate more resources to R & D in general, and specifically develop a national technology policy, including a policy of innovation as well as of innovation diffusion. Today, Norway devotes only 1.7 per cent of GNP to R & D, while Sweden spends 2.5 per cent and Japan and the US around 2.8 per cent (Frogner, 1987; Larsen, 1987).

Definition of Concepts

Before proceeding with the presentation of the theoretical framework it is necessary to try to give a more precise definition of the concepts 'flexible specialization' and 'regional systems of production'. The concept of flexible specialization has over time gradually obtained a content so broad as to cover almost all forms of production that cannot be characterized as 'Fordism' in its typical assembly line version (e.g. flexible manufacturing systems) (Piore and Sabel, 1984). Thus, the reality the concept is supposed to grasp has become so heterogeneous that it loses both its analytical and descriptive potential. Furthermore, such a 'chaotic conception' (Sayer, 1984) provides an incorrect understanding of industrial development, the most notable example of which is when it is claimed that flexible specialization/flexible production methods is the new hegemonic model of industrialization replacing the old mass production methods. (Storper and Scott 1988). As Sayer correctly pints out, the flexible production method of 'just-in-time' has enabled the mass-production industries of Japan to stay extremely competitive and profitable (Sayer 1988). The 'just-in-time' system of production as such can be described as a vertically disintegrated and spatially integrated subcontracting system, where the flexibility and cost efficiency of the system (through stock control) is provided by the smaller subcontracting firms lower down in the production hierarchy (Sheard, 1983).

Another problem connected with the conceptualization of flexible specialization is the historical perspective of this form of production presented by Sabel and Zeitlin (1985). They argue that there are many examples of viable local economies based on flexible specialization in craft production within industrial districts in England, Germany and France in the nineteenth century, representing historical alternatives to mass production, and that their eventual defeat was due to national economic policy strongly supported by large, monopoly capitalist firms rather than based on strict intra-firm/industry productivity arguments. This author believes this 'political technology' line of arguing to be wrong both theoretically and empirically. What is characteristic about flexible specialization is neither the flexibility nor the specialization, but the combination of both aspects (Hull Kristensen, 1984), which is attained through minimizing the readjustment time, i.e. the ability to shift promptly from one product configuration to another (Storper and Scott, 1988), by re-programming computerized production equipment. This combination is a result of technological development (process innovations) rather than regression to more simple techniques of craft production (technological retardation), and which could not have taken place without the **scientifically** based transformation of the production technique (i.e. **technological** development) of the mass production industry in the second part of the nineteenth century and the beginning of the twentieth (Asheim 1985). Thus, it cannot be correct to identify today's flexible specialization with craft production in the past when the present means of labour is that advanced. Therefore, Piore and Sabel's argument that flexible specialization represents a revival of craft forms of production has to be wrong (Piore and Sabel, 1984). It would be more correct to refer to it as a **new form of manufacture**, (but) with advanced technology (means of labour) producing relative surplus value with an efficiency which earlier (in principle) could only be attained by machinofacture (large industry). Theoretically, this reflects the important difference between the **formal** subordination of the labour process of craft production and traditional manufacture on the one hand, and the **real** subordination of the labour process of flexible specialization today (due to technological development) on the other hand. Once more, then, Piore and Sabel are wrong when they claim that 'the advent of the computer restores human control over the production process; machinery again is subordinated to the operator' (Piore and Sabel, 1984, 261).

Empirically the crux of the matter is the significant difference between the production of **relative** surplus value today and the production of **absolute** surplus value of yesterday's craft production and manufacture, which would not have been able to provide a large enough material surplus to finance the creation of the modern welfare state (Asheim, 1984). This is convincingly demonstrated by Williams et al's reference to the production of the Ford model T, stressing that the new 'process innovations had an overwhelming economic advantage over the craft methods of production'... which... 'suggests there never was a choice because there was not a viable craft alternative to mass production of complex consumer durables' (Williams et al, 1987, 420). They go on to conclude that

> 'whatever new technology does offer... it is not going to inaugurate a new era of flexible specialization which restores those craft methods of production that lost out at the first divide. New generations of computer controlled equipment may deliver a more varied output but they do not restore an economic system based on redeployable productive resources and low fixed costs. That is a world which we have lost'. (Williams et al, 1987, 433).

In contrast, a sound conceptualization of flexible specialization should take as its point of departure Brusco's typology of small firms, which differentiates between various types on the basis of market relations, technological level, degree of skill of the labour force and relationships between the firms, ending up with the following result (Brusco, 1986):

- the traditional artisan firm;
- the dependent sub-contracting firm;
- the independent small firm of industrial districts.

There are three important aspects of this typology, which are useful when making a more precise definition of flexible specialization.

(a) Firstly, it distinguishes between various types of small firms. While the first two categories represent traditional types of small firms (i.e. the second could, for example be a subcontractor of a Japanese just-in-time system, and the first an example of former craft production), the third category comprises the new, advanced and innovative small-firm sector of, for example, Emilia Romagna, based on a model of flexible specialization.

(b) Secondly, the typology distinguishes between the types of small firms on the basis of some important characteristics **internal** to the firm (i.e. market, technology, skill).

(c) Thirdly, it distinguishes on the basis of factors **external** to the firm (i.e. the relationships between firms).

Thus, a definition of **flexible specialization** must contain (a) **intra-firm** characteristics (production in small or medium-sized firms of specialized (semi-customized/customized) products with (advanced) flexible, multi-use equipment and skilled workers, and selling in a national and international competitive market) as well as (b) **inter-firm** characteristics of **the regional system of production** (industrial district) where the firm is located (relationships between the independent small firms within a formal (functional) and informal network of economic, technical and social character exploiting external agglomeration economies associated with geographical concentration) in order to be descriptively precise and analytically useful.

Theoretical Famework

On necessary conditions

The most important task for the abstract part of the research is to identify the necessary regional and national conditions for the existence and diffusion of flexible specialization and regional systems of production. This can be outlined as shown in Figure 3.1.

Studies have shown that **social-cultural factors** such as common values in a community (i.e. collective identity and consciousness) serve as references for the attitude and behaviour of people as well as 'examples' which demonstrate that the role of 'independent' and entrepreneur is a possibility open to most persons (Edvardsen, 1988; Gustavsson, 1985). In some Scandinavian studies the concept of 'life-modes' have been used to explain how the so-called 'independent life-mode' of the self-employed in specific regional contexts - in contrast to the life-mode of the working class - stimulates entrepreneurial activity and growth of the small-firm sector (Hjalager and Lindgaard, 1984). Moreover, in Great Britain and Italy the presence or absence of specific socio-cultural factors favourable to successful small firm enterprise seem to play an important role in certain regions during the process of industrial restructuring (Massey, 1983; Morgan and Sayer, 1985; Brusco, 1986; Fua, 1983; Piore and Sabel, 1984).

Another important specific regional condition required for flexible specialization

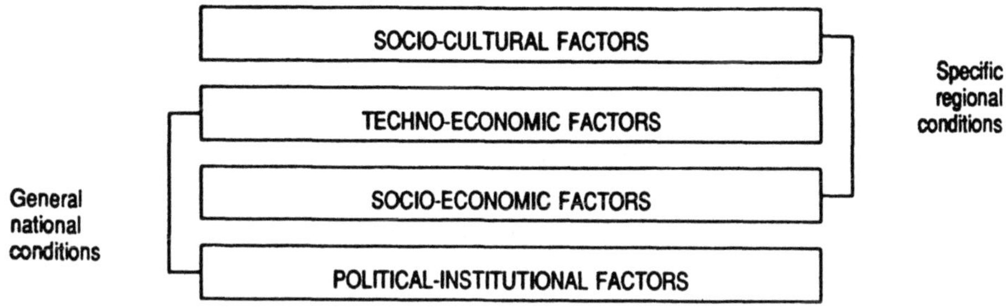

Figure 3.1: Necessary conditions for the existence and diffusion of flexible specialization and regional systems of production.

is the **socio-economic factors**, which comprise the relations of production (where class relations are understood as 'Klasse für sich') of regions. A central determinant of the socio-economic factors is the qualitative and quantitative aspects of the regional industrial structure; i.e. their impact on the social structures within regions dominated by central **or** peripheral firms and workers (Asheim, 1984). Of importance in the Norwegian context, among other things, is the fact that mass production never became the dominant method of production, as is seen very clearly from the present industrial structure, which is characterized by very small firms compared to what is found internationally. With considerable regional variations this can constrain as well as enable the process of innovation diffusion.

A significant general national condition is the **techno-economic factors** understood as the relationship between economy and technology manifested as the actual and potential technological and innovative capacity of the country, which means the capacity of private and public R & D organizations to adopt and develop radical innovations and new technological systems, and eventually to promote changes in the 'techno-economic paradigm' (Freeman and Perez, 1986). In this connection incremental innovations are not considered to be part of the general national conditions, as they primarily take place within the individual firms at the regional level.

And lastly there is the **political-institutional factors** referring to the development of the welfare state, a high standard of educational system and public policy within the fields of industry and technology, which together represent the general strategic national conditions for generating and establishing creative and innovative milieus for industrial renewal based on flexible specialization within regional systems of production.

However, as the main task of this paper is to develop a theoretical framework for analysing the process of innovation diffusion to and among small firms, i.e. the processes of **hierarchical** and **epidemic** diffusion respectively, all of these factors are not of equal importance and relevance. According to Freeman and Perez (1986) conditions for the diffusion of innovations in the context of technological change and economic development are to a large extent related to changes in the **social-institutional** subsystem, thus making the techno-economic factors (i.e. technological and innovative capacity) less relevant in an analysis of innovation diffusion. As an example they emphasize that

> 'the full realization of the productivity gains which can be achieved as a result of information technology depends on the diffusion of the new paradigm throughout the economy. This in turn will be possible only as a result of many social and institutional changes, which will involve interrelated organizational and technical innovations, as well as a large increase in new skills and a transformation of the existing capital stock'. (Freeman and Perez, 1986, 26).

This has also been stressed in a recent book review by Martin, who argues that it is

> 'particularly pertinent at the regional level: to the extent that the inherited socio-institutional structure of accumulation varies geographically then the attractiveness and ease of adjustment of different regions to new technology will also vary. It is within this perspective, of spatial uneveness in and differential geographical inertia of the socio-institutional environment, rather than a narrowly defined micro-economic or behavioural context, that questions of regional variations in "entrepreneurship", "innovation potential" and technological development might more usefully be situated'. (Martin, 1987, 255).

Referring to the conceptualization in Figure 3.1, the **social-institutional subsystem** is constituted by (a) the **socio-economic** factors and (b) the **political-institutional** factors. In addition, Figure 3.1 also stresses the importance of **socio-cultural** factors in explaining innovation diffusion in the context of small (and medium sized) firms in industrial districts.

On agency-structure and lifeworld-system

The next step in the process of conceptualization is to elaborate these concepts further in order to give them a more precise **theoretical** content. Important analytical dimensions in this theoretical construction will be Giddens' concepts 'agency - structure' (Giddens, 1984) and Habermas' concepts 'lifeworld - system' (Habermas, 1981). Habermas defines **lifeworld** as the spheres of society where the interaction between people must be symbolically structured, i.e. based on the communicative rationality of discussion (communicative action). **System** on the contrary includes the two subsystems economy and politics, i.e. the spheres for strategic, instrumental rationality (Mjøset, 1983). While lifeworld as an arena is based on **social** integration, i.e. 'reciprocity between actors in context of co-presence' (Giddens, 1984, 28), the system is dominated by **system** integration, i.e. 'reciprocity between actors or collectivities across extended time-space' (Giddens 1984, 28). **Agency**, according to Giddens, refers to doing, i.e. 'not to the intentions people have in doing things but to their capability of doing those things in the first place', and '**structure** is not to be equated with constraint but is always both constraining and enabling'. (Giddens, 1984, 9, 25). The relation between agency and structure is described in this way:

> 'The constitution of agents and structures are not two independently given sets of phenomena, a dualism, but represent a duality. According to the notion of the duality of structure, the structural properties of social systems are both medium and outcome of the practices they recursively organize' (Giddens, 1984, 25).

When placing the original concepts concerning the necessary conditions for the existence and diffusion of flexible specialization and regional systems of production within this theoretical context, it can be illustrated as in Figure 3.2. In this connection

> 'it must be remembered that merely noting these necessary conditions is not sufficient to explain how they are met, if they are. To do this we must refer back to the level of actions. This movement also illustrates the interdependency of structure and agency' (Sayer, 1984, 103).

The process of innovation diffusion can be constrained by at least two sets of structural constraints leading to either **the fallacy of division** or **the fallacy of composition**. 'The fallacy of division' refers to a suboptimal situation, where lack of co-operation leads to results which are poorer than the attainable ones (i.e. the so-called prisoner's dilemma). Thus, the solution can be rational to individual capitals in the short run, but unfavourable to a collective of capitals (e.g. to the small-firm sector in an industrial district). 'Fallacy of composition' refers to the production of **contre finalité** or **unintended effects** (i.e. not wanted and not expected consequences) of human action. Contre finalité is produced when agents act according to presumptions made on each other which due to the necessary, internal relations (i.e. structures) of the capitalist mode of production cannot at the same time be true (Asheim, 1986). These situations occur as

> 'the invisibility of structure to common-sense thinking leads to an underestimation of the **interdependence** of positions. ...This is the assumption that, in all cases, what is possible for an individual must be

possible for all individuals simultaneously' (Sayer, 1984, 86).

This is illustrated in Figure 3.3. Position I in Figures 3.2 and 3.3 (the numbers correspond in the two figures) is of less relevance in the context of innovation diffusion as instrumental activity of the system, and will not be dealt with explicitly. However, to give a brief example of an activity that belongs to this arena, invention, i.e. the ideal, non-instrumental creative activity of the researcher/inventor, could be mentioned.

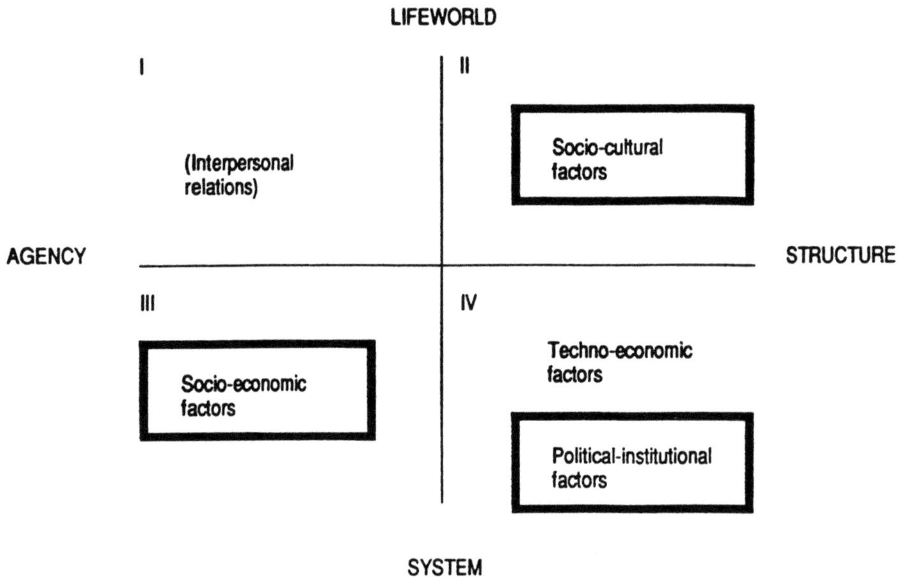

Figure 3.2 : Analytical dimensions in the study of innovation diffusion.

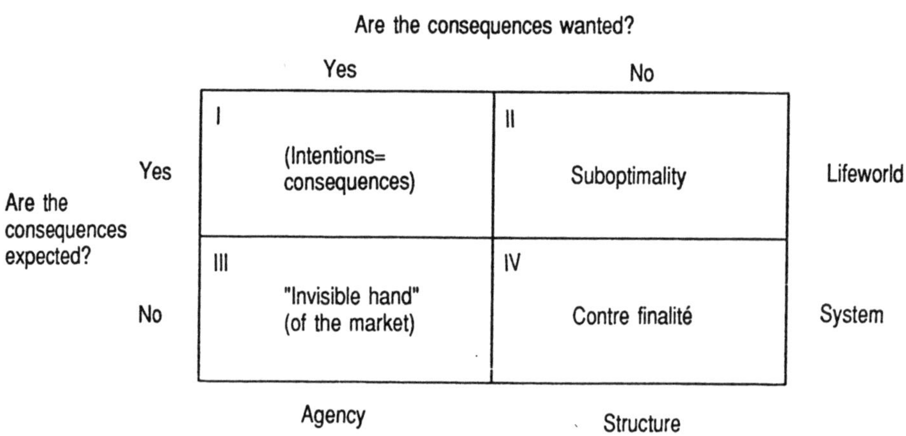

Figure 3.3 : Problem generation mechanisms in social action

Position III refers to the function of the market as a co-ordinator, in which typical agents are the entrepreneurs of small firm owners in their role as atomistic actors. According to Adam Smith's 'invisible hand' (or in a modern version: 'a great IBM-machine in the sky' (Gould and Ferguson, 1980, 410)) the market (as a stable, self-regulating system in a free market economy) is supposed to have only positive consequences, including the unexpected ones. But as history has shown this notion is of a pure ideological character. Market 'co-ordination' results in market failure as well as market imperfections, thus needing the positions II and IV respectively to be dealt with analytically.

On the social dimension to regional industrial structure

As indicated earlier, one important aspect of the socio-economic factors is the social dimension of the regional industrial structure. In our context the existence of an industrial structure dominated by small (and medium sized) firms is of interest to the process of diffusion, as there are signs indicating that such an industrial structure enables optimal balance between competition and co-operation to operate. This characterizes some of the most successful examples of regional systems of production based on flexible specialization in the 'Third Italy'. (Brusco, 1986; Garofoli, 1982; Piore and Sabel, 1984). This co-operation may take the form of regional innovation centres, such as CITER (Centro Informazione Tessile dell' Emilia Romagna) in Carpi, and TESA (technical co-operation) in Sandnes, Jaeren which has existed since 1957 to strengthen the competitiveness and the capacity of the restructuring of the member firms by promoting technological competence (and which has resulted in Jaeren having one third of the industrial robots in Norway with only about three per cent of total employment in manufacturing industry), or more informally through levelling out periods of peak and slump among competitors (Camagni and Rabellotti, 1988; Finne, 1985; Piore and Sabel, 1984). Camagni and Rabellotti consider what they call 'systems areas', as

> 'a sort of co-operative agreement on a territorial basis, featuring high stability and easy information transfers. What most uniquely differentiates a system area from the more generic inter-company agreement is the former's relevance of historical, cultural and environmental factors versus the latter's straightforward functional integration; in the system area, the relations are bred by a co-operative habit and by geographical closeness, and are mainly managed through informal personal contacts' (Camagni and Rabellotti, 1988, 16).

However, this indicates that in addition to socio-economic factors such as the existence of entrepreneurs and the social impact of a small-scale industrial structure, co-operation in regional systems of production seems to be dependent on viable, local socio-cultural factors in order to function properly (Position II). According to Piore and Sabel this represents a violation of

> 'one of the assumptions of classical political economy: that the economy is separate from society. ... By contrast, ... it is hard to tell where society (in the form of family and school ties or community celebrations of ethnic and political identity) ends, and where economic organization begins' (Piore and Sabel, 1984, 275).

As mentioned earlier, studies from several countries have pointed out the importance of specific, regional socio-cultural factors in stimulating entrepreneurship and co-operation. Two out of four crucial factors identified by Piore and Sabel in explaining the development of the innovative system of small firms in Italy, i.e. the Italian extended family and the view of artisan work as a distinct type of economic activity (Piore and Sabel, 1984, 227) can be categorized as socio-cultural factors, and the same is true for Brusco's observation 'that the "Third Italy" areas correspond to the diffusion of métayage (share-cropping) agricultural areas'. (Brusco, 1986, 195). Fua also stresses this point when stating that 'the continuity with local history and traditions is doubly valuable. It is something positive in itself, and it is a source of strength in achieving other positive results. Hence the system's merits' (Fua, 1983, 376). Corresponding socio-cultural factors have been identified in regions in Sweden (e.g. Smaland) and Norway (e.g. Sunnmore), which are characterized by this type of entrepreneurial activity (Gustafsson, 1986; Edvardsen, 1988).

Such socio-cultural factors are rooted in the traditional sector of society, i.e. reminiscences of pre-capitalist forms of production (e.g. simple commodity production) (Asheim, 1988b), and can be understood as **culture** interpreted as **tradition** (i.e. life-modes, norms, common value system). Thus, socio-cultural structures are important constituent elements of the lifeworld. However, when remembering that historical contingent factors represent conditions which can enable as well as constrain economic and social development, in which contexts are socio-cultural factors 'functional' to the formation and growth of regional systems of production? The studies referred to seem to indicate that where traditional structures still play a substantial role and have retained influence **in spite of** an advanced process of modernization, they can be reinforced and expanded for new progressive purposes, and thus not impede economic and social development (Berger, 1980). Piore and Sabel argue that the traditional, socio-cultural factors of familiarism and the exemption of 'artisan' firms from certain requirements of the labour laws protected the 'infant' firms, and facilitated capital accumulation in the early phases of industrial development (Piore and Sabel, 1984). And there are relatively few examples of such artisanal firms staying small (i.e. with fewer than 15 employees) just to avoid losing these advantages (Fua, 1983). One conditional factor behind this successful development is, according to Fua, that 'transplants and changes are not so drastic as to shatter the existing social pattern, thereby squandering valuable resources' (Fua, 1983, 355). If the changes on the contrary are that drastic, what is occurring is referred to by Habermas as the colonization of the lifeworld by the system (Habermas, 1981), i.e., the reorganization and instrumentalization of the lifeworld to become part of the system (Mjøset, 1983).

A tentative answer to the question of which mechanisms ensure the socio-cultural factors of the lifeworld to be system functional, may be found in the concept of **'loyalty'** (Hirschman 1970). According to Hirschman, the informal barrier of loyalty is particularly 'functional' to an organization when it ensures that the

> 'potentially most influential... members will stay on longer than they would ordinarily, in the hope or, rather, reasoned expectation that improvement or reform can be achieved "from within". Thus loyalty, far from being irrational, can serve the socially useful purpose of preventing deterioration from becoming cumulative, as it so often does when there is no barrier to exit' (Hirschman, 1970, 79).

Hirschman elaborates this further and maintains that

> 'in the choice between voice and exit, voice will often lose out, not necessarily because it would be less effective than exit, but because its effectiveness depends on the **discovery** of **new** ways of exerting influence and pressure towards recovery. However "easy" such a discovery may look in retrospect the chances for it are likely to be heavily discounted in **ex ante** estimates, for creativity always comes as a surprise. Loyalty then helps to redress the balance by raising the cost of exit'. (Hirschman, 1970, 80).

Hirschman's line of reasoning can be applied to formal regional institutions such as CITER and TESA as well as to more informal networks between small firms in industrial districts. Thus, it may offer an explanatory framework within which the conditions for the optimal balance between competition and co-operation found in these districts can be analysed. The most essential channel for information and diffusion of new technology among small and medium sized firms is other firms (epidemic diffusion), thus emphasizing the importance of interaction between firms within formal and informal networks in avoiding suboptimal solutions and situations with reference to innovation diffusion. In this way technological development within one firm stimulates development in the other firms of the industrial district (Hull Kristensen, 1984). This plus sum game is conditioned upon the absence of behaviour causing fallacy of division. In addition such regional systems of production in their capacity of innovative milieu can be situated higher up in the central place system than the size of population should indicate, which increases the chances of firms being early adopters in a process of hierarchical diffusion (Vatne, 1987).

However, loyalty has to be reproduced, and important conditions in this reproduction are **proximity in space** and **stability over time**. These conditions refer especially to the question of ownership and control of firms within a regional system of production. Increasing external ownership and control of firms in the form of branch plants or subsidiaries is believed to seriously weaken the effect of loyalty. Another factor of importance to the reproduction of loyalty may be found on the cultural/ideological level, which can be illustrated by two regional examples. In Emilia Romagna in particular, small firms were often run by socialist and communist artisans, and many of the skilled workers who left or were expelled from the large-scale industry and became independent subcontractors were members of the Italian Communist Party. Thus, employers were often the ex-workmates and the political co-militants of their employees, and these ties possibly created a feeling of equality and solidarity which was never found in the large-scale industry (Piore and Sabel, 1984). In the northwestern part of Southern Norway (Sunnmore) it was not communism but Christianity that dominated, as the region is one of the strongholds of the Christian lay movement in Norway (Edvardsen, 1988). But the effect on people's perception of equality and solidarity in the local communities might be identical; thus promoting co-operation and industry in the same way as in Emilia Romagna.

As pointed out earlier, lifeworld as an arena of socio-cultural factors as structuring principles is based on social integration 'on the level of face-to-face interaction' (Giddens, 1984, 28) and governed by common cultural norms and values. But if either a) the co-presence of actors disappears due to increased **time-space distanciation** of the system of production, or b) the traditional culture disintegrates through **the system's colonization of the lifeworld**, system integration - referring to connections with actors across extended time-space (Giddens, 1984) - has to replace social integration in order to ensure reproduction of the system. In addition to these cases, represented by increased external control and dominance by the international commercial cultural industry respectively, another two situations provide the basis for system integration. The first case relates to c) regions without socio-cultural factors functional to system reproduction, i.e. factors which do not develop **loyalty** among members of the community. Examples of such localities can either be regions where traditional social structures have retarded development (i.e. economic and social dualism), which is the case in parts of Northern Norway (e.g. Finnmark) and Southern Italy (e.g. Sicily) (Asheim, 1988b), or regions characterized by a 'working-class life-mode', due to the dominance of heavy industries typically located in one-company towns (e.g. in South Wales, where on the contrary **solidarity** is manifest) (Massey, 1983; Morgan and Sayer, 1985). The last case refers to d) the contradictions of capitalism, i.e. within capital and between capital labour, which produce the basically unintended effects of the capitalist mode of production manifested as overaccumulation and class struggle (where class is understood as 'Klasse an sich'). But as already emphasized, techno-economic factors, which on an abstract level of analysis constitutes the 'logic' of the system causing these contradictions (Asheim, 1985), is of less relevance with respect to analysing innovation diffusion. Thus, we concentrate on the remaining cases of historical, contingent situations where system integration by means of political-institutional factors is made necessary by the process of development (position IV).

Policy Implications for Regional Systems of Production

In accordance with the argument of Freeman and Perez, such an approach should focus on 'the institutional factors affecting the behaviour of both users and suppliers of innovations and on the enabling role of policies which might accelerate or retard diffusion' (Freeman and Perez, 1986, 26-27).

One problem confronting the independent small firm sector is the possible reduction of innovative capacity both with respect to new products and the adaptation of new processes to local needs (incremental innovations). In the absence of loyalty in the community (due to either factors a), b) or c)) the environment within which small firms make decisions about investments in new technology is structured by international competitive markets, conditioning behaviour which potentially leads to the fallacy of composition. This can be manifested in two ways, either as unsuccessful investments in expensive new technology due to all firms behaving in the same way and thus deteriorating market prospects; or on the contrary, as neglect of investments in new technology and an attempt to survive by means of cutting production costs, i.e. reducing wages and using inferior materials, which can develop into a functional incapacity of the system for further technological development. Eventually both groups of firms will lose in the competition with other more cost efficient and productive firms. An important task to be solved through system integration is therefore to ensure or maintain an optimal technological dynamism in the regional systems of production, so as to avoid technological stagnation or retardation as well as overinvestment.

Thus, what is needed is a strategic public policy to control the development of technology, i.e. a **technology policy** - comprising an **innovation** policy as well as a **diffusion** policy - defining **what** should be produced (product innovations) and **how** it should be produced (process innovations). This refers to the (direct) demand of the public sector for new and advanced products as well as to the (indirect) public support of product and process innovations, together with bringing about knowledge and information in order to stimulate diffusion, adoption and adaptation of innovations among small firms. Part of such policies must also apply to existing well-functioning regional systems of production so as to encourage the creation of regional institutions to ensure continual technological development (Piore and Sabel,

1984).

In order to accomplish the establishment of regional systems of production as creative and innovative milieus for industrial restructuring and renewal, a policy based on a modified growth pole concept may be pursued (Perroux, 1970). The success of such a policy would depend on whether it is possible to call **planned** growth poles based on small and medium sized firms into being, i.e. to create functional integrated systems of production - e.g. in the form of (geographically) expanded industrial districts (Camagni and Rabellotti, 1988) - with high innovative capacity, which can ensure the diffusion of new technology throughout the economy to exploit maximum productivity gains (Haraldsen, 1987).

A policy for technology will render necessary interventions in the sphere of production based on a broad conception of the goals of societal development. This strategy would increase the possibilities of preventing the unintended effects of technological change such as structural unemployment, regional imbalances and a high degree of dependence on foreign import of technology. Consequently, this would imply a change in the national economic policy to include a **qualitative** dimension (i.e. policy of technology), in addition to the **quantitative** dimension of Keynesianism (i.e. policy of demand regulation) (Asheim, 1986).

But if the main strategy of implementing a technology policy, with special reference to innovation diffusion, is to establish and build up regional systems of production based on flexible specialization, what will then be the chances for these systems to be extended throughout the economy so as to provide a national basis for reindustrialization?

The view expressed here is that flexible specialization can only work as a general model of industrialization either in small countries with a small-scale industrial structure, and which have a tradition for state intervention together with favourable socio-institutional prerequisites, i.e. democratic traditions, egalitarian social structures, a good educational standard and high per capita income (e.g. the Scandinavian countries); or for a limited number of regions with supporting socio-cultural structures within larger countries (e.g. 'The Third Italy'). In large countries without the necessary socio-institutional prerequisites referred to above on a national level, it is quite difficult to imagine that these regional islands of prosperity, with their specific form of production, could expand to dominate the whole of the national economy (Asheim, 1988b; Williams et al, 1987).

A strategic public policy for technology would increase the role of the superstructure to become a more decisive factor in technological development. This change would imply a modification of the capitalist relations of production as it weakens the control of the means of production by capital, and so the control of technological development. This can also be regarded as a renewal of the mixed economy where 'the mixture proportion' of the public and the private sectors is qualitatively different as compared to the present situation (Asheim, 1986).

References

Asheim, B.T. (1984) Informal economy, small firm development and spatial structures in Italy, paper presented to the 24th European Congress of the RSA, Milan. Published (in Norwegian) in Nordisk Samhällsgeografisk Tidskrift, No.2, 1985, 34-47.

Asheim, B.T. (1985) Capital accumulation, technological development and the spatial division of labour: A framework for analysis, Norsk Geogr. Tidsskr., 39, 87-97.

Asheim, B.T. (1986) The Public Sector, Information Technology and Regional Development, in Guesnier, B. (ed.) Développement Local et Décentralization, Editions Regionales Europèennes S.A., Poitier, 81-102.

Asheim, B.T. (1987) Radikal samfunnsgeografi - regional, romlig eller 'realist'? Nordisk Samhällsgeografisk Tidskrift, 5, 9-16.

Asheim, B.T. (1988a) Macro-geography: Technical change and capitalist development, Norsk Geogr. Tidsskrift, 42, (2-3), 175-78.

Asheim, B.T. (1988b) Peripheral industrialisation and capitalist development: a macro-geographical comparison of Norway and Italy, in Linge, G. (ed.), Peripheralization and Industrial Change: Impacts on Nations, Regions, Firms and People, Croom Helm, London, 53-71.

Berger, S. (1980) The traditional sector in France and Italy, in Berger, S.and M. Piore (eds.), Dualism and discontinuity in industrial societies. Cambridge University Press, 88-131.

Brusco, S. (1986) Small firms and industrial districts: the experience of Italy, in Keeble, D. and E. Wever (eds.) New firms and regional development in Europe, Croom Helm, London, 184-202.

Camagni, R. and Rabellotti, R. (1988) Technology, innovation and industrial structure in the textile industry in Italy, paper presented to the Conference on 'The application of new technologies in existing industries: prospects for regional industrial regeneration and employment in Europe', The University of Newcastle upon Tyne, 23-25 March.

Edvardsen, T.E. (1988) Lokal kultur og småforetaksvirksomhet i Norge og Italia. Nordisk Samhällsgeografisk Tidskrift, 7, 49-56.

Finne, G. (1985) TESA-gruppen, in Johnstad, T. (ed.), Kompetanse- og innovasjonssentra, Report No.53, Institute of Industrial Economics, Bergen, 85-90.

Freeman, C., and Perez, C. (1986) The Diffusion of Technical Innovations and Changes of Techno-economic Paradigm, paper presented to the Conference on Innovation Diffusion, Venice.

Frogner, T. (1987) En industristrategi for Norge mot år 2000, in Aalbu, H. (ed.), Kompetansespredning som vekststrategi, Norske samfunnsgeografers forening - Skrifter, 15, Oslo, 154-71.

Fua, G. (1983) Rural industrialization in later developed countries: the case of northeast and central Italy, Banca Nazionale del Lavoro Quarterly Review, 147, 351-77.

Garofoli, G. (1982) Areas of specialized production and small firms in Europe, paper presented to the Conference on National and Regional Development in the Mediterranean basin, Durham.

Giddens, A. (1984) The Constitution of Society: Outline of the Theory of Structuration, Polity Press, Cambridge.

Gould, J.P. and Ferguson, C.E. (1980) Microeconomic Theory, Richard Irwin Inc., Homewood, Illinois.

Gustafsson, B. (1986) Kultur för entreprenörskap, in Johannisson, B., O. Spilling (eds.), Lokal naeringsutvikling. Entreprenørskap og nettverksstrategier i noen norske og svenske kommuner, Universitetsforlaget, Tromsø, 79-96.

Habermas, J. (1981) Theorie des kommunikativen Handelns, M.:Suhrkamp, Frankfurt.

Haraldsen, T. (1987) Vekstpol - en teoretisk begrepsdroftelse, Nordisk Samhällsgeografisk Tidskrift, 6, 20-25.

Hirschman, A.O. (1970) Exit, Voice and Loyalty. Responses to Decline in Firms, Organizations and States, Harvard University Press, Cambridge, Mass.

Hjalager, A.M. and Lindgaard, G. (1984) Livsformer og lokale erhvervspolitiske initiativer, NordREFO.

Hull Kristensen, P. (1984) Dansk produktionskultur ved industrimodellernes skillevej, paper presented to the Conference on Technology - Society initiatives, Nyborg.

Larsen, K.A. (1987) Norge må reindustrialiseres, Forskningspolitikk, 4, 8-9.

Martin, R. (1987) Book review of 'Technological Change, Industrial Restructuring and Regional Development' (edited by Amin, A. and J. Goddard), IBG Transactions, 12(2), 254-56.

Massey, D. (1983) Industrial Restructuring as Class Restructuring: Production Decentralization and Local Uniqueness, Regional Studies, 17, 73-89.

Mjøset, L. (1983) Jürgen Habermas' reformulering av den klassiske sosiologiens bekymringer, Sosiologi i dag, 2-3, 3-27.

Morgan, K. and Sayer, A. (1985) A 'modern' industry in a 'mature' region: the remaking of management-labour relations, International Journal of Urban and Regional Research, 9(3), 383-403.

Perroux, F. (1970) Note on the Concept of 'Growth Poles', in McKee, D., R. Dean and Leahy, N. (eds.) Regional Economics: Theory and Practice, The Free Press, New York, 93-103.

Piore, M. and Sabel, C. (1984) The Second Industrial Divide: Possibilities for Prosperity, Basic Books, New York.

Sabel, C. and Zeitlin, J. (1985) Historical alternatives to mass production: politics, markets and technology in nineteenth century industrialization, Past and Present, 108, 133-76.

Sabel, C. (1982) Work and Politics. The division of labour in industry, Cambridge University Press, Cambridge.

Sayer, A. (1984) Method in Social Science: A Realist Approach, Hutchinson, London.

Sayer, A. (1988) Post-fordism in question, paper presented to the Industrial Specialty Group, AAG's Annual Meeting, Phoenix, Arizona.

Sheard, P. (1983) Auto-production systems in Japan: organizational and location features, Australian Geographical Studies, 21, 49-68.

Storper, M. and Scott, A. (1989) The geographical foundations and social regulation of flexible production complexes, in Dear, M. and Wolch, J. (eds.) The Power of Geography, Winchester, Mass.

Vatne, E. (1987) Teknologisk utvikling, arbeidsdeling og lokalisering av naeringsvirksomhet. Perspektiver for den regionale utvikling, Report No.93., 'Institute of Industrial Economics, Bergen.

Williams, K. et al (1987) The End of Mass Production? (review of Piore and Sabel's 'The Second Industrial Divide'), Economy and Society, 16(3), 405-39.

4. TECHNOLOGY TRANSFER, TECHNOLOGICAL CAPABILITY AND INDUSTRIAL STRUCTURE IN DEVELOPING COUNTRIES

Sunwoong Kim

There has been a long tradition in the economics of growth and development that the growth of per capita income (or output) is mainly determined by the accumulation of physical capital. In the view of early students such as Ragnar Nurkse (1953) and Paul Rosenstein-Rodan (1943), underdevelopment is basically a result of lack of capital stock. In this tradition, the celebrated Harrod-Domar model (Harrod, 1939 and Domar, 1946) explains economic growth by the accumulation of capital stock and by assuming that the capital-output ratio is constant. Thus, the development programme should accelerate capital formation by encouraging savings.

Solow's (1956) neoclassical refinement of the model was motivated by the razor-edge stability of the Harrod-Domar model. As he allowed labour-capital substitution by assuming a neoclassical aggregate production function, the model produces globally stable steady state equilibrium growth. The major prediction of the model is that initial capital-labour ratios or short run disturbances do not have any impact on the long run balanced growth at the natural rate, which is determined by the savings rate and technology.

The most influential growth model in relation to developing countries is probably the dual-economy model of Lewis (1954). Although the working of the model hinges on the assumption that wage rates in developing countries stay fixed, the growth of the economy is basically fuelled by large savings generated by capitalists in the modern sector generate. The modern sector expands by the accumulation of capital, and the expansion is supported by the shift of labour from the low productivity traditional sector to the high productivity modern sector. Implicitly assumed here is that high productivity technology in the modern sector is available without any cost. Beyond that, technology has very little to do with economic growth.

However, Solow's (1957) empirical implementation of the aggregated production function model in the US economy generated a very surprising result that upsets the ubiquitous capital fundamentalism:

> 'Gross output per hour of work in the US economy doubled between 1909 and 1949; and some seven-eighths of that increase could be attributed to

"technical change in the broadest sense" and only the remaining eighth could be attributed to conventional increase in capital intensity'. (Solow, 1988, p.313).

This so-called large residual, the component not explained by the growth of factor inputs and attributed to technological improvement, has generated 'growth accounting' which decomposes the residual into a number of different constituents. More notably,

'Taking the period from 1929 to 1982 and smoothing away the business cycle, he (Denison) finds that real non-residential business output increased at an average rate of 3.1 percent per year. ... Denison (1985) estimates that a quarter of it can be attributed to the increased educational qualifications of the average worker. The growth of "capital" accounts for 12 percent of the growth of output; ... Then Denison imputes 11 percent of total growth to "improved allocation of resources" (by which he means such things as the movement of labour from low-productivity agriculture to higher productivity industry). Another 11 percent goes to "economies of scale" (but this must be a very insecure imputation). Finally, 34 percent of recorded growth is credited to the growth of knowledge or technological progress in the narrow sense. ... Thus technology remains the dominant engine of growth, with human capital in second place'. (Solow, 1988, pp.313-4).

Certainly technology can take many different forms : machinery, human capital, books, reports, blueprints, patents, organizational structure, standard operating procedures, and so forth. First of all, more efficient technologies can be incorporated into the design of new capital goods. Solow (1988) in particular thinks that

'much technological progress, maybe most of it, could find its way into actual production only with the use of new and different capital equipment'.

However, the empirical studies show that the embodiment idea is not really noticeable, as shown in Denison (1985).

This leads to the hypothesis that technologies embodied in workers and organizations are more important. Indeed this is the key observation of this chapter. Without having a better terminology, we shall call it technological capability (TC). In this paper, TC is viewed as a **stock** of knowledge pertaining to transforming inputs into output embodied in workers and organizations. TC as a distinct category separated from technology per se was identified by Slater (1960), who distinguishes technical knowledge and techniques of production. Hayami and Ruttan (1971) distinguish the transfer of material, design and capability in the discussion of international agricultural technology transfer.

The knowledge and experience embodied in workers have been identified as human capital for a long time (see Becker 1964 for an elaboration on the subject). The idea of social capital or institutional capital is relatively new. It refers to the know-how used to combine human skills and physical capital into systems of producing goods and services to satisfy human needs. Hall and Johnson (1970) identified system-specific, firm-specific, and some general types of institutional capital. Coleman (1988) argued that social capital is an important determinant of human capital formation in an analysis of high school drop-outs' behaviour in the US. Martin Fransman categorizes TC by many different aspects:

> '...these capabilities involved the following kinds of activities : The search for available alternative technologies and mastering of the technology, that is its successful use in the transforming of inputs into outputs; the adaptation of the technology in order to suit specific production conditions; the further development of technology as the result of minor innovations; the institutionalized search for more important innovations with the development of R & D facilities; the conducting of basic research'. (Fransman and King, 1984, p.10).

More specifically, Westphal et al. (1985) classify TC into three broad areas: production, investment and innovation. According to them,

> 'The first capability is for operating productive facilities, the second is for expanding capacity and establishing new productive facilities, and the third is reflected in (technical or x-) efficiency in the ability to adopt operations to changing market circumstances. Proficiency in investment capacity is reflected in project costs and in the ability to tailor project designs to suit the special circumstances of the investment. Proficiency in innovation capability is reflected in the ability to develop technologies that are less costly and more effective'. (Westphal et al., 1985, pp.172-173).

If technological innovation is important and if technological knowledge embodied in workers and organization is more important than that embodied in capital goods, we have to re-think the process of economic growth and international technological transfer. First of all, the development process should not be thought of as the process of accumulation of capital stock. Rather it is the process of technological improvement. As indicated above, if TC is more important that capital goods in embodiment of technology, then more adequate attention should be given to the accumulation of TC.

Moreover, if the major source of economic growth through technological innovation comes mainly from the enhancement of TC, then the study of technology transfer should also focus on those aspects as well. It is unfortunate, however, that TC has not been paid adequate attention in the technology transfer literature until recently. For example, according to the Feldman-Mahalonobis-Domar (FMD) model (see Taylor, 1979, for detail), in order to achieve high rates of economic growth, an economy should not invest in capital goods that produce consumption goods but in capital goods that produce other capital goods. This model has been very influential in shaping economic development plans of large closed economies such as the Soviet Union, India and China. Their experience suggests that large investment in the capital goods sector only is not enough for rapid economic growth. Investment in TC is also necessary. Conversely, the incredible reconstruction and consequent rapid catch-up of West Germany and Japan after World War II also suggest that much of their TC was intact even after the large loss of capital stock during the War (see Hall and Johnson, 1970).

It is easy to recognize that the accumulation of TC is both costly and time consuming. Certainly, TC is complementary to capital goods and disembodied technological knowledge. But, both capital goods (machines) and TC (workers and organizations) are necessary in the actual production process, and conscientious investment decisions and planning consideration should be given to both aspects. Before going into the economic consequences of such explicit considerations of TC, it would be useful to point out several special characteristics of TC. TC is very different from technologies embodied in capital goods. First of all, it is less visible. Also, since it is embodied in individuals or groups of individuals of human beings, it is very difficult to claim exclusionary property rights. Therefore TC has a strong public goods character. There are several reasons for this. Workers switch firms. When a worker quits, the firm cannot successfully undo the technology embodied in the worker. Although the patent system authorizes a property right on an innovation, it is far from perfect in protecting the inventor's exclusive claim. Certain technical details have to be disclosed in the application process, and others can invent around the patent in order to achieve similar results. There are usually time limitations for the protection of property rights, after which it becomes a public good. Organizational structure and standard operational procedures can be easily imitated by other firms (see Pack and Westphal, 1986, for a survey).

Although TC creates externalities **within** an economy, it is very difficult to transfer **between** different economies. The accumulation of TC is primarily a local effort. Importing foreign personnel is more difficult than importing machines, since the workers usually have to be compensated for living abroad. If there are differences in language and/or culture, the diffusion of TC is difficult even with the importation of personnel. Sending workers abroad to acquire technology is usually not perfect. Many developing countries have experienced a brain-drain by sending their students and professional workers to developed countries. Moreover, as Richard Nelson (1987) has eloquently put it, the receiver of technology is bound to have less information than the originator. Thus, technology transfer involves a relatively quick process of importing technology embodied in capital goods and a relatively slow process of acquiring technology embodied in workers and organizations.

In order to illustrate the point, let us consider the debate on appropriate technology. It has been argued that capital intensive technologies in a labour-abundant developing country are not appropriate. The typical criticism is two-fold:

- Firms in developing countries choose capital intensive technologies because of factor price distortion favouring capital usage through credit rationing and/or favouritism toward large firms;

- Even if factor prices are 'correct' the adoption of capital intensive technologies is inequitable, because they do not generate a great deal of employment.

The usual counter argument is that although capital intensive technologies may be inappropriate in the short run they are appropriate in the long run because of their ability to increase productivity. As the argument goes, the increased output will generate more savings (investment) and thus more employment in the long run. But the discussion can be put in a different perspective by considering TC explicitly.

It would not be safe to assume that technologies are available for a continuously varying combination of factors. In reality, there are usually several technologies with different factor requirements with corresponding levels of TC required to use such technologies. The major production decision of the firm in a developing country with low TC is not the choice of a technology with continuously varying capital-labour ratio. Rather, it is the choice between an indigenous technology with lower productivity or a foreign technology with higher productivity. For expositional convenience, let us assume that the firm has only three alternative technologies: the indigenous craft technology with low capital-labour ratio, a turnkey contract of a multi-national corporation with high capital-labour ratio, and the intermediate technology with medium capital-labour ratio. If the firm adopts the latter technology, it appears that the firm chose a capital intensive technology. However, it is critical to note that both technologies involve a similar level of TC. Although the turnkey project may be an embodiment of high technology, the level of TC needed to operate the plant would be very low. The cost and time required for training workers and setting up operating procedures for the new plant may be minimal. If the firm chooses the latter technology, this just reflects the fact that it is more profitable. Why did it not choose the more labour intensive intermediate technology? The choice of technology is simply not feasible because of its TC. For example, licensing the technology and constructing the plant with local talent requires a higher level of TC. Although the choice of such schemes is more labour-intensive and thus more 'appropriate', it is not feasible for firms with the low TC constraint.

Thus, this model predicts that the firm in a country with low TC is likely to choose a technology with low local TC required. Typically, this scheme involves more embodied capital goods and more capital-intensive technologies. But the level of TC may increase over time, through the enlargement of the formal education system and the quality standards imposed by foreign buyers. Westphal et al. (1985) describe the process of the accumulation of TC in the South Korean case. Hence a NIC (Newly Industrialized Country) which has accumulated a substantial amount of TC may want to choose a more labour intensive technology with higher TC. Thus, the paradox of 'inappropriate technologies' can be explained easily by considering TC explicitly.

In the last few paragraphs, it has been argued that the real choice of the firm in developing countries is not capital-intensive versus labour intensive technologies but the level of profitability and technology transfer mechanism with varying levels of TC. Let us be more systematic and consider Figure 4.1. Its horizontal axis is capital goods (K) and its vertical axis is TC. In the short run, TC of a firm (or a country) is fixed at TCo. Through the investment, TC can be increased in the long run. But in the short run, the technological choice is restricted to those whose TC requirement is below TCo. Various mechanisms of technology transfer can be imposed on Figure 4.1 also. The ratio of required capital over require TC would be highest in Direct Foreign Investment (DFI). Turnkey contracts, licensing, and domestic development will have lower ratios in decreasing order. In our earlier example, the profit is maximised at the point C, in which the most profitable technology transfer mechanism is a turnkey contract. The economic trade-off here is that technology embodied in machine versus TC. In some cases, high capital-ratio technology is more profitable even when the labour is relatively cheaper.

A second example of the usefulness of TC is in the debate over the choice between indigenous development and the importing foreign technology. Some developing countries pursued self-reliant economic policies regarding technology. India is a good example of a country that has been favouring indigenous development over the importation of foreign technologies until very recently. But too much investment in TC and too little investment in K would yield lower productivity growth. By putting more resources into the accumulation of TC, India has moved from point A in Figure 4.1 to point D over the last twenty-five years. During this time period it could have invested more in imported hardware and less in the development of TC which might yield point E, which is more efficient than point D. See Lall (1985) for more descriptions of the Indian experience.

Having stressed the importance and usefulness of the TC concept, we now turn to a simple two-period model to derive a number of propositions regarding investment for TC and industrial structure. The choice of technology transfer is viewed as the

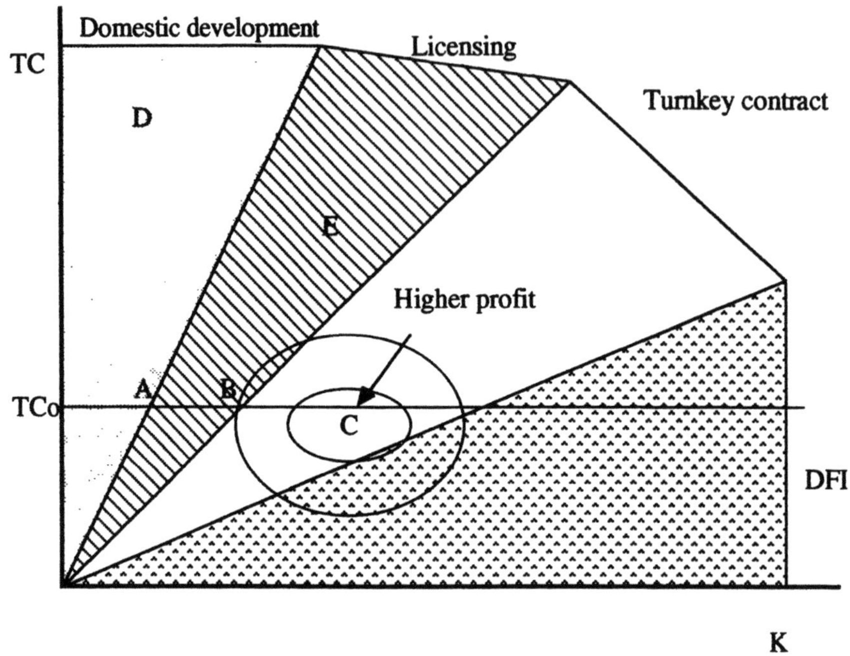

Figure 4.1: Trade-off between TC and Embodied Capital Goods.

choice of a relative share of TC versus K. There are two periods in the model. Production is carried out in both periods. In order to capture the slowness of TC accumulation, it is assumed that the investment in TC can only improve productivity in the next period. Since the second period is the last, investment for the TC will be made only in the first period. The total production cost of the firm i is a function of its output (q_i), its TC (x_i), and the aggregated level of TC (Σx_i) in the economy. We shall assume that the cost function c(.) has the following characteristics:

$$D_1\ c(q,\ x_i, \Sigma x_i) > 0,\ D_{11}\ c(q,\ x_i, \Sigma x_i) > 0$$

$$D_2\ c(q,\ x_i, \Sigma x_i) < 0,\ D_{12}\ c(q,\ x_i, \Sigma x_i) < 0$$

$$D_3\ c(q,\ x_i, \Sigma x_i) < 0$$

where Dj represents a partial derivative with respect to the jth argument. The profit of the firm i (Πi) will be:

$$\Pi i = p_1 q_{1i} - c(q_{1i}, \bar{x}_i, \Sigma \bar{x}_i) - f(x_i) + \beta(p_2 q_{2i} - c(q_{2i}, \bar{x}_i + x_i, \Sigma(\bar{x}_i + x_i)))$$

where p_j is the price of output in period j, $1-1/\beta\,(0<\beta<1)$ is the discount rate, f(.) is the cost of making an investment of TC, and $\bar{x}_i$ is the level of TC of the firm i in the first period, which is taken to be exogenous.

Proposition 1: In the competitive market structure, firms investment in TC less than the socially optimal level.

This level is a straightforward consequence of the positive externality assumption regarding TC. Since firms do not take into account the productivity increase due to the increase of overall TC, they tend to under-invest in TC. In the competitive market structure, therefore, the government should play a role in order to achieve productive efficiency. A subsidy in TC or a government sponsored consortium for sharing technological information can achieve the efficient allocation. The government should put more emphasis on the activities which have strong externality effects such as an information-gathering agency or job training.

Proposition 2: In the competitive market structure, $dTCi^*/d\beta>0$, where TCi^* is the equilibrium level of TC investment.

In words, firms with lower discount rates will invest more in TC. Since accumulation of TC is a time-consuming process, firms with higher discount rates will find the investment in TC is relatively less attractive than devoting resources to the current production. In a regime of credit rationing, small firms that have less credit-worthiness are likely to have high discount rates, and consequently will be less willing to make an investment in TC. Thus, we would observe that innovative technological transfers will be made initially by large firms. A similar proposition can be proved in the monopolistic and oligopolistic market structure. Davies (1979) contains some empirical evidence that large firms are more likely to innovate. However, he gave different reasons for the relationship such as scale, profitability, and so forth. This does not imply government policies favour large firms. On the contrary, the provision of credit for small businesses can be a viable policy option. Entrepreneurship and vitality of such firms should be encouraged as they are often the major source of employment and local economic development.

Proposition 3: Let q_{ji}^* be the equilibrium level of output by firm i in period j. If the demand stays the same (either price or demand curve is the same in both periods), then $q_{1i}^*<q_{2i}^*$.

A higher level of TC results in lower costs. Given the same demand schedule (or the same price), firms will produce more as the marginal cost declines. If the good is an importable good, an increase in TC will increase domestic production and reduce the import. If the good is exportable, the reduction of marginal cost will enhance the international competitiveness. Thus, exports will increase. The increase of the firm's competitiveness results not only from its own increased TC but also from the aggregated level of TC available in the local area. Such externality will generate synergic increase of the productivity among different firms in the related industry.

Proposition 4 (Schumpeterian trade-off): The competitive market structure will yield marginal cost pricing. However, since firms will under-invest in the competitive market, the competitive market will not be efficient in the long run. On the contrary, the monopolistic firm will not under-invest in TC since there is no externality problem. However, the monopolistic firm will reduce output and raise prices so that market outcomes will not be efficient.

Monopolistic market structure is one way of internalising the externality in TC investment. In order to achieve efficiency a regulated monopoly such as public enterprise may be a viable solution. An equally viable policy is to encourage TC investment in the competitive market structure. The choice of market structure would depend on many factors such as the openness and the size of the economy. See Kamien and Schwartz (1982) for a detailed discussion on the Schumpeter hypothesis.

Proposition 5: If demand is expected to increase more, then investment in TC will be higher.

The firm's demand can be increased by two factors:

- Holding total demand constant, the firm's market share may increase;
- Holding market share constant, total demand may increase.

Consider the following example of monopolistic market structure in which:

$c(q, TC) = mq/TC$,

$f(TC) = n\ TC$,

$p = a - bq$.

Profit maximising monopolist will produce $q = (a-m)/2b$. The optimal level of TC is $\sqrt{\frac{\beta m(a-m)}{2bn}}$. Thus, it is clear that $\frac{dTC}{da} > 0$. Proposition 5 suggests that firms expecting growth are more likely to invest in TC. And, firms in a growing industry are more likely to invest in TC.

To summarize, it has been found that technological innovations are the most important determinants of economic growth. These persistent empirical findings are quite contradictory to the common belief among economists that capital accumulation is the most important factor. Moreover, technological innovations embodied in workers and organizations are more important determinants of the rate of economic development than those in capital goods. In order to consider such phenomenon, the concept of technological capabilities (TC) was proposed. TC can be regarded as a stock of technological knowledge embodied in organizations as well as workers. Because of the mobility and opportunities to observe the other agents in the local area, TC can be diffused quite quickly among locals. This relatively quick local diffusion process is highly contrasted with the slow and costly diffusion between different economies. The accumulation of TC is time-consuming and occurs mainly through local effort. Importing TC is not usually as effective as importing capital goods.

Technological improvements in developing countries have two aspects: accumulation of TC and accumulation of capital goods. The level of TC is a serious bottleneck for a developing country in the short run as it is time consuming to accumulate the local TC. However, optimal technology transfer policy should properly consider the long term benefit of the investment in TC. Thus, the major trade-off in the short run is how to balance the two aspects of technological improvement. Because of the externality problem, industrial structure is also an important determinant of the accumulation of TC. Monopolistic market structure can internalise the externality problem. But the monopoly can result in higher prices unless there is enough foreign competition. The government can stimulate economic growth by proper policies on industrial structure and TC investment.

Acknowledgement

The author would like to thank Harriet Newberger for her comments and suggestions.

References

Becker, G. (1964) Human Capital, National Bureau of Economic Research, New York.

Coleman, J.S. (1988) Social Capital in the Creation of Human Capital, American Journal of Sociology 94, 95-120.

Davies, S. (1979) The Diffusion of Process Innovations, Cambridge University Press, Cambridge.

Denison, E. (1985) Trends in American Economic Growth, 1929-1982, The Brookings Institution, Washington.

Domar, E.D. (1946) Capital Expansion, Rate of Growth and Employment, Econometrica 14, 137-147.

Fransman, M. (1986) Technology and Economic Development, Westview Press, Boulder, Colorado.

Fransman, M. and King, K. (eds.) (1984) Technological Capability in the Third World, Mcmillan, London.

Hall, G.R. and Johnson, R.E. (1970) Transfers of United States Aerospace Technology to Japan, in Vernon, R. (ed.) The Technology Factor in International Trade, National Bureau of Economic Research, New York.

Harrod, R.F. (1939) An Essay on Dynamic Theory, Economic Journal, 49, 14-33.

Hayami, Y. and Ruttan, V.W. (1971) Agricultural Development: An International Perspective, John Hopkins Press, Baltimore.

Kamien, M.I. and Schwartz, N.L. (1982) Market Structure and Innovation, Cambridge University Press, Cambridge,

Lall, S. (1985) Trade in Technology by a Slowly Industrializing Country: India, in Rosenberg, N. and Frischtak, C. (eds.), International Technology Transfer, Praeger, New York.

Lewis, W.A. (1954) Economic Development with Unlimited Supplies of Labour, Manchester School of Economics and Social Studies 22, 139-191.

Nelson, R.R. (1987) Understanding Technical Change as an Evolutionary Process, North Holland., Amsterdam.

Nurkse, R. (1953) Problems of Capital Formation in Underdeveloped Countries, Oxford University Press, Oxford.

Slater, W.E.G. (1960) Productivity and Technical Change, Cambridge University Press, Cambridge.

Solow, R. (1956) A Contribution to the Theory of Economic Growth, Quarterly Journal of Economics 70, 65-94.

Solow, R. (1957) Technical Change and the Aggregated Production Function, Review of Economics and Statistics 39, 312-320.

Solow, R. (1988) Growth Theory and After, American Economic Review 78, 307-317.

Pack, H. and Westphal, L.E. (1986) Industrial Strategy and Technological Change: Theory versus Reality, Journal of Development Economics 22, 87-128.

Rosenstein-Rodan, P. (1943) Problems of Industrialization of Eastern and South-eastern Europe, Economic Journal, 53, 202-211.

Taylor, L.J. (1979) Macro Models for Developing Countries, McGraw-Hill, New York.

Westphal, L.E., Kim, L. and Dahlman, C.J. (1985) Reflections on the Republic of Korea's Acquisition of Technological Capability, in Rosenberg, N. and Frischtak, C. (eds.) International Technology Transfer, Praeger, New York.

5. THE ORGANIZATION VARIABLE IN DEVELOPING COUNTRIES INDUSTRIAL DEVELOPMENT

Roberta Rabellotti

Introduction

In the neoclassical literature, a competitive market is assumed to be the optimal organizational structure characterized by firms that are all pretty much the same - small, price takers, rational and in competition with many other similar firms.

The obvious question is why do other organizational forms (e.g. multinationals or small family enterprises) exist if competitive firms are optimal? According to neoclassical economics, the only possible explanation for the existence of other forms of organization is technological indivisibility. Evidence suggests however that large and multinational firms do not always adopt indivisible technologies. For example, vertically integrated textile firms, where the production cycle is split up among several sequential production phases (fibres preparation, spinning, etc.), can each operate with quite distinctive technological features unexplained by the indivisibility arguments.

Therefore, in order to explain this organizational variety we must seek out alternative explanations. We suggest that one way forward is to integrate the concept of the firm as performing a function of production with the concept of the firm as performing a function of governance. This notion is based upon the premise that unequal access to information, the existence of uncertainty and the not-perfectly-rational behaviour of the agents hinders the functioning of the neoclassical market. In these circumstances, transactions have a cost since markets are costly and prices are imperfect. The objective of the firm performing a function of governance is therefore the minimisation of transaction costs. According to this approach, the costs of internal co-ordination of economic activities, the costs of production and the costs of purchasing the productive factors on the market should be **jointly** taken into account in order to determine the size and the structure of a firm.

In this paper we want to regard firms, markets and mixed modes as **alternative means of organization** and to consider the allocation of economic activity among them as a **decision variable**. The advantage of this approach seems to us the possibility of identifying and defining a wide range of organizational structures, which are heterogeneous, within a homogeneous analytical framework.

In the next section of this chapter we present the main reasons for the existence of a continuum of organizational forms, passing from the competitive market, to intermediate structures and finally to vertically integrated firms. In contrast to the traditional industrial organization approach adopted by Williamson, (1967, 1975, 1986) we shall emphasize the role of the intermediate forms, which have only recently stopped being considered as second best alternatives to 'make or buy' solutions.

In the following section, we put forward a few hypotheses about the different effects on technology, products, learning, etc., in the Least Developed Countries (LDCs), of a very wide variety of organizational forms, instead of the classic alternative between 'make or buy'. The basic idea, derived from analysis of the few contributions which exist on this topic, is the relevance of new forms of co-operative agreements for the development of local industrial capabilities. In this paper empirical testing has not been attempted because suitable data are not currently available. The paper therefore concludes with a call for stronger empirical evidence to test the proposed approach.

The Framework

The 'make or buy' alternative

Arguments for the existence of other economic institutions besides the market can be traced back to Coase (1937), who studied the firm as an organizational form of economic activities, which provided an alternative to the market. He stated that 'the distinguishing mark of the firm is the supersession of the price mechanism' (Coase, 1937), in other words, he defined the firm as a way of organizing production in which the market system is suppressed and resource allocation is accomplished instead by authority and direction.

Williamson (1967, 1975) provides a critical development of Coase's analysis to illustrate the way in which the activities undertaken within the economic system are divided up between firms. According to him, the task of the hierarchy is to transmit orders down to the lowest level, (and to receive information back), but only a proportion of such orders can actually be transmitted at each stage in the hierarchy. The reasons for 'control loss' are the distortion that occurs in communicating across successive hierarchical levels and the loss of detail which is due to the acquisition of additional information in the process.

If high organization costs can lead to a situation of hierarchy failure, we must also take account of the costs of market failure. According to Williamson, the market fails as a mode of co-ordination, allocation and monitoring of productive inputs because; firstly, it is costly to measure the dimensions of the good or service transacted; secondly, it is costly to enforce the terms of exchange, and finally, the measurements are very often imperfect. All these factors cause an increase in transaction costs. Therefore, according to transaction-cost economics, the trade-off between the transaction and organization costs is the main explanation of organizational variety in the economic system. The market approach is most suitable when certainty prevails, the items involved are not specific and transactions are random; whereas a hierarchical organization works more effectively in situations of uncertainty, non-standardized products and when more recurrent and regular transactions take place (Silva, 1985).

The choice between firm and market governance structure is therefore determined by transaction costs which can vary depending on:

- the specific item being exchanged;
- the frequency with which they recur;
- the uncertainty to which they are subject; and finally,
- the degree to which they are supported by investments in durable, transaction-specific assets (Williamson, 1986).

Technology delimits the feasible governance set: economising on transaction costs is accomplished by effecting a discriminating match between transactions and governance structures within the feasible set.

If we assume the production costs are fixed, the organization structure should be comprised of: the market, in the case of very high costs of organization and very low costs of transactions; or the vertically integrated firm, in the case of very high costs of transaction and very low costs of organization.

Moreover, firms and markets are only two possibilities, indeed polar extremes, on a governance continuum that includes joint ventures, franchising, and a variety of complex contracting forms (Richardson, 1972). The boundaries between markets and enterprises are becoming increasingly difficult to identify, spawning a wide range of intermediate, non-market, non-hierarchy situations. As a result new forms of organization as quasi-firm or quasi-market structures become more and more prevalent. These networks of quasi-integrated relationships organize production processes within protected markets and long-term contracts, when neither integrated firm nor market are suitable (Antonelli, 1987). The trend towards these intermediate forms materializes in more co-operative agreements and in new forms of inter-firm linkages, namely setting up co-operative relations of a 'make together' or 'buy together' type, as opposed to the traditional 'make or buy' trade-off.

Between market and hierarchy: new forms of inter-firm linkages

According to the kind of transaction costs approach used by Williamson (presented in the previous section), intermediate forms of organization are possible, but they represent a second-best solution in the cases of market and hierarchy failure.

In this section, we want instead to define the intermediate strategies as possible long-term optimal solutions not only according to the minimisation of transaction or organization costs, but also according to the characteristics of products, technologies, economic and institutional environments as well as the size and structure of the firm, just like the market and hierarchy alternatives. As Richardson (1972) said:

> '... what confronts us is a continuum passing from transactions, such as those organized on commodity, intermediate areas in which there are linkages of traditional connection and goodwill, and finally to those complex and inter-locking clusters, groups and alliances which represent co-operation fully and formally developed'. (Richardson, 1972, p.887)

According to Richardson, firms involve an indefinitely large number of activities which have to be carried out by organizations with appropriate capabilities. The existence of scale economies and the requirement of specialized capabilities explain the diffusion of co-operation agreements to co-ordinate complementary, but dissimilar, activities.

The diffusion of these forms of agreement is a new phenomenon. Nevertheless, we believe that there is not a sharp differentiation from traditional forms of behaviour, but rather a continuum of complex behaviour with a prevalence of non-traditional forms of co-operation. As has been rightly stated (Chesnais, 1987), it is fairly rare that all the new elements, in economic and legal terms, of the new forms of agreement (subdivision of risk, effective sharing in management, pooling of technical and financial resources) are found clearly in all cases simultaneously. The sharing of risks and benefits is often unequal, and traditional aspects of decentralization of production, production licenses, agreement on division of markets in the respective countries, are often found combined in the agreement.

Regarding type of agreements reached between organizations of both equity and non-equity kind, we can identify the following (Oman, 1980):

- **technological co-operation**: by this we mean the transfer of technology between companies. The technology may be leased or transmitted, may include know-how, trademarks, patents and may also call for the training of local personnel;

- **product co-operation**: this concerns agreements for developing new products;

- **participation**: this is taken as an acquisition of a minority equity interest;

- **control**: this is taken as the acquisition of the majority equity interest;

- **joint venture**: the creation of a new company, while the contracting companies maintain their independence;

- **commercial agreement**: a traditional agreement for distributing and marketing products;

- **turnkey operation**: this is taken as a setup of plants and equipment by the supplier in the host country. It sometimes also calls for the extensive training of local personnel and for assistance in the commercial phase;

- **management contracts**: an arrangement under which operational control of an enterprise is vested by contract in a separate enterprise which performs the necessary managerial function;

- **subcontracting**: this normally involves a principal placing orders with a subcontractor to produce components or assemble finished products with inputs provided by the principal.

There are several reasons indicated in the literature to explain the growth of these intermediate forms of organization which differ from the minimisation of transaction or organization costs in the model posited by Williamson. Firstly, firms look for complementary assets; technological, commercial, financial or productive. It is important to emphasize the complementarity aspect: the existence of know-how within the firm is in fact a necessary condition for the success of the agreement. When the 'local' capability is very scarce, the transfer of knowledge is usually very low (as is often the case in LDCs).

Secondly, another reason for the growth of intermediate forms of organization can be the adoption of a technological 'catching up' strategy. In order to reduce the technological gap, it can be easier, quicker and more efficient to find the innovations on the market rather than to develop them oneself. If imitation is relatively easy, then imitators may enter the market, modifying the product and finding a market niche for their products (Teece, 1986).

Thirdly, firms can co-operate in order to reduce risk and costs of new activities. This is particularly true in the case of new research activities with an unknown probability of success.

Fourthly, firms tend to widen their market through diversification of products which can be achieved through co-operation. Furthermore they can acquire knowledge about different markets from local partners.

Fifthly, the achievement of economies of scale (Mariti and Smiley, 1983) and economies of scope (Teece, 1980) can be a reason for an agreement.

Finally, co-operative agreements can be induced by national regulations in order to promote co-operation between local firms and foreign investors and among local firms.

To conclude this section, we emphasize two further points concerning co-operative agreements: a negative aspect and a positive one. On the negative side, co-operative agreements save transaction and organization costs, but they generate integration costs. The integration between very different organizations can be very costly and can create problems of communication. Finally, and on the positive side, agreements between firms are very often asymmetrical, but very often perform a learning role where the minority partner can learn, gain experience, and self-confidence from the majority partner, which may finally lead to independence. In our framework we take into account the possibility of the evolution in any possible direction of the relationships between firms.

Different Forms of Organization in LDCs

Make, buy, or make together?

Within the theoretical framework set up in the first part of this chapter, we want to explore the effects of different forms of industrial organization in LDCs on technology, products, learning, etc.. The hypotheses put forward below are not meant to refer to any particular typology of developing countries. We know that within the developing world, a very wide range of situations exists, but at this stage of the work we simply have in mind a very generalised LDC in which industry is trying to take off.

We will firstly discuss the two alternative situations stressed by Williamson - make or buy - and then show why, in our view, these two organizational forms are not the two optimal solutions in LDCs. The increasing diffusion of new forms of inter-firm linkages seems to us to provide particularly helpful ways of enabling locally-based industrial development in LDCs to push forward.

According to Williamson, as noted above, the reasons for market failure are bounded rationality, opportunism, uncertainty/complexity of the environment and the small number of economic actors. In LDCs all these conditions play a particularly significant role.

Bounded rationality in LDCs is highly visible owing to the lack of information and the low capability in these countries to process it. Information is: (a) very often unevenly distributed among the actors; (b) often acquired only at very high cost; (c) almost always incomplete, in the sense that any actor's information set is smaller than the total information set in existence; (d) very often to some extent implicit rather than explicit; and finally (e) acquired, assimilated, stored and recalled in ways that require the operation of complex organizational practices; processes which seldom occur in LDCs outside multinational or, sometimes, large national firms (Fransman, 1986).

Opportunism is a very common problem in LDCs. In order to realize individual gains through transactions, rent-seeking behaviour is widespread. Flows of information are also very often distorted, because of individual and personal interests.

Uncertainty/Complexity of environment refers to the problems of collecting information. We have already stressed the very strict national or local regulations for industrial intervention which are sometimes implemented in LDCs, but additionally, foreign investors' limited knowledge of the local economic, social and political mechanisms which exist.

The existence of only a small number of economic actors leads to a limited availability of services, particularly of advanced services such as managerial, marketing, advertising, engineering, technological, and informational services, a limited availability of skills, particularly technological and managerial skills, together with limited competition within industrial sectors.

From what has been said so far we can conclude that LDCs' markets exhibit gross imperfections which compel their actors to resort to other means of achieving the allocation of resources.

The vertically integrated firm is the alternative suggested by Williamson in the case of market failure. In LDCs the imperfections of the markets frequently favour a hierarchical solution to the allocation problems by internalization of markets within multinational firms. In this case a whole bundle of resources, consisting of capital, technological skills and a wide range of rights, usually covering rights to produce, market and develop products, is transferred inside the firm rather than through external markets. The internalization permits effective parental control, avoids market uncertainties, and guarantees standardized quality control of products (Buckley, 1983). Nevertheless, multinational firms can be attracted by some intermediate forms of organization, such as joint ventures with local partners, because this can ease access to the local market and the resources of the host country. There is also the possibility of using already established infrastructures such as distribution networks, and of taking advantage of the lower wages paid in local firms. The acquisition of general knowledge of political and administrative mechanisms and the possibility of political preferment can also induce multinationals to enter into joint ventures with local firms.

As might by expected, local vertically integrated firms represent another possible alternative in the case of market failure. However, to undertake every single stage of the production cycle involves very high costs, in order to set up new facilities or to take over other independent firms. Moreover, vertically integrated firms are characterized by low flexibility, in response to market fluctuations and peak demands, without maintaining redundant capacity during off-peak periods. They also tend to be characterized by higher wages and a more unionized workforce than is the smaller firm. The search for inter-firm linkages, in contrast, can also be explained by the limited availability of certain skills or expertise within the firm and by the necessity to acquire additional information and managerial and technological knowledge.

A second solution to the imperfections of the market is represented by the wide array of in-between or intermediate situations when, in the case of partial-market failure, we have extra-market linkages which by-pass pure market relations, but do not involve complete internalization. It is important to emphasize that these may involve inter-firm linkages between local and foreign firms as well as between local firms themselves.

A schematic view of the organizational continuum is illustrated in Table 5.1. The columns represent a continuous spectrum of the various organizational forms available to firms, ranging from a full vertical integration (make) to the market (buy), across several intermediate accord-based forms (make together). The entries in the cells of Table 5.1 indicate the main effects of these organizational forms, which we wish to explore in the context of developing countries.

Table 5.1

The Organizational Continuum

ORGANIZATIONAL FORMS EFFECTS	VERTICAL INTEGRATED FIRM	EQUITY AGREEMENTS ie. JOINT VENTURES	TECHNOLOGY COOPERATION	PRODUCT COOPERATION	COMM. AGREEMENTS	MANAGEMENT CONTRACTS	SUB-CONTRACTS	TURNKEY CONTRACTS	BUY MARKET
Sharing and reduction of costs	NO	YES	YES ie.Egypt Mead, 84	YES	YES	YES	POSSIBLE	POSSIBLE ie. India Lall, 80	NO
Appropriate technologies	NO	POSSIBLE ie. LDC's SMEs, Lall, 84	YES					YES	NO
Appropriate products	NO	POSSIBLE, ie. LDC's SMEs, Lall, 84	YES					YES	NO
Exploitation of local capabilities	LOW	YES ie. Korea UN, 87	YES	YES	YES ie. Chile Tunisia, Watanabe, 71	YES	YES	YES	LOW
Managerial learning	NO	YES				YES ie.Cyprus IDS, 89 India Nagaraj, 84	YES ie.India Lall, 80	YES	LOW
Technological learning	LOW	YES	YES ie.Cyprus IDS, 88 India Nagaraj, 84	YES ie. India Nagaraj, 84			YES ie.India Lall, 80	YES	LOW
Bargaining power	NO	YES in the long term					YES		NO

* BLANKS: NOT RELEVANT

A first reason to enter into new forms of inter-firm linkages is to minimise the cost of investment by paying for only those specific factors or combination of factors really needed. In the case of foreign investment, for instance, it is possible to 'unpack' the set of factors traditionally incorporated in the direct investment package - financial capital, embodied or disembodied technology, management skills, access to world markets, etc. (Oman, 1980) - and examine them separately. For example, it could be cheaper to handle certain of these activities by hiring outside firms or producers to undertake them through subcontracting agreements.

From the technological point of view, extra-firm linkages give access to a considerable amount of technical know-how. There are reasons to believe that these new forms of co-operation often favour the adoption of more appropriate technologies than the traditional methods of technological transfer through foreign direct investments. For instance, in the case of the transfer of technology from small and medium size enterprises (SMEs) of developed countries to LDCs, the technologies are usually found to be closer to the economic conditions and needs of developing countries than those of large corporations. This type of firm tends to transfer techniques of production which, generally, are relatively more labour intensive than those of larger enterprises. In addition, small and medium sized enterprises seem more likely to transfer unpatented know-how and also their latest know-how than their larger counterparts. They also have a propensity to transfer technology with less emphasis on package and to accept payments and contract periods which are not so oppressive for LDCs (Herbolzheler and Ouane, 1985).

Small and medium sized firms also tend to transfer technologies for the production of intermediate or standardized products, which are appropriate for the local markets and the income levels of most developing countries. In fact, the evidence suggest that SMEs in general tend to be specialized in undifferentiated products where price is the main element in competitiveness. It could also be hypothesized that SMEs' industrial affiliates established in developing countries would be relatively more inclined to source their supplies locally rather than depending on the imports from the parent company, which is the typical behaviour of multinationals.

The emergence of Third World multinationals (LDCs' MNCs) is also characterized by the transfer of small scale technology which tends to be relatively labour-intensive, often embodied in second-hand equipment and more flexible to use than modern, large-scale, more specialized machinery coming from the first world (Lall, 1983). Third World multinationals often play an important intermediary role, by exporting technology that was originally imported from industrially more advanced countries and then adapted through experience to the needs or requirements of less advanced countries. Moreover, Third World multinationals mainly produce labour-intensive and standardized products, such as textiles, footwear and electrical instruments. In certain cases, and in comparison, Third World firms may develop differentiated products which compete with the branded products of a developed country. The competitive elements can be small innovations adapting existing products to a particular market, or in bringing market policy closely in line with the demands of local buyers.

More generally, we can argue that the new forms of co-operative agreements are based to a greater extent on the exploitation of local industrial capabilities than the traditional vertically integrated firm. Firms may recognize that they do not have the skills, either technical or managerial, to do all the work themselves and may find it easiest to look for them on the local market. The external environment plays therefore a very important role in ensuring the success or otherwise of co-operative agreements. The development of a local industrial capability is therefore crucial in this process and depends on many inter-related factors such as accessibility to information, receptiveness to innovation and 'business climate', presence of production services, availability of skilled labour and management capability, efficiency of public administration and a general social consensus on the goals of any technological strategy.

Industrial capability, therefore, involves a very broad range of factors, but not all their functions need to be performed at local or regional level, or even within the country in question. Successful industrial development depends on a long-term effort to build systematically on externally generated physical, organizational and social inputs and internally on accumulated experience. Industrial capability can therefore be defined as the capability to undertake the entrepreneurial, managerial, productive and technological functions, and selection of the right mix between external and internal inputs. In the case of foreign investment, co-operative agreements involving a local partner therefore tend to fit into the local vocations productives, or typical sectors of the local industry. This association implies the existence of local technological know-how (very important in the phase of assimilation of the new techniques transferred) and a knowledge of the market, which is mutually beneficial and allows commercial and other types of co-operation to flourish.

The process of assimilating and reproducing technology and new products, both from local firms and from abroad, in fact requires acquiring firms to solve numerous problems, the answers to which are not always given by the seller of the technology. The assimilation and reproduction of technology, therefore, involves a process of technological change, however minor. Technical effort is needed not just to change a given technology or to introduce a new one, but also to master it in order to make it operative and efficient in a new environment. This mastery of technology requires participation in the learning process. The sharing of risks and responsibilities, implied by co-operative agreements, has a very important effect in promoting learning among local partners. In the case of subcontracting, for instance, parent firms encourage their subcontractors to adopt the most efficient and up-to-date manufacturing methods. This often raises the optimum scale of production to a level exceeding the parent firm's own requirements. In such cases, subcontractors can start to supply the same components to the parent firm's competitors, increasing their degree of independence (Watanabe, 1971, 1972). In other cases, subcontractors can introduce small innovations to their products in order to sell them in a different market niche. Co-operative agreements are not limited to technology transfer but also allow a relevant transfer of entrepreneurial and managerial skills, promoting an additional learning effect within the partner firm's management.

The question is whether learning increases the degree of effective control and bargaining power (Barba Navaretti, 1987): particularly relevant when the partnership involves inward foreign investment. These new forms of foreign investment, by definition, involve co-ownership or even outright ownership of the investment project or enterprise by the host-country partners. But does this trend towards change in ownership patterns through co-operative agreements mean that a corresponding shift in effective control from foreign investors to host countries is taking place? Control in this sense means the exercise of decision-making powers in such vital matters as budget, expansion and development programmes, appointment of top management, pricing, marketing, re-organization and procurement of equipment (Oman, 1980). A proper test of the degree to which host countries attain effective control through the new forms of co-operation is whether they effectively transfer the power to make or influence the critical decisions on these specific matters to the host country. It seems reasonable to expect an effective increase in bargaining power and control in the key functions in the longer term, due to the learning effect and to the strengthening of the local industrial capability.

From what has been said so far, it is possible to conclude that the main advantages of the introduction of the new forms of inter-firm linkages for LDCs' firms are the minimisation of costs of investments, the adoption of appropriate technologies, the development of appropriate products, the exploitation of local capabilities, the learning effect and finally an increase in bargaining power and control over the local industry. In the next section we will analyze some empirical studies illustrating these arguments.

Some empirical evidence

It is hard to find empirical evidence to support these speculations' about the main effects of the diffusion of the new co-operative agreements among firms located in LDCs. The adoption of these forms of organization is quite recent and research has not yet begun, with the exception of a few case studies. Nevertheless, the fragmentary evidence available seems to confirm the relevance and general direction of the effects of co-operation discussed above.

A very interesting analysis of inter-firm linkages in two truck manufacturing companies in India has been carried out by Lall (1980). He found significant linkages between both firms and their suppliers. Both firms helped to launch several new suppliers in order to obtain components more cheaply than they could manufacture them in-house. They offered their suppliers technical, informational and managerial assistance as well as the assurance of a stable market. Furthermore, they offered their assistance in procurement of raw materials and provision of grants and loans.

The importance attributed to the growth of subcontracting in India is also illustrated by the establishment of a large network of Small Industry Service Institutes for technical assistance, dissemination of information, training, market research, etc.. Since the development of small industries and subcontracting are State objectives, a number of promotional agencies have come into existence offering a wide range of specialised services; State financial corporations, in general, provide long-term loans and risk capital; State industrial development corporations supply raw materials, technical and managerial assistance (Nagaraj, 1984).

In Chile and Tunisia, too, subcontracting has become a State issue. Institutes for the promotion of subcontracting very similar to the French bourses de sous-traitance have been established. Their objectives are:

- to increase productivity by obtaining larger orders;
- to rationalise investments;
- to secure as much subcontracting work in the region as possible;
- to co-ordinate the resources of individual firms to meet large orders.

They also provide an information, managerial and technical guidance service (Watanabe, 1971).

An example of the development of service centres to provide mechanized and specialized services is described by Mead (1984) with regard to small furniture manufacturers in Cairo. These small producers generally own no machinery, but instead arrange to have their wood cut, stopped, drilled or planed at woodworking machine shops scattered throughout the city.

The necessity to organise on a collective basis the supply of technological, financial, informational and managerial services has also been emphasized in Cyprus (IDS, 1988). The establishment of associations of industrialists to carry out collective services in collaboration with public centres for technical and financial assistance is considered a necessary support to an industrialization model based on small scale enterprises. The aim is to support small firms mainly in the food, clothing, footwear, furniture and metal-working sectors, often working as subcontractors to foreign firms. With the implementation of this network of assistance, Cyprus' firms should increase their independence and their capability to carry out their own production for their own markets.

All the previous examples show a lively interest in small firms at the State level. The common pattern is the emphasis on the development of extra-market linkages between firms and between firms and national or local agencies. The common idea is that an industrial tissue of small and medium sized firms, specializing in the different production phases or in differentiated products of the same industry, will prove to be highly efficient in promoting very rapid imitation rates in process and organizational innovations (Camagni and Rabellotti, 1988). The cases of the so-called system areas, mainly located in the Third Italy (the North-Eastern and Central part of the country), like Prato (textiles), Sassuolo (ceramics), Udine (chairs), Ancona (furniture and shoes) and many others, are very significant examples of this kind of spatial organization of economic activities. An additional type of firm is sometimes found in these areas: producers of machinery necessary for the production of the system area's specialized commodity.

Two further characteristics of this kind of industrial structure are spatial concentration and internal cohesion within local society. The Italian system areas are characterized by a high frequency of informal agreements and low transaction costs as a result of constant contact-making, information availability and regular relationships. In LDCs the strongest cases of these forces at work arise in situations involving ethnic, racial or tribal groups within an alien society (for example, Chinese in Vietnam and Hausa in Western Nigeria) (Mead, 1984). Informal networks of information keep participants well informed about people, markets and new business opportunities. The accepted rules of behaviour are widely agreed upon in the community, and are enforced by a network of sanctions, both positive and negative, economic and non-economic. The system is well suited to minimise both transaction costs and the risk of opportunistic behaviour.

In contrast, an oligopolistic structure may be highly innovative within itself, but it seldom provokes spin-off effects into the external environment, as the main information and service flows take place through internal channels.

Another interesting example of a developing country that has made selective use of foreign direct investment, together with a variety of other means of securing access to foreign technology, is the Republic of Korea. After a considerable influx of foreign (mainly Japanese) direct investments in the 1960s, the Korean Government adopted a more restrictive policy towards such investments in 1973. The measures established were aimed at promoting joint ventures between local and foreign firms as the favoured form of foreign investment in the economy. By 1980, over 60 per cent of total foreign direct investment in Korea was in 50:50 and minority foreign-owned joint venture firms. Technology was also acquired through a variety of non-equity channels such as machinery imports, turnkey contracts, the copying of foreign products, and through specifications and advice provided by customers located abroad (UN, 1987).

The low degree of reliance upon the local suppliers of foreign multinationals is confirmed by a study on the electronics industry in Singapore (Fong and Lim, 1977). According to this study, American firms were found to buy less than 10 per cent of their material inputs locally because they were 'locked in' to the world-wide supply network of the vertically integrated parent company. On the other hand, European firms had a higher degree of reliance on local supplies (between 40 per cent and 50 per cent of their inputs) because European sourced inputs are more expensive.

Finally, it is interesting to sketch the main results of a study on new forms of investment in LDCs made in India (Lall, 1984). Firstly, the local firms undertaking the new forms of co-operation are generally larger than the standard Indian firm, most are long-established and are among the most dynamic and progressive firms in their industrial sector. Moreover, the pressure for import substitution has forced Indian firms to generate or adopt new technology in order to adapt products, raw materials and processes to meet the increasingly sophisticated needs of the market. Finally, the Indian intervention is not limited to low-technology, small-scale, labour-intensive sectors, like textiles or food processing, but also takes place in more advanced, capital- and skill-intensive sectors, producing sophisticated differentiated products like chemicals and machine tools.

Concluding Remarks

The choice between different forms of industrial organization must be taken into account in the formulation of industrial development strategies for LDCs. The establishment of a vertically integrated firm, the creation of a joint-venture or the adoption of a technological agreement have different effects on industrial development.

On the one hand, the industrial structure of LDCs requires managerial, technological, commercial and financial assistance. On the other, LDCs have to absorb industrial knowledge in order to develop a local industrial capability. This leads to the necessity of adopting methods of transfer which an effective impact on local economic systems and the exploitation of existing local capabilities.

In countries characterized by a lack of physical and human capital, co-operation between firms is particularly important in order to increase the endowment of resources and to favour their exchange.

We have seen that co-operation among firms and between firms and other institutions can take many different forms and have different impacts on the economic system. It is important to know how the different forms of organization are related to the capacity of the system to develop and to its dynamism, in order to enable firms to overcome their limitations by moving into new markets or more complex production patterns, which overcome their own internal managerial and technological constraints.

In order to assess these impacts properly there is a need for further research and the collection of relevant data. The collection of this data will not be easy because co-operation agreements involve not only technological and financial flows, but also other issues which are very difficult to quantify, such as learning, bargaining power, development of local industrial capabilities, etc. One suggested way forward is through case studies which will enable us to verify the validity of the proposed framework for analysis.

Acknowledgements

I wish to thank Frances Stewart for her supervision and invaluable suggestions. I have also benefited greatly from the comments of Carlo Pietrobelli and the participants in the seminar held in the Institute of Economics and Statistics, Oxford, England in May 1988.

References

Antonelli, C. (1987) Dall'economia industriale all'organizzazione industriale, Economia Politica, August, (2), 277-320.

Barba Navaretti, G. (1987) Control Sharing Foreign Investments in Developing Countries: the Long Term Transfer of Capabilities, M.Phil Thesis, Oxford University.

Buckley, P.J. (1983) New forms of international cooperation: a survey of the literature with special reference to North-South technology transfer,

Aussenwirtchaft, 38, 195-222.

Camagni, R. and Rabellotti, R. (1988) Technology, Innovation and Industrial Structure in the Textile Industry in Italy, paper presented at the Conference on 'The application of new technologies in existing industries: prospects for regional industrial regeneration and employment in Europe', University of Newcastle upon Tyne, 23rd-25th March.

Chesnais, F. (1987) Les accords de cooperation technologiques et les choix des enterprises europeennes: le cas des industries de haute technologie dans une contexte mondial de turbolence economique, Europrospective, Atelier 3, cooperations technologiques europeennes.

Coase, R.H. (1937) The Nature of the Firm, Economica, 4, 386-405.

Fong, E.P. and Lim, L. (1977) The electronics industry in Singapore, Economic Research Centre, University of Singapore.

Fransman, M. (1986) Technology and Economic Development, Wheatsheaf Books, Brighton.

Herbolzheimer, E. and Ouane, H. (1985) The transfer of technology to developing countries by small and medium-sized enterprises of developing countries, Trade and Development, 6, 131-148.

Institute of Development Studies (1988) Cyprus industrial and technology strategy, reports for UNDP, Brighton.

Lall, S. (1980) Vertical Inter-Firm Linkages in LDCs: An Empirical Study, Oxford Bulletin of Economics and Statistics, 42, 2, 203-226.

Lall, S. (ed.) (1983) The new multinationals, John Wiley, London.

Lall, S. (1984) New Forms of Investment in Developing Countries by India, OECD Working Paper, Paris.

Mariti, P. and Smiley, H. (1983) Co-operative agreements and the organization of industry, Journal of Industrial Economics, 31(4), 437-451.

Mead, D.C. (1984) Of contracts and subcontracts: small firms in vertical dis-integrated production/distribution systems in LDCs, World Development, 12, 1095-1106.

Nagaraj, R. (1984) Sub-contracting in Indian Manufacturing Industries, Economic and Political Weekly, 19, 1435-1453.

Oman, C. (1980) The New Forms of Investment in Developing Countries a-state-of-the-art, OECD Working Paper, Paris.

Richardson, G.B. (1972) The Organization of Industry, The Economic Journal, 82 883-896.

Silva, F. (1985) Qualcosa di nuovo sulla teoria dell'impresa?, Economia Politica, April, (1), 95-133.

Teece, D.J. (1980) Economies of Scope and the Scope of the Enterprise, Journal of Economic Behaviour and Organization, 1, 223-247.

Teece, D.J. (1986) Profiting from technological innovation: Implications for integration, collaboration, licensing and public policy, Research Policy, 15, 285-305.

United Nations Centre of Transnational Corporations (1987) Transnational Corporations and Technology Transfer: Effects and Policy Issues, United Nations, New York.

Watanabe, S. (1971) Subcontracting, Industrialization and Employment Creation, International Labour Review, 104, 51-76.

Watanabe, S. (1972) International Subcontracting, Employment and Skill Promotion, International Labour Review, 105, 425-449.

Williamson, O.E. (1967) Hierarchical control and optimum firm size, The Journal of Political Economy, 75, 123-138.

Williamson, O.E. (1975) Markets and Hierarchies: Analysis and Antitrust Implications, The Free Press, New York.

Williamson, O.E. (1986) Vertical integration and related variations on a transaction-cost economics theme, in Stiglitz, J.E. and Mathewson, G.F. (1986) New Developments in the Analysis of Market Structure, MacMillan Press, London.

6. THE IMPLICATIONS OF MONOPSONY FOR REGIONAL TECHNOLOGICAL CHANGE

Michael Bradfield

Introduction

Much of the economic literature, including that dealing with technological change, implicitly assumes competitive markets. To the extent that the literature introduces market imperfections, it focuses on the product market, largely ignoring factor markets. For example, Scherer (1980) devotes less than two per cent of his study of market structure to the analysis of monopsony. With respect to the impact of product market imperfections on technological change, the academic community is divided. The followers of Schumpeter argue that the scale advantages of oligopoly are required to support research and to reap its benefits before competitors have a chance to imitate innovations. Others argue that oligopoly power reduces the incentive to innovate because the excess profits generated by market power reduce the incentive for efficiency. This paper considers the impact of monopsony on the incentive to adopt technological change, focusing on process rather than product innovation.

Under the assumption of competitive factor markets, regional differences in factor rewards are attributed to differences in productivity, often because of assumed lags in the adoption of new technologies in peripheral regions (Batra and Scully, 1972; Macdonald 1985, Vol III p. 203). The policies recommended are either to move labour to where the wage (and therefore the assumed productivity) is higher, or to stimulate the more rapid adoption of technological change to improve productivity. But do these policies apply equally to monopsonistic markets? This is an important issue if a substantial portion of the labour force faces a monopsony. In Canada, it has been noted that more than one quarter of the population of the Atlantic region (the poorest region) lives in small fishing communities (Macdonald 1985, Vol.II p. 454) and that 37 per cent of the non-urbanized population lives in single industry communities (Department of Regional Economic Expansion, undated, p. 35).

While monopsony is often discussed with respect to labour markets (Brusco and Sabel, 1981, p. 105), it applies to other inputs as well. In Canada, it is common for competitive primary producers in a region (e.g. in fish, food, and forestry products) to have only one processor to which to sell. In many areas, a single pulp mill will have a spatial monopsony over the small woodlot owners which supply it. For instance, the Irving company's three pulp mills and 25 sawmills control an area of the province of New Brunswick and the state of Maine that is larger than the Netherlands. Similarly, fishermen may have only one fish processor near their home port. Farmers often buy their planting seed, fertilizer, and equipment on credit from the major processor in their area and sign a contract to sell their output only to that processor. In western Canada, the National Farmers Union spent years challenging Kraft Food's monopsony over dairy products. However, monopsony is not limited to rural or isolated communities. In some industries, an employers' association may bargain collectively, as is the case for crews on Great Lakes shipping, for stevedores in many Canadian ports, and for forest workers in British Columbia.

This paper compares competitive and monopsony factor markets, and shows that market structures can determine long run regional income differentials in otherwise identical regions. It also shows that the monopsonistic long run equilibrium implies profit maximizing lags in the adoption of technology by the monopsonist. The rest of the paper looks at the effects of technological change from the perspective of the competitive suppliers in a monopsonistic market for a factor of production. Section one establishes a model of long run monopsony equilibrium. Section two shows that changes in productivity may increase regional disparities and be less beneficial to monopsonized suppliers than to firms supplying to a competitive factor market. Section three looks at the implications of the factor intensity of technological change for regional linkages. Section four provides a summary and policy conclusions, including a brief discussion on the impact of monopsony on product innovation.

The Long Run Monopsony Equilibrium

To establish the long run equilibrium conditions for a monopsony, we will start with a model of a perfectly competitive industry producing product X with factor inputs of capital and factor F. The individual firm is competitive in all markets - for capital, for F, and for the product. Thus the prices of inputs and of the output are exogenous to the firm. It is also assumed that the regional industry is competitive in the capital and product markets so that variations in the entire regional industry's demand will not affect the price of capital and variations in the industry's output of X will not affect the price of X. It is only in the factor market for F that the regional industry is a sufficiently large influence that variations in demand will affect its price level, R. These assumptions are necessary to isolate the effects of a monopsony in one market (for factor F) and also are consistent with a situation in which factor F is less mobile than either capital or the final product.

The equilibrium achieved after a market becomes a monopsony is slightly more complicated than conventional monopsony analysis predicts and affects the monopsonist's timing for the introduction of technological change. If we have an industry which is buying an important input, factor F, in a competitive factor market, the factor return would be R_E with demand and supply D and S, respectively, in Figure 6.1.a. In the conventional analysis, if a monopsony took over the firms in the market, the upward sloping supply curve would generate the marginal factor cost curve, MC, and the monopsony would set an initial factor price of R_M and hire F_M units of inputs. Under monopsony there is a decline in both the return to the factor and the amount employed. Thus two otherwise identical regions would have different factor rewards and incomes if one region has a competitive factor market while that factor faces a monopsonist in the other region.

This equilibrium determined by the intersection at B of the demand and marginal factor cost curves is a short run equilibrium, taking the initial capital stock as given. It implies that the monopsonist has a lower input of factor F with the existing input of capital, and therefore suggests a decline in the monopsony's output. However, the use of isoquant-isocost analysis (Figure 6.1.b) shows that the monopsonist can increase output in the long run, to X_2. This is due to the monopsonist's concave isocost line, KF_{mon}.

The monopsonist's isocost is concave because the marginal cost of the input is rising as the monopsonist hires more of the factor F. Assuming each plant is identical, the monopsonist will divide output equally among all its plants to equate the marginal cost of X in each. As the monopsonist moves from hiring all capital (OK) to a little of F, the monopsonist pays a price for F substantially below either R_E or R_M. This means that for a given initial sacrifice of a small amount of capital the monopsonist can hire much more labour than the competitive firm can. Thus as one moves away from 0K, the monopsonist's isocost line lies beyond the competitive firm's isocost line, KF. However, as more and more of the factor F is hired, its marginal cost rises and therefore the slope of KF_{mon} becomes progressively steeper - the monopsonist can hire smaller and smaller amounts of F as it gives up more and more capital, i.e. as it moves up the region's supply curve for factor F. If the monopsonist were to hire the competitive level of input, F_e, it would pay the competitive price, R_E, and therefore face the same average cost for the factor, even though the marginal cost of the factor would be much higher. The higher marginal cost of F is indicated by the steeper slope of the monopsonist's isocost at point a in Figure 6.1.b. Thus the monopsonist's isocost intersects the competitive firm's isocost line at point a, the equilibrium point for each of the competitive plants.

The concave isocost curve makes sense when it is remembered that the slope of the isocost curve reflects the ratio of the marginal costs of the two inputs. For a

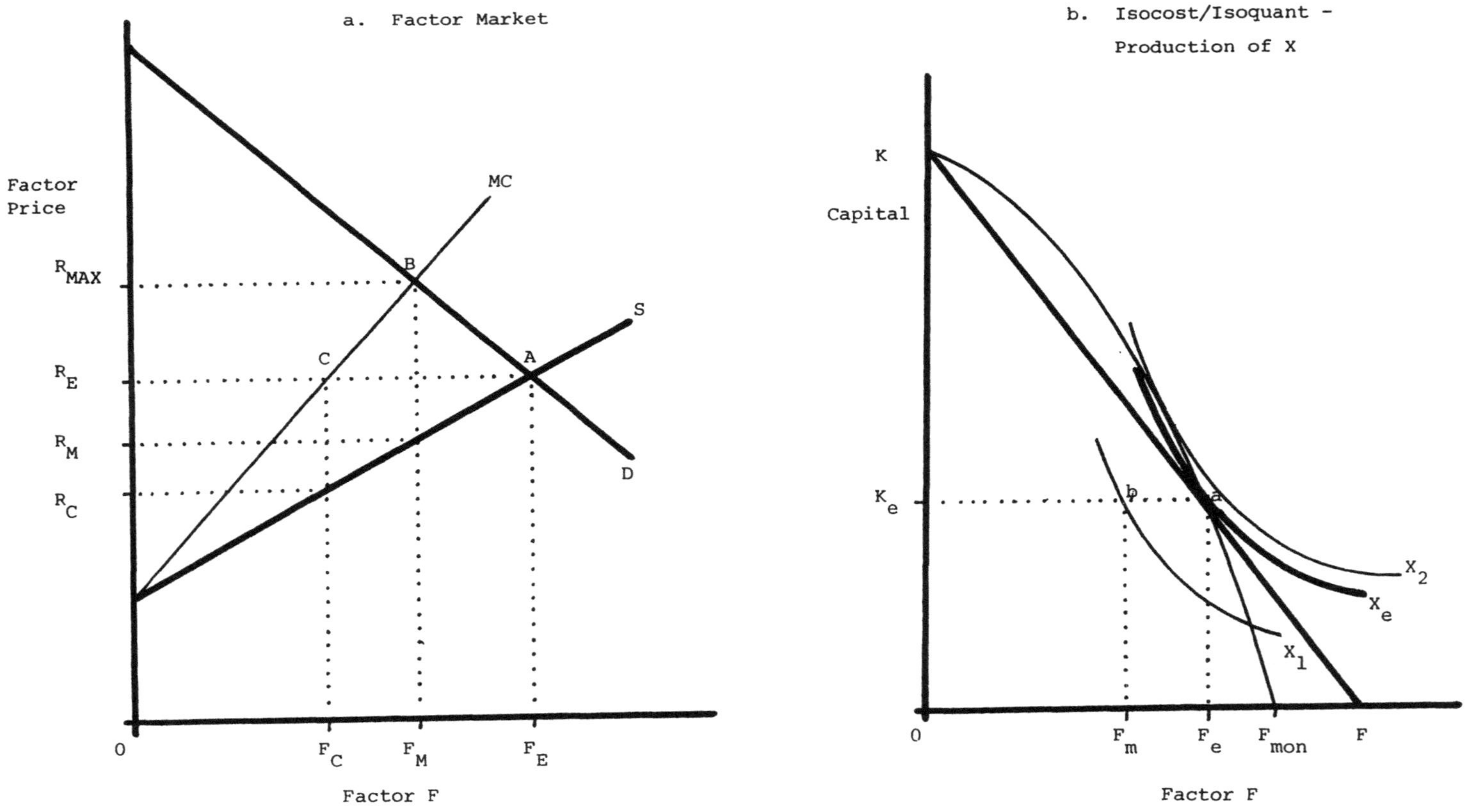

Figure 6.1: A Competitive versus a Monopsonistic Factor Market

firm operating only in competitive factor markets, as much or as little of a factor can be purchased without affecting its price. Thus the marginal cost of the factor is its price or average cost. Therefore a firm purchasing inputs competitively has an isocost curve whose slope is equal to the (constant) ratio of factor prices, because factor prices equal the marginal cost of the factors. However, the monopsonist, as the industry decision maker, faces the rising supply curve of factor F and recognizes that an increase in demand for F will drive its price up. Thus the monopsonist's marginal cost is not equal to its price and the ratio of the factor prices no longer determines the slope of the isocost curve. With a rising marginal cost of factor F, the slope of the monopsonist's isocost curve is increasing as more purchases of F drive up its price and its marginal cost.

Given the concavity of its isocost through point a, the monopsonist's long run optimum appears to be producing the higher level of output, X_2. However, if there are initially increasing then decreasing returns to scale, the long run average total cost curve would be U shaped. The competitive output X_e would be the optimal scale - the minimum long run average total cost output. Therefore the monopsonist would produce quantity X_e at point b', with F'_m units of input F, in Figure 6.2.b. Thus, the conventional level of inputs at F_M is a short run level, representing an output such as X_1 in Figure 6.1.b. In the long run, the monopsonist can decrease its expenditure level to $K'F'_{mon}$ and still produce output Xe. It is not accidental that the shift to the long run monopsony isocost curve, $K'F'_{mon}$, is depicted as a vertical shift downward, not a parallel shift inward from KL_{mon}. OK'/OK represents the reduction in the expenditure necessary to produce output X_e (and therefore indicates the amount of monopsony profits, KK'/OK, as a percentage of sales). Although each of the monopsonist's plants is spending less, they collectively face the original supply curve of labour. Therefore the initial sacrifice of a unit of capital can hire the same amount of factor F, whether the move is from OK or OK', as long as the monopsonist is obtaining financial and real capital in competitive markets (cf. Winch, 1960, p. 633).

In this model, the monopsonist's long run equilibrium involves an increase in capital inputs, as is suggested by Canadian data (Bradfield, 1988, p. 103). Since a factor's marginal physical product is a function of the capital/factor ratio, this increased input of capital implies that the marginal revenue product of factor F is increased for each level of input of F. This means that the 'demand' curve for F must move above the short run competitive demand, D, for instance to D_M. Thus the long run monopsonistic equilibrium will be determined by the intersection of the MC curve with D_M, at B'. Since the monopsonist still hires less of F than the competitive firms ($F_e > F'_m$), the long run monopsonistic factor price will be lower than the competitive equilibrium ($R_E > R'_M$), but above R_M. The monopsonist's plants obviously have a higher capital/factor ratio at b' than did the competitive

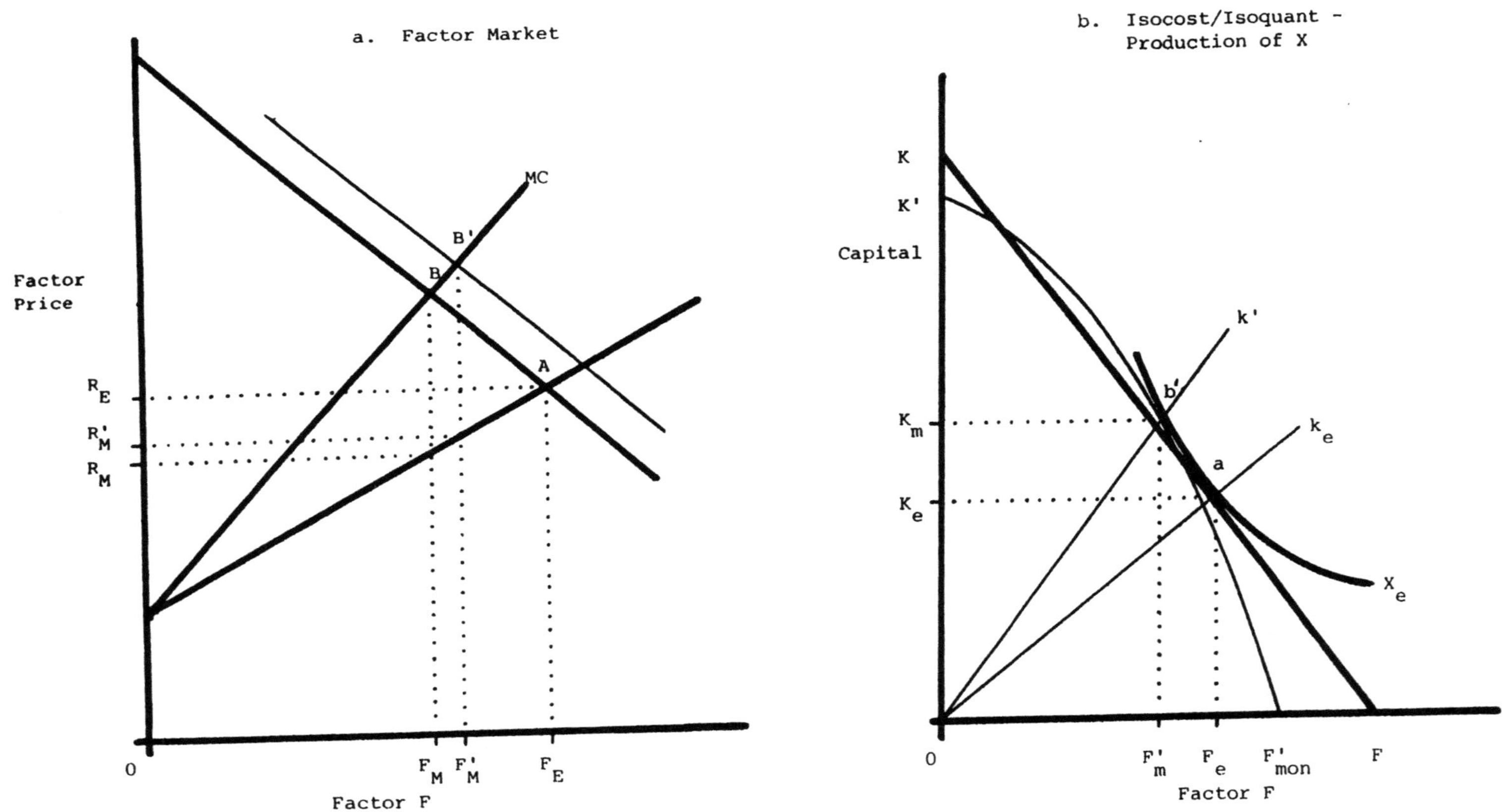

Figure 6.2: Competitive versus Monopsonistic Factor Market - Long Run

firms at a. The capital input has increased to K_m and the factor F input decreased. The tangency at b' means that the ratio of capital's and factor F's marginal products equals the ratio of their marginal costs. The higher capital/factor ratio ($k' > k_e$) implies a fall in the marginal product of capital and disequilibrium in the capital market if the cost of capital remains unchanged. That is, if the regional monopsonist is a competitive purchaser of capital and a competitive seller of product X, the competitive capital/F ratio, k_e, is the equilibrium capital/F which sets the value of the marginal physical product of both capital and factor F. The exogenous product price means that the equilibrium factor marginal revenue products are also determined. This would mean that the long run monopsony marginal revenue product curve for F would reflect the original capital/F ratio, k_e, and marginal revenue product of factor F which equals R_E. The long run factor marginal revenue product would be a horizontal line at RE and the monopsonist's equilibrium point would be where MC = R_E, i.e., at point C, which implies an input of factor F of F_C, in Figure 6.1.a., and a factor reward of R_C, below R_M.

How then do we rationalize the equilibrium at B'? It may be that the monopsonist is a profit satisficer rather than maximizer. This may reflect the social pressures felt by the major employer in the region who may have status as a 'pillar of the community'. To maintain that status, the monopsonist may feel it necessary to constrain profit maximization. Alternatively, operating at a higher than optimal scale may be a profit maximizing strategy to maintain monopsony control of factor F - if too many suppliers are cut loose, other demanders may enter the market. Finally, the capital/factor ratio k' need not be a disequilibrium if movement along the isoquant means adopting a more capital intensive production function. This possibility is developed below. The monopsony long run equilibrium, with a higher capital input than found in the comparable competitive market, can now be shown to lead to a slower adoption of new techniques by the monopsonist.

If technological change is embodied in a new capital stock, the decision to adopt it requires a comparison of the minimum average variable cost of the existing technique (since its capital costs are fixed) with the minimum average total cost of the new technique (Bradfield, 1988, pp. 217-220). If the firm wishes to switch to the new technique, it must buy the capital in which the new technique is embodied. However, to continue to use the old technique requires only variable inputs; the capital is already in place. The only relevance of the cost of the existing capital is its resale value - if the firm decides to switch to the new technique, how much can it sell off its old capital for? This resale value is then a cost of producing the first unit of output with the old capital and therefore is a part of the variable cost of producing that first unit. In this way, the value of the old capital stock affects the firm's variable costs and its decision as to when to switch to the new technique. As the old capital stock wears out, it can be expected that its maintenance, operating, and parts replacement costs will rise, raising the variable cost of its use. Eventually, these variable costs will become high enough that they can be expected to exceed the minimum average total cost of the new technique. When this occurs, the firm will switch to the new technology by replacing the old capital stock.

For the monopsony scenario of Figure 6.2, there are two factors which tend to push the monopsonist's marginal and variable costs below those of the competitive buyer. Over some range of hiring the input F, the monopsonist has a lower marginal cost for the factor (MC $< R_E$ until more than F_C is hired in Figure 6.1.a). Secondly, the greater capital inputs which raise the marginal physical product of factor F cause a corresponding fall in the marginal cost of the product X since the marginal cost of X is the marginal cost of the variable factor F divided by its marginal product. Ceteris paribus, the monopsonist will have the lower marginal cost of production with the old technology and therefore a profit maximizing incentive to delay the switch to the new technology, relative to competitive firms producing under otherwise identical conditions. Thus the market structure will not only generate regional disparities in factor returns and in incomes, it will also cause a lag in the introduction of technological change for the production of X for the low income region. However, the lag is not a determinant of the income differential but is a function of the market structure which caused the regional disparities in factor prices and incomes in the first place.

The lower return to input F may be mistaken as an indication of the lower productivity of the suppliers of the factor. This is obviously untrue since the 'firm' (worker, fisherman, small woodlot owner, farmer, etc.) which can still supply in the long run at the lower return must have lower costs than the firms in the other region receiving R_E. Alternatively, the competitive suppliers to the monopsonist may be receiving a lower return on their labour and investment relative to those supplying to the competitive factor market for F. Similarly, suppliers of factors to the monopsonized producers of F may also suffer lower returns because the lower factor price, R'_M, lowers their marginal revenue product. For these conditions to persist, there must be factor immobilities between, and perhaps within, regions.

The fundamental point is that the monopsonist has a lower incentive to adopt new production techniques because of the monopsonist's lower marginal cost with the existing technique. In the model developed in Figure 6.2.b this lower marginal cost reflects the lower-than-competitive factor cost and the higher-than-competitive capital/factor ratio. The lower marginal cost also holds true for the 'pure' monopsony case where the long run marginal revenue product of the factor F is a horizontal line at R_E. In this case, the monopsony factor payment would be R_c, below R_M, and the original capital/factor ratio maintained in the long run. Thus the monopsonist's marginal cost would be decreased by this lower factor cost ($R_c < R_M < R'_m < R_E$)

although not by a change in the capital/factor ratio. Thus, it appears that the monopsony long run equilibrium will consistently imply a diminished incentive for production technique innovation. The only exception to this would be if the long run equilibrium in Figure 6.2 itself involved a change in production technique.

If the tangency at b' in Figure 6.2.b. means that there has been a switch in technique to a more capital intensive technique ($k' > k_e$), this could imply an equilibrium in the capital market. The more capital intensive technique requires a higher capital/factor ratio at existing capital costs to maintain the equilibrium in the capital market between the cost of a unit of capital and its marginal revenue product. Thus k' need not be inconsistent with fixed capital prices to the monopsonist. To be a superior technique with higher fixed (capital) costs, the new technique must also imply lower variable costs. If newer techniques are generally capital intensive, this equilibrium interpretation of Figure 6.2.b implies a paradox. The formation of the monopsony leads to the adoption of a (newer) capital intensive technique. However the effects of this new technique and of monopsony (sub-competitive) factor payments mean decreased marginal costs and again, less pressure to have any further innovation in production technique. The paradox is that monopsony may initially stimulate a change in production technology and then retard further change.

Having concluded that the monopsonist has a profit maximizing incentive to delay technological change, relative to a competitive firm in a comparable market, let us now look at the effects of technological change occurring in the other side of the monopsony market.

'Short Run' Productivity Changes

Let us initially examine the effects of a productivity increase which affects marginal but not fixed costs. Often this is not the result of technological change, but occurs from changes in supply conditions for the firms supplying factor F. For example, it could be the effect of weather changes or variations in fish stocks which affect the cost of harvesting. We will examine the impact in the monopsony compared to the similar competitive markets of Figures 6.1 and 6.2. As a 'pure' form of technological change, this type of marginal cost shift could occur from an organizational change which kept the same capital stock, but re-arranged the assignment of labour, as with Henry Ford's shift to the assembly line. Any technological change which made variable inputs more efficient would lower marginal costs. This would include changes in labour's health or attitudes. Ironically, these are now being achieved by moving away from the monotony of the assembly line to work teams in which workers have greater participation, responsibility, and diversity in their jobs.

Assume that the marginal cost for each unit of production falls by a constant amount, aa', in Figure 6.3.b. The drop in the marginal cost by aa' means that the average total cost will fall by the same amount. The minimum average total cost level of output will not change and the market supply curve will also shift by aa', down to S' (Figure 6.3.a). The price in the competitive market for the factor will fall to R'_E; in the monopsonistic market, the factor reward will fall to R''_M.

Figure 6.3.a has been drawn so that in the competitive factor market the supply curve has shifted along the inelastic part of the demand curve and therefore total revenue to the suppliers of factor F has fallen. This is the classic short run 'farm' problem. However, this fall in total revenues does not mean that the suppliers are worse off in the short run. Although total revenues have fallen, profits have risen. Even though the price paid to the competitive supplier has dropped to R'_E, the downward sloping demand curve means that the price has not fallen by as much as costs have, i.e., $R_E R'_E < aa'$. Thus there is a short run competitive profit despite falling total revenue.

Similarly for the suppliers to the monopsony, the fall in price to R''_M will not equal the fall in average total cost, so that there will be short run profits attained. However, there are important differences, vis-a-vis the competitive solution, in terms of the absolute price fall (and its effects on relative regional incomes), of total incomes and of short run profits.

Let us begin with the fall in the price received from the monopsonist. With the same fall in marginal costs from the same increase in productivity as experienced by the firms supplying to the competitive factor market, the suppliers in the monopsony will face a larger drop in price. In Figure 6.3.a, B'HF and B"H'F' are similar triangles, as are the triangles AHE and A'H'E'. Since the supply curve has shifted down to S', we know that B"F' > B'F (the intersection points are further along MC and S' than along MC and S respectively). With B"F' > B'F, it is also true that B"H' > B'H and that H'F' > HF. With Dm parallel to D, H'A' > HA and E'H' > EH. But $EH = R_E - R'_M$ and $E'H' = R'_E - R''_M$. Thus $R'_E - R''_M > R_E - R'_M$.

Therefore the absolute difference in rewards to the supplier of factor F has increased between the two regions despite the fact that the only difference between them is in their factor market structures. This effect is even stronger if we compare relative incomes. The price of the factor has declined in both regions, but in absolute terms has fallen more to the suppliers to the monopsonist. Since the suppliers to the monopsonist were receiving the lower price before the technological

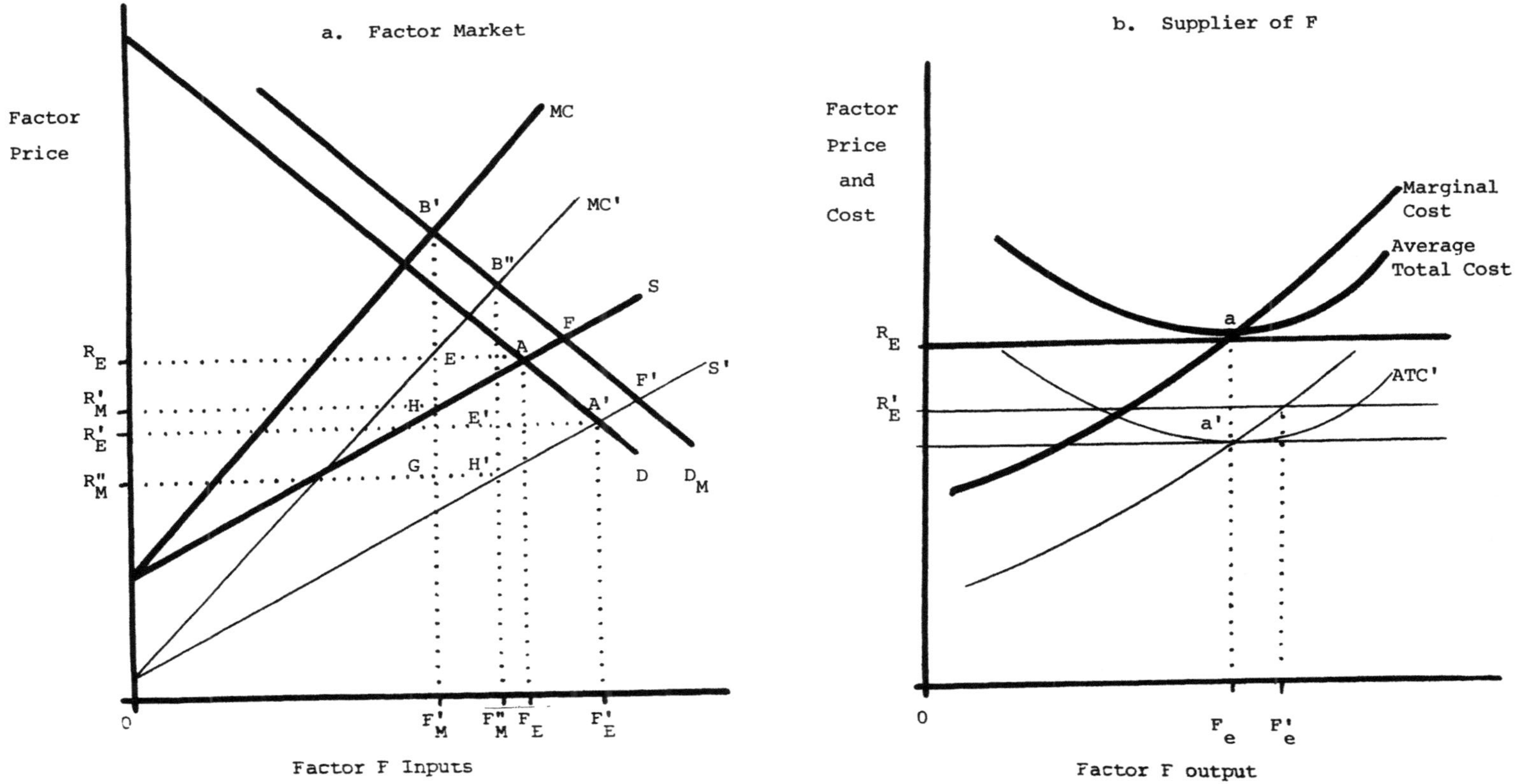

Figure 6.3: Increased Productivity/Lower Marginal Cost of Producing Factor F

change, the relative factor price differential has been increased, $R'_E/R''_M > R_E/R'_M$.

The suppliers to the monopsonist have experienced the same decline in costs but a greater decline in price relative to the suppliers in the competitive factor market. Thus the short term profits of the former must be smaller. The former must share more of the productivity gains with the monopsony buyer. This is true even though the monopsonist may be buying the factor inputs in the elastic part of the demand curve for the factor, Dm. In the initial monopsony equilibrium, the suppliers receive $OR'_M HF'_M$ for their efforts. After their marginal costs fell and the industry supply curve has shifted to S', they receive $OR''_M H'F''_M$. It may be that $F'_M GH'F''_M < R''_M R'_M HG$. It is the elasticity of supply, not of demand, that is crucial to the monopsony equilibrium. The result is that total revenue may fall for the suppliers of F even though they are supplying in the elastic part of the monopsonist's demand curve.

This lower reward to the firms supplying the monopsonist puts those firms in a position similar to the producers in a competitive market where the supply intersects the demand in the inelastic part of the demand curve. If all firms incur costs to expand supply, everyone will be worse off and some must leave the industry. In this respect, there is no reward for adopting technological change. Nonetheless, there is an incentive to do so for the individual firm. If the individual firm does not adopt the new technology as others do, the decline in price will mean losses with the old technology. The producer must then compare the falling price with their minimum average variable cost with the old technology. When the price is forced below the minimum average variable cost, the producer must make an either/or long run decision, either go out of business or adopt the new technology, knowing that some firms must ultimately leave the industry.

If this decrease in the marginal cost were both costless and permanent, then the short run excess profits would attract new suppliers of the Factor F until the price fell by the drop in average total costs, aa'. This would be true in both regions so their long run price changes would be identical. The absolute difference in regional factor rewards would not change. However, the lower original reward in the monopsonistic market would mean a higher percentage fall in the factor reward in the long run as well as the short run. Thus, absolute regional differentials would not change but percentage differentials would be widened by a given technological change.

We have assumed a constant shift in the marginal cost curve, but if the marginal cost were to decrease very little for low levels of output and much more for higher levels, the effect would be to flatten the new industry supply curve. The more elastic the supply curve at any point, the closer the monopsony solution will be to the competitive solution in terms of factor price. The steeper the new supply curve, the more the factor payments will diverge between the competitive and monopsony markets as supply shifts. In addition, there will be effects on the degree of linkage with the local economy. The following section suggests that the factor intensity of technological change may affect the shift in the marginal cost curve and therefore affect the regional impacts in the different market structures.

Factor Intensity of Technological Change

Capital Intensive

Technological change is often capital intensive - fishermen buy fish finding sonar, farmers and woodlot owners invest in more sophisticated harvesting equipment. When technological change is capital intensive, the capital/labour ratio increases at existing factor prices. Thus the tangency of the isoquant for the new ("Y") production technique with the isocost will occur closer to K on KL, at a point such as a' in Figure 6.4.a. The Y technique is more capital intensive than the initial technique since the Y technique involves a higher capital/labour ratio, i.e. $k_y = K_y/L_y > K_e/L_e = k_e$. For the new technique to be superior, it must be true that $F_{Y1} > F_e$. Since more money is being spent on capital, less is spent on labour to attain output level Y1.

Of course, the optimal scale of plant for the new technique need not involve the same level of total expenditures as the old technique. It may well involve a larger scale, in terms of increased capital inputs at the original expenditure level and an increase in the level of expenditures to take full advantage of available economies of scale. If this were the case, the isocost line would move outward to a tangency with whatever isoquant represents the new optimal scale of output. One would normally not anticipate the reverse, a reduction of the scale of expenditures with the new, capital intensive technique, although this is obviously possible.

Figure 6.4.b illustrates the effect of technological change on the firm's cost curves, assuming the level of expenditure is constant or increases, and ignoring changes in the price of labour from the decline in its demand which might accompany the shift to the new technology. Since more is being spent on capital, for the new technique to be more efficient there must be an even greater decline in expenditures on the variable input, labour. Thus the higher average fixed costs at low levels of output must eventually be more than offset by lower marginal costs, generating shifts in the marginal and average total costs indicated by MC_y and ATC_y, respectively, in Figure 6.4.b.

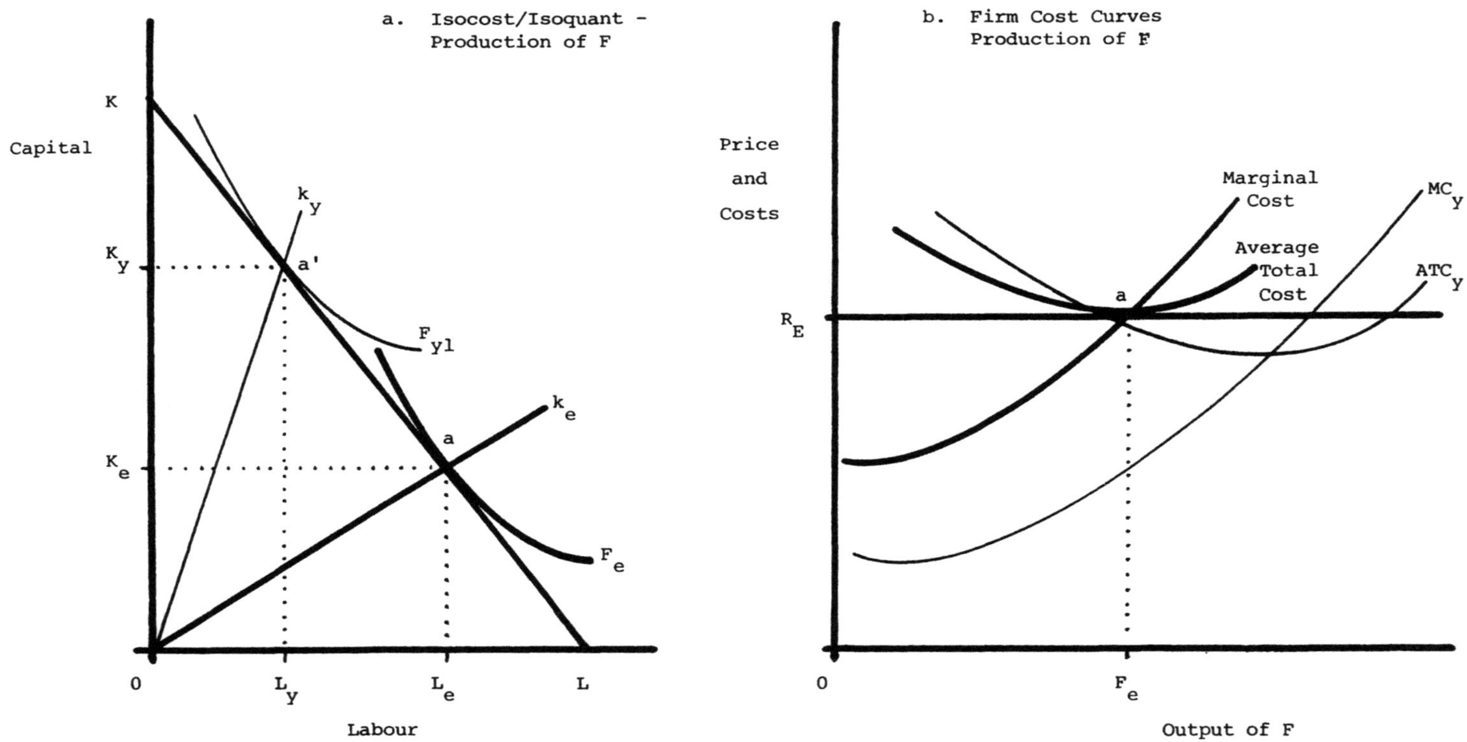

Figure 6.4: Capital Intensive Technological Change in the Production of Factor F

While the assumption of capital intensive technical change implies a downward shift in the average total cost, we do not know the nature of the shift in the marginal cost. It may shift down by a constant amount, as discussed in Figure 6.3, or by varying amounts, as indicated in Figure 6.4.b. Even if the marginal cost falls by a constant amount, this no longer means that the average total cost will fall by the same amount, because of the increase in fixed (capital) costs. The adoption of the new technique would shift the marginal cost and the industry supply curve and thus lead to lower prices, as before. However, the minimum average cost will not fall as much as marginal costs, because of the added fixed costs. As firms supplying to a monopsonist will see prices fall more than firms supplying in a competitive market (Figure 6.3.a), the rewards for innovation are lower. There is therefore less incentive to take the risks of innovation for firms supplying to a monopsony.

Moreover, a given technological change may have different effects in the two types of market. Except for technological change which reduces the slope of the marginal cost curve significantly, we would expect the capital intensive change to cut prices more in the monopsony market. This may mean that the suppliers there would find prices below the new average total cost if every supplier adopted the new technique. Because of the greater fall in the price of factor F in the monopsony case, this effect is more likely and stronger than in the competitive market. Thus capital intensive technological change is likely to require that some firms leave the industry as others adopt the new production technique. The proportion of firms leaving the industry will be greater in the monopsony than in the competitive market, because of the greater drop in prices to the monopsony.

The exit of firms represents a trade-off or dilemma. The more the new technique reduces the slope of the marginal cost curve, the closer will be the price changes generated in the two regions, i.e., the less likely technical change is to increase regional disparities in the incomes of the suppliers of factor F (and of the suppliers of inputs to the producers of F). On the other hand, the reduction in the slope of the marginal cost curve will mean, ceteris paribus, a greater rightward shift in the marginal cost and in the supply curve of factor F. This will then mean a larger displacement of firms supplying the factor F.

Capital intensive technological change has another effect which may appear neutral with respect to market structure, but may not be. Obviously, capital intensive production techniques mean a reduction in the use of other inputs, the variable inputs such as labour or raw materials. This will be true regardless of the market structure in the two regions. However, it may well be that the regions facing more monopsonistic markets are the peripheral regions, the low wage regions. This is certainly the case in Canada (Department of Regional Economic Expansion, undated). The high wage region is often the capital goods producing region. In the Canadian case, with roughly one-third of Canada's population, Ontario (the high wage region) produces almost half of Canada's manufacturing output (Bradfield, 1988, p. 117). We may therefore have a situation where the regions with the predominantly competitive factor markets are the regions which produce the capital goods. For them, the capital intensive technological change means a shift in local demand, away from labour and raw materials and toward capital goods produced in the region. Thus the factors displaced by the decline in variable inputs may be offset by those employed in the capital goods industries (although they need not be the same factors). The effect of technological change may therefore be a change in the industrial mix of the regional economy without an increase in net unemployment in the competitive high wage region.

The results will be different in the low wage region with a higher proportion of monopsonistic factor markets and a lower share of the capital goods industries. The capital intensive change means a reduction in locally supplied variable inputs and an increase in the importation of capital goods. Local linkages decline and the region becomes more dependent on outside capital. Unemployment in the region goes up and imports increase, but there is not an automatic link to increase the region's exports.

Thus technological change may have several negative impacts on a low wage region, particularly if it is also characterized by a higher proportion of monopsonistic markets. It may cause a greater exit of firms than in the competitive region, a sharper drop in factor prices and incomes, a deterioration in the region's balance of trade, and an increase in unemployment. Each of these may be interpreted in a competitive market context as more evidence that the low incomes of the lagging region are simply market signals of lower efficiency in those regions. However, we obtain these regional disparities even though we initially assumed identical conditions in the factor markets, except with respect to market structure.

Labour Intensive Technological Change

If there is labour or raw material rather than capital intensive technological change, some of the differential regional impacts will be avoided or decreased. With increased use of variable inputs and a reduction in capital inputs, we might anticipate a shift in costs curves such as represented in Figure 6.5. We assume that the 'scale effects' of the change in production function are such that the optimal output will be higher than F_e, the original equilibrium output of F. As before, it is not clear where the new marginal cost curve will lie relative to the original, but we have shown it as higher for smaller units of output. Nonetheless, if the new technique is superior, average total cost with it must be lower and

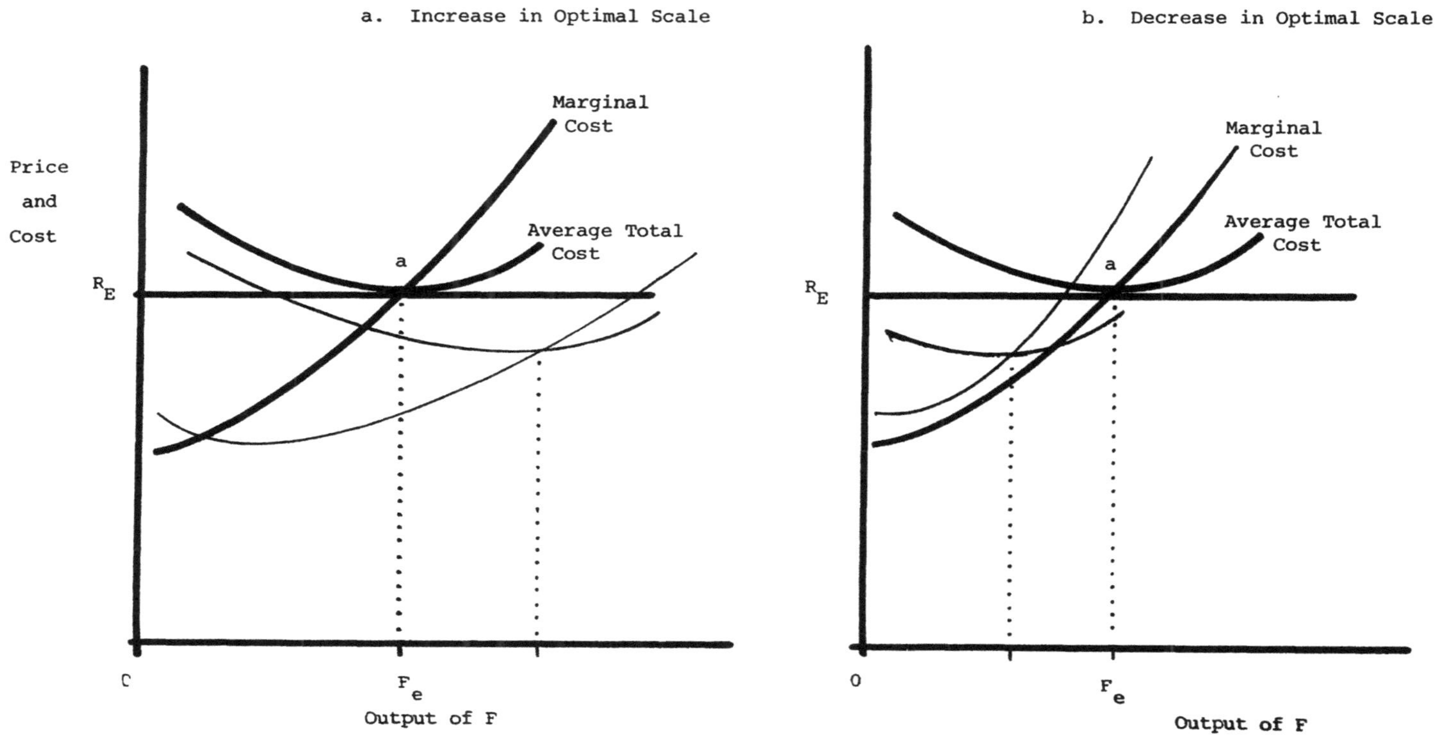

Figure 6.5: Labour Intensive Technological Change in the Production of Factor F

therefore, at least for higher units of output, the new marginal cost will lie below the old. It seems likely that labour intensive technological change will reduce the slope of the marginal cost curve since the share of production going to labour must increase. If the labour intensive technique involves a lower optimal scale than F_e, the effect on the marginal cost curve might be more like that depicted in Figure 6.5.b. The discussion will assume that the shifts of Figure 6.5.a are relevant.

If the shift of the marginal cost with the more labour intensive production technique generates a reduction in the slope of the marginal cost curve, this would offset the tendency of technological change to generate a larger increase in regional disparities with a monopsonistic market in the low wage region. Moreover, the increased proportion of variable inputs means that the low wage region will increase its local linkages and decrease its reliance on imports, under the conditions discussed above. On the other hand, it would also lead to a greater increase in the supply from each firm and therefore require that a greater number of firms exit the industry. Thus the impact of capital versus labour (or other factor) intensive technological change not only varies with the regions' industry mix but also their factor market structures.

Summary and Policy Conclusions

Several useful conclusions and directions have been developed in this paper. Two regions may be identical with respect to everything except factor market structure, but this single difference can generate regional differences in incomes. Moreover, the long run equilibrium of the monopsonist implies a lower marginal cost, delaying the introduction of new technologies by the monopsonist.

In terms of the suppliers to the monopsonist, the situation leads to differential effects from technological change. The same increase in productivity will lead to smaller benefits to the supplier in the monopsonistic market compared to those supplying in a comparable competitive market. The factor intensity of technological change also has a differential effect on the monopsonized region, as it is likely to make the industry supply curve more elastic if it is not capital intensive. A more elastic supply curve will reduce the differential in factor prices between the competitive and monopsonistic markets, reducing regional disparities. On the other hand, a capital intensive technical change will often tie the lagging region more strongly to the developed region, exacerbating both its dependence and its unemployment.

While some would argue that science and technological change are independent of spatial bias, many models conclude that innovation is selective. Both urban hierarchy (Richardson, 1975, p. 51) and centre-periphery approaches (Friedmann, 1975, pp. 278ff.) suggest that it is the centre which invents and innovates. Since the centre is likely to have high wages and easier access to capital (Richardson, 1973, p. 42), the firms there may prefer capital intensive inventions. Governments often show a bias toward 'high tech' inventions, even in low wage, primary producing regions where this paper suggests high tech is least appropriate. They apparently see high tech as a way of breaking with the past, not recognizing that it may be a very expensive way to create an enclave economy.

The emphasis on high tech may also broaden the applicability of the monopsony analysis. To attract a high tech firm may require major expenditures on human capital. For instance, to convince Pratt and Whitney to locate an engine parts plant in Nova Scotia required more than $40 million in subsidies, including at least $17 million for worker training (Bradfield, 1987). To the extent that this training is firm specific, Pratt and Whitney is the only firm demanding those skills and thereby exercises monopsony power. Thus human capital expenditures by firms or governments may create monopsony power even in an urban setting. This may be a much more important area for monopsony analysis in Europe than is the resource and spatially isolated labour model of monopsony which applies to Canada and even the U.S. (Wheat, 1986, p. 646).

This paper suggests that government policy should do more than be wary of funding high tech research or industries. It is common for governments to wish to stimulate the adoption of new technologies and to implement information and loan programmes to provide access to financial and real capital. While access to funds is often a problem for small businesses and primary producers, enabling them to borrow money for capital intensive techniques may create instability and unemployment in peripheral regions. If small primary producers have low incomes, they may be assisted far more effectively by destroying the monopsony power which is holding factor prices down. Loan Boards may be less useful than marketing boards in raising the incomes of primary producers.

It might appear that a marketing board for primary inputs or a union for workers facing a monopsonist could increase input rewards beyond the competitive level of R_E in Figure 6.1.a. The normal interpretation of 'bilateral monopoly' is that there is a range of possible factor prices, between R_M and R_{MAX} in Figure 6.6.a, in which the factor price can rise without a decline in factor employment. Where the actual factor price is set is then a function of relative bargaining strength. However, it should be clear that the monopsonist cannot pay above R_E. The negotiated factor price restores a fixed marginal cost to the monopsonist, equal to the negotiated price. This means the isocost curve becomes a straight line over the range of factor input before the factor supply curve becomes a constraint. If the

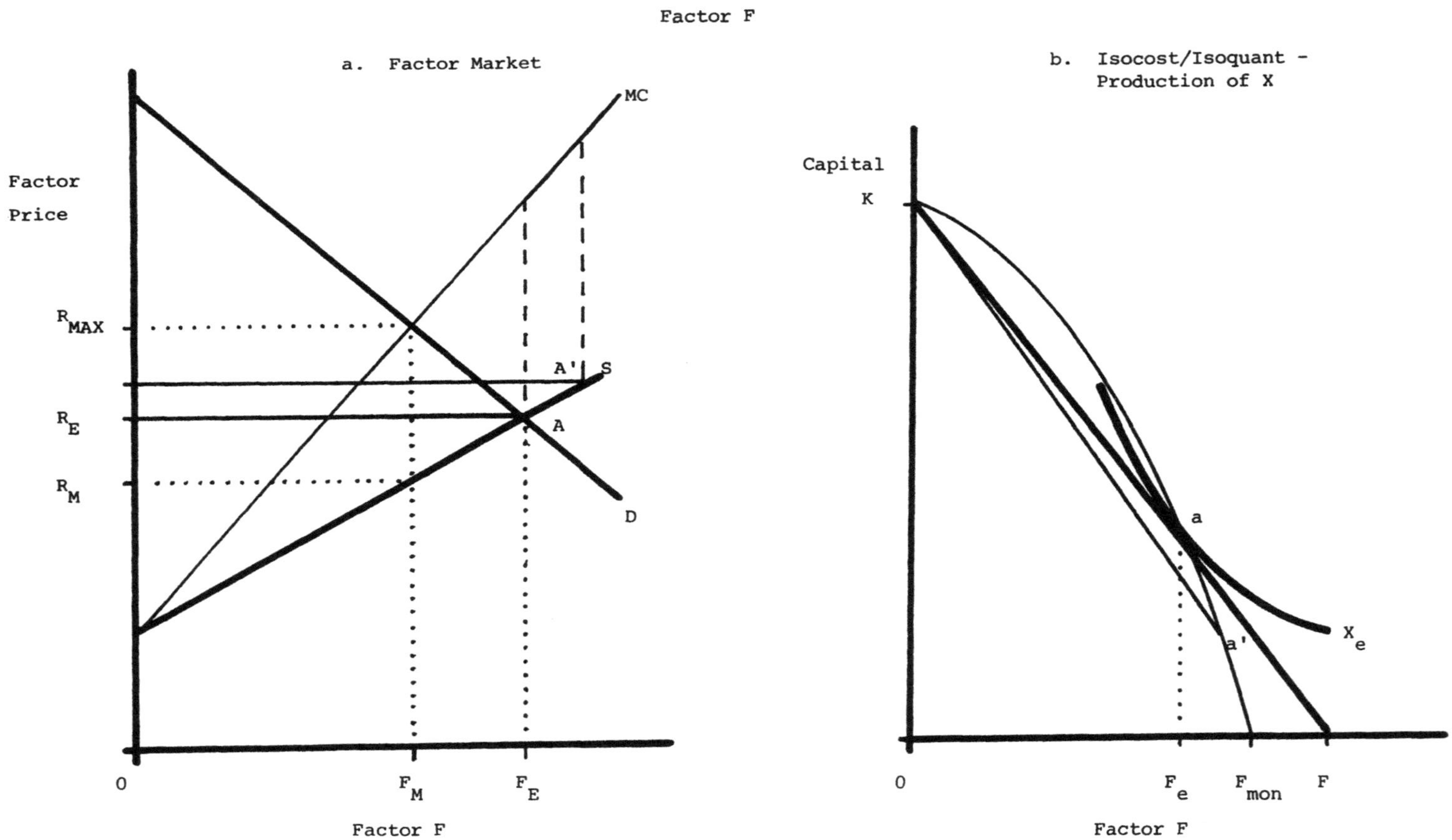

Figure 6.6: "Bilateral Monopoly" in the Market for Factor F

negotiated price were set at R_E, the monopsonist's plants would face a kinked isocost curve. Moving away from OK, it would have the same slope as KF until point a in Figure 6.1.b. After F_E units of F are employed, no more units of F are available at the negotiated price RE. If the monopsonist wishes to employ more than F_E (more than Fe in each plant), the price of F must increase and the marginal cost beyond F_E jumps to the old MC (Figure 6.6.a.) and the monopsonist's isocost would then be KaF_{mon}. The bilateral monopoly with a negotiated price of R_E removes all monopsony profits.

However, if the negotiated price were above R_E, the monopsonist's straight line isocost segment would lie inside KF and intersect KF_{mon} at a higher input of F than point F_e. In this case, the monopsonist would be unable to compete with the competitive firm as it would be impossible to produce X_e for the same expenditure of firms buying F in a competitive factor market. Thus providing countervailing power would permit an increase in factor F's reward and total income up to $R_E F_E$, but this is the upper limit. In the long run a marketing board or union could not raise the factor price above R_E without destroying the industry in that region. Nonetheless, the result of the bilateral monopoly situation can unequivocally improve the welfare of the suppliers of factor F. This is not a certain result with technological change.

Of course, this is not to take a Luddite view of technological change. It is simply to warn that there may be market phenomena which explain or contribute to regional disparities in both incomes and the speed of innovation. Policies which assume perfectly competitive markets may compound the problems of lagging regions. They may also lead both economists and policy makers to interpret reality in such a way that it is the victims who are blamed for regional problems.

The focus of technological change in this paper has been on changes in the production process, not on changes in the products themselves although product change has attracted considerable attention in the literature. The analysis here would suggest that factor market imperfections lead to less ambiguous results in terms of the incentive to innovate. Since the monopsonist faces lower factor costs, the monopsonist's marginal costs are lower than the competitive firm's marginal costs. Ceteris paribus, the monopsonist would have less incentive to innovate and would therefore take longer to adopt technological change affecting the production process. This conclusion is independent of the Schumpeterian considerations of ability to perform R & D or to protect the advantages of innovation long enough to cover costs and make excess profits from it.

When it comes to product innovations, this paper is generally neutral. The model assumes that the monopsonist faces an exogenously determined price for his/her product, i.e. the monopsonist in the factor market is a competitive firm in the product market and subject to the incentives and constraints of the competitive firm. However, it might be argued that being a monopsonist nonetheless gives the firm some advantage when it comes to product innovation. If the Schumpeterian position is correct - and not everyone believes it is (Scherer, 1984, p. 198) - that excess profits are needed to finance the R & D activities which generate both process and product innovation, this model would suggest that the monopsony might be more likely to innovate in the product market. Although lacking excess profits from its activities as a competitive supplier of a product, the monopsony does generate excess profits from the market power exercised in its factor market(s). It is possible that the monopsonist could use these excess (monopsony) profits to develop product innovations which could provide at least a temporary advantage in the competitive product market. If the reaction time of the competitors lacking monopsony profits is long enough, the monopsonist may well feel that the investment in R & D is worthwhile even though the advantage in the product market is temporary. Of course, the hope that the advantage may be built upon over time may lead the monopsonist to make the R and D investment, anticipating a long run advantage over the other firms in the product market. On the other hand, the existence of monopsony profits could lead to complacency in the product market since the firm's monopsony power generates excess profits already. Thus extending the monopsony model to product innovations leads to the same ambiguities already in the literature.

References

Batra, R. and Scully, G.W. (1972) Technical Progress, Economic Growth, and the North-South Wage Differential, Journal of Regional Science, 12, 375-386.

Bradfield, M. (1988) Regional Economics: Analysis and Policy in Canada, Toronto McGraw-Hill Ryerson.

Bradfield, M. (1987) Making the Economy Boom: Is Militarization the Answer for the Maritimes?, This Magazine, 21 (4), 25-27.

Brusco S. and Sabel, C. (1981) Artisan Production and Economic Growth, in F. Wilkinson (ed.), The Dynamics of Labour Market Segmentation, Academic Press, London,99-113.

Department of Regional Economic Expansion, undated Occasional Paper, Single-Industry Communities, DREE, Ottawa.

Macdonald, Donald S., et al. (1985) Report of the Royal Commission on the Economic Union and Development Prospects for Canada, Minister of Supply and Services, Ottawa.

Richardson, W. (1973) The Economics of Urban Size, Saxon House, Farnborough.

Scherer, F.M. (1980) Industrial Market Structure and Economic Performance, Houghton Mifflin Co., Second Edition, Boston.

Scherer, F.M. (1984) Innovation and Growth: Schumpeterian Perspectives,

The MIT Press, Cambridge, Mass.

Richardson, W. (1975) Regional Development Policy and Planning in Spain, Saxon House, Farnborough.

Wheat, L. F. (1986) The Determinants of 1963-77 Regional Manufacturing Growth: Why the South and West Grow, Journal of Regional Science, 26 (4) 635-660.

Winch, D. M. (1960) Input and Output Determination Under Imperfect Competition, The Canadian Journal of Economics and Political Science, 26 (4), 631-638.

7. DIFFERENTIATION, ENTRY, INNOVATION IN THE SYSTEM OF CITIES

Tomaso Pompili

Introduction

The theoretical analysis of urban systems, and of the appearance and diffusion of innovations within them, has essentially followed two approaches: on one hand, there are developments of the traditional, static paradigm grounded in the German tradition of Christaller (1933), Loesch (1940) and others; on the other hand, there are dynamic studies, essentially based on the biological-evolutionary paradigm (e.g. Allen and Sanglier, 1979 and 1981; Sonis, 1983; Dendrinos and Mullally, 1985). The latter tend to consider the old approach, linked to mainstream economic theory, as a spent force, whereas it is the contention here that the German tradition of spatial economics needs revitalization rather than dereliction, re-designing it on the basis of the state of the art in industrial economics.

Therefore, after recalling the essential traits of the Christaller-Loesch inter-urban location model, the goals of this paper are: (i) to give a spatial interpretation of selected recent advances in the theory of differentiated competition, (ii) to show how urban innovation can be conceptualized on the basis of developments in industrial economics, (iii) to present recent models, building also on this perspective, among others. Recent important developments in the theory of competition on product differentiation, gained by means of non-cooperative game theory, are surveyed in the next section, where an interpretation of these theoretical developments as applied to an urban context is presented; entry, pre-emptive investments and cost-reducing innovations are then explored, and outlined in a metropolitan context; finally, a link between these applications and the recent family of SOUDY models, which highlights the role of innovation as the driving force of urban development, is traced.

Classical Models of the Urban System

General equilibrium theory was formulated a century ago (Walras, 1874; Pareto, 1896); afterwards, there have been great advances in the study of the time process by which equilibrium is achieved in Anglo-Saxon countries, whereas scant attention has been paid to the space issue. In Germany, instead, the Lausanne school's results, anticipated locally (Menger, 1871), found an environment sensitive to this issue. In this section the classical models by Christaller (1933) and Loesch (1940) are presented in their essentials (see also Beavon, 1977), bearing in mind the similarity of the subject with non-cooperative game theory of horizontal differentiation (Gabszewicz et al., 1986).

Central places

Christaller's basic assumptions are that there is a two-dimensional uniform transportation surface and that distance is costly to overcome, therefore total transport costs should be minimized. This is equivalent to maximizing the number of central places, which in turn requires an equilateral 60-degree lattice pattern. There is perfect information. On the supply side, ubiquity of all inputs is assumed, together with homogenous commodities, a single production technology, zero fixed costs (absence of technical barriers to entry), zero relocation costs, implicitly constant average costs and mill pricing. Therefore, **producers maximize their profits** under a regime of perfect competition. On the demand side, the population of consumers is regularly and continuously distributed in terms of its size, purchasing power and preferences. Using as a benchmark a reservation price related to their preferences, consumers decide whether to purchase the commodity, after which they choose the supplier by minimizing total costs i.e. their own transport costs: in sum **consumers maximize their utility**.

For each commodity defined by these parameters (Gabszewicz et al., 1986), a **maximum demand range** appears, where local demand for the commodity declines to zero, i.e. where total demand of that supply is maximum, and a **minimum supply threshold** can be identified at the distance where total demand ensures that the producer is not making losses (minimum viable market size). Space endows each producer with some monopoly power within its own potential market, hence with ability to make excess profits, but this is countervailed by competitors, the more so the wider the gap between threshold and range. In fact excess profits induce additional producers to enter the industry, so that in equilibrium the whole territory is supplied (maximum production) through **a pattern of non-overlapping market areas** and no producer is making excess profits.

In Christaller's model, commodities basically differ because of the unit distance cost, which determines the range, and they can be ordered according to this range: this is in itself sufficient cause for a **hierarchical structure of centres** to emerge (Berry and Garrison, 1958). In a top-down approach, Christaller identifies a metropolis, as the centre supplying those commodities whose range is such that only one supply point is needed for all the territory (the highest-order commodities).

Additional functional ranks will be created whenever a commodity is reached whose range is such that the previous rank does not have enough centres to supply the whole territory, and each new rank will include all the previous centres to minimize total transport costs, while the new centres are located at maximum distance from the existing centres and from each other. Thus, the centres of each functional rank perform all the activities performed by centres belonging to lower ranks, which for some commodities are part of the market areas of the higher-order centres. Thus, a pattern of **intra-activity dispersion** and **inter-activity agglomeration** is displayed, which is regular both horizontally and vertically.

Economics of location

Loesch classifies and develops Christaller's model, while maintaining all his basic assumptions. Loesch also maintains the definitions of demand range and supply threshold, so that Christaller's final result of non-overlapping market areas for each commodity, with supply of the whole territory and zero excess profits, is therefore shared by Loesch.

Where Christaller and Loesch differ is in the **derivation of the hierarchy and pattern of activities and centres**. If commodities can be differentiated by their own range, each one of them should be characterized by its own supply network, so that no commodity may provide excess profits in equilibrium. Following a bottom-up approach contrary to Christaller's, the various market networks are superimposed, with the less restrictive constraint that every supply point must be located in one point of the basic lattice.

There are two important points where Loesch is closer than Christaller to empirical findings: first, the Loeschian **city-size distribution is more continuous** than Christaller's stepped hierarchy, and second, and more important, **centres supplying the same number of commodities may perform totally different activities** and, conversely, the same commodity may be supplied from centres of different 'size' (i.e. with a different number of activities). Although intra-activity dispersion and inter-activity agglomeration are still observed, **it is now difficult to define a hierarchy of functional ranks, or even of city sizes**, as directly related to the number and not to the 'quality' of activities. This result implies the idea of two-way influences among centres, hence the 'grid' pattern recently put forward by Dematteis (1985).

The greater number of centres involved in Loesch's design of supply networks requires their alignment along major transportation axes in order not to jeopardize the minimization of distances travelled to supply points. This is achieved by adopting some **assignment criteria** whenever a network has more than one feasible positioning. In rank order these involve maximizing: first, agglomeration among centres (returns to 'regional' scale), second, agglomeration among activities within a centre (returns to urban scale), third, functional 'proximity' of activities within a centre (synergies among activities of a similar order).

A Spatial View of Competition

Monopolistic competition, according to Chamberlin's (1933) classical treatment, is characterised by long-run zero profits and excess capacity (too many firms); long-run entry is an essential element of monopolistic competition. However, conjectural rules on the opponent's behaviour are crucial to the conclusions: Cournot's rule produces a quantity equilibrium associated with profits, whereas Bertrand's produces a price equilibrium where price equals average cost. What happens with entry in the case of homogenous goods? If the strategy is a quantity one (Cournot), competition is restored by entry; if the strategy is a price one (Bertrand), competition takes place already with only two firms. Recent important developments in the theory of imperfect competition, gained by means of non-cooperative game theory, are surveyed in this part, grouped around the themes of horizontal and vertical product differentiation.

An inter-urban model of cities competing for one activity (good market on the demand side or factor market on the supply side) is presented here, as an interpretation of the theoretical advances on 1930s knowledge (the one employed by Christaller and Loesch), surveyed in this section. The urban hierarchy can thus be interpreted both in terms of output demand market areas and in terms of localized external economies and input supply market areas.

Horizontal differentiation

By horizontal product differentiation (HPD) we mean a market structure in which **goods differ in variety and consumers differ in preferences**: this is the traditional form of differentiation.

Hotelling (1929) concluded that competition among sellers leads to no differentiation i.e. an homogeneous good in each industry/market, if firms compete to maximize market share (and profits). Recently, d'Aspremont et al. (1979) have proved that Hotelling's principle of minimum differentiation is invalid: product differentiation must be an important component of oligopolistic competition. After Lerner and Singer (1939) and d'Aspremont et al. (1979), de Palma et al. (1985) show that the principle of minimum differentiation (agglomeration of producers) does hold when products and consumers are sufficiently heterogeneous. In fact one firm leaving the others makes a greater gain the smaller the heterogeneity of consumers, i.e. the more consumers are affected by price rather than preference. As for the Nash equilibrium (NE) in prices, heterogeneity hinders price undercutting; the combination of location and price choices into a two-stage model suggests that no regular spacing of producers may occur.

With HPD, seeking a Bertrand-Nash equilibrium (i.e. a NE in prices), each firm has a positive market share and a price exceeding unit variable costs, and at least covering sunk costs. Increasing market size (e.g. by duplication) allows entry of an unbounded number of firms, each earning revenue exceeding variable cost, so that at equilibrium the structure becomes more and more fragmented: this is the so-called '**fragmentation property**' (Sutton, 1986).

Also **spatial competition** analyses are relevant here (e.g. Capozza and van Order, 1982), in that a group of consumers can be modelled as if they inhabited a Hotelling Loesch location model (Horstmann and Slivinski, 1985): spatial competition is inherently different from non-spatial perfect competition, which is in fact the limit case with zero fixed costs and/or zero transport costs (Bowes, 1984) i.e. with homogeneous goods.

The usual determinants are price, transport rate and distance on the demand side; fixed cost, marginal cost and quantity on the supply side. As a result, fewer firms and lower prices (equal to marginal cost) would be more efficient than the equilibrium values i.e. would minimise costs producing a higher quantity, even taking into account the transport rate. With costless relocation, these results are qualitatively the same with any conjectural assumptions on the behaviour of firms, although welfare, i.e. the sum of consumer surplus and producer profits per capita, is quantitatively different. All conjectures lead to positive supplier's profit in long run equilibrium. Notice that equilibrium under spatial competition, just like in HPD, need not involve zero profits, but simply negative expected profits for entrants (Bowes, 1984).

According to Scholer (1985), the type of entry is crucial to the results: with sequential market entry, i.e. with no relocation, there is no zero profit condition for long-run equilibrium. If entry is continuous, profits are zero, but not if entry is sequential (Hay, 1976; Rothschild, 1976; Prescott and Visscher, 1977).

Economic models wherein consumers are spatially separated from production and distribution outlets invariably show imperfectly competitive equilibria, characterized by prices in excess of marginal costs, welfare losses and excess capacity. However, a few authors cast doubts on these conclusions. Benson (1984) argues that, even though free entry in Loeschian spatial competition leads to Chamberlinian tangency, it is a second best solution. In fact, there are not too many inefficiently small firms in equilibrium, in that the difference between price and marginal cost just covers social cost i.e. additional marginal cost of output resulting from multiple location (increased proximity benefit to consumers versus sunk cost of establishing an alternative location). More fundamentally, de Serpa (1986) argues that partial equilibrium analysis cannot adequately address this problem because of interactive effects of product market, factor market and land market. The conclusion is that a market equilibrium, apparently monopolistically competitive but actually purely competitive, is consistent with Pareto optimality, even though firms have some control on prices and operate under increasing returns.

Vertical differentiation

By vertical product differentiation (VPD) we mean a market structure in which **goods differ in quality and consumers differ in incomes** (Gabszewicz and Thisse, 1979). By a better, higher-quality product we mean that all consumers would prefer it, if all products were sold at the same price (Sutton, 1986).

The conclusions that can be reached are the following (Gabszewicz and Thisse, 1979):

- non-cooperative market structure is fully characterized by relative product quality and income distribution;
- income differentiation plays a central role to sustain product differentiation, even more so if cost differences are considered.

This can be construed as the search for **perfect equilibrium** (Selten, 1975) in a three-stage non-cooperative game, where the choices are on entry (with positive entry cost), quality and price (Shaked and Sutton, 1982). At the price stage, we find that the top quality product enjoys a positive market share, that if a product has no market so do all lower quality ones, and that there is a unique price pair forming a NE point for any two distinct levels of product quality (with greater income to higher quality). The quality-price game has a perfect equilibrium in pure strategies with distinct qualities and net revenues greater than zero, whereas with homogeneous products Bertrand competition would drive prices to marginal costs. In this case, price competition drives prices down so that, within limits given by income distribution, at most the top k products have a positive market share; therefore, with more than k potential entrants, only the first k have a positive payoff and hence operate.

With VPD, therefore, the fragmentation property need not hold good, depending on consumers' willingness to pay for quality improvements ('preferences') and the increase in unit variable costs associated with such improvements ('technology'). Brands of successively higher quality are purchased by consumers in successively higher income bands, while they have identical preferences. If quality price equals quality cost we can identify a different preferred quality with each income level: this is a correspondence with the locational model of HPD and the fragmentation property holds good, provided that quality cost increases steeply enough (Sutton, 1986).

Otherwise, all consumers may prefer the highest quality and there exists an upper bound to the number of firms offering quality differentiated products which can co-exist at a non-cooperative NE in prices with positive market shares and a price exceeding unit variable cost. The bound, independent of the qualities on offer, depends only on preference and income distributions and produces a 'natural oligopoly': there is no limiting process towards competition, i.e. the fragmentation tendency fails, and the **'finiteness property'** holds true instead, proved by Gabszewicz and Thisse (1980), with zero costs, and by Shaked and Sutton (1983), with positive costs.

The result produced by this mechanism is that price competition between some group of high quality products drives their prices down to a level at which even the poorest consumer prefers to buy one of these goods at its equilibrium price, rather than buy any of the excluded (lower quality) goods at a price sufficient to cover unit variable cost; so the property relates to the last stage of the above mentioned game (Sutton, 1986).

Differentiation is maintained in the case of changes in product quality: the assumption that entrants are of higher quality produces the result that, after the upper bound is reached, each new entrant of increasing quality must simply displace the (hitherto viable) lowest quality existing firm; further entries entail exits, yet reduce prices to competitive ones (Gabszewicz and Thisse, 1980).

The same result holds in a **market with VPD and HPD**: there is an incentive to escape horizontal crowding in a vertical direction because consumers may operate a trade-off between quality and variety. Again, convergence to a fragmented market structure depends on technology and preferences.

Sufficient conditions for a natural oligopoly are that unit variable cost increases slowly with quality increases and that the rate of increase in fixed costs associated with quality changes is bounded above for all qualities (Gabszewicz and Thisse, 1982). The mechanism is quite general, in that it does not depend on types of entry, competition, firm, and so on, holding good with respect to alternative specifications of the oligopoly model (Shaked and Sutton, 1987).

A demand-side approach to space

The above results can be employed to sketch an inter-urban model of cities competing for the location of a one good industry with complete freedom of entry, considering the cases of a homogeneous good, of a horizontally differentiated good and of a vertically differentiated good.

A **homogeneous good** in space, as is well known, behaves like a horizontally differentiated good: consumers, homogeneous in income and preferences but not in location, choose whether and where to purchase it according to its price and to the cost of overcoming distance between them and the supply point (that it be overcome by the consumer (service model) or by the good (manufacturing model) is irrelevant). Consequently, as the good is homogeneous, price and transport cost are the same for all producers; therefore, each consumer will purchase from the nearest supplier and market areas are mutually exclusive. The number of producers is as high as demand conditions allow, according to the fragmentation property.

From an urban, rather than industrial, viewpoint, this is the traditional Christaller - Loesch model. For each activity, the different centres housing it have an exclusive market area and increases in demand (in its spatial density) lead to increases in the number of producing centres, while market area sizes shrink in geographical but not in economic terms (surplus profits tend to zero), and to increases in consumers' welfare, because average distance to the supply point decreases. Thus, it can be said that the function filters down to lower urban ranks and in this sense it gets 'banalized'.

A **horizontally differentiated good** in space behaves like a 'doubly' horizontally differentiated good: consumers, homogeneous in income but not in preferences and location, choose whether and where to purchase it according not only to its price and transport cost but also to the cost (i.e. reduced utility) of overcoming 'distance' between the preferred variety and the one under consideration. Consequently, as by definition the first two elements are the same for all the varieties of the good, the effect of the subjective element is that supply does not necessarily come from the nearest producer: market areas, possibly defined by probabilistic rules, overlap. In this case, too, we can observe fragmentation in the number of suppliers, although heterogeneity of goods encourages their spatial agglomeration, in that price competition among suppliers matters less relative to the correspondence with consumers' preferences.

From an urban, rather than industrial, viewpoint, this leads to the formation of dense metropolitan areas from which to serve the whole territory: the location-variety trade-off is profitable to the supplier only within limits. Increases in demand lead to increases in the number of varieties supplied and 'subjective distance' falls; hence consumers' welfare rises, but increased variety makes transport cost less crucial to consumer choice and therefore we may experience greater spatial agglomeration and correspondingly lower welfare. If the number of potential varieties is finite, at a given point in time the conditions of the previous model will re-appear, with market fragmentation among various competing suppliers located in different centres.

A **vertically differentiated good** in space behaves like a VD-cum-HD good: consumers, homogeneous in preferences but not in income and location, choose whether and where to purchase it according not only to transport cost but also to its price, which is no longer the same for all producers. There is an incentive to producers to escape horizontal (spatial) competition in a vertical (quality) direction, but the finiteness property defines a natural oligopoly, provided that cost differentials do not fully reflect quality differentials. From a spatial viewpoint, markets do overlap and may even coincide.

From an urban, rather than industrial, viewpoint, this leads again to the formation of dense metropolitan areas from which to serve the whole territory. However, in this case, increases in demand lead to upward shifts of quality and therefore welfare increases. This may happen both with the appearance of the new activity in a new centre and with the downward re-location of activities in the existing centres to make room for the new activity in the largest centre. In any case, neither the number of supply centres nor the shape of their locational distribution vary: thus the individual identities of supply centres may change, but not the general form of the system.

A supply-side approach to space

The above paragraphs are paralleled here by the outline of an inter-urban model of cities competing, with complete freedom of entry, to obtain the use of a spatially localized input, considering the cases of a homogeneous input, of a horizontally differentiated input and of a vertically differentiated input. A **homogeneous input** (say, labour) in space corresponds, just like the homogeneous good, to a horizontally differentiated good: the wage is the same for all localized employers, homogeneous in purchasing power and technical needs, and labour chooses whether and where to supply itself according to its wage and to the cost of overcoming distance between itself and workplace location (commuting). Consequently, each worker will work for the nearest employer, labour market areas are mutually exclusive and a number of local labour markets appear as labour demand conditions allow, according to the fragmentation property (increasing with labour demand size i.e. with expected production profits).

In an urban approach, for any input the different centres employing it have an exclusive market area and increases in supply (in its spatial density) lead to increases in the number of employers, while market area sizes shrink in geographical but not in economic terms (quasi-rents on wages tend to zero), and to increases in workers' welfare, because average distance to the workplace decreases. Thus, it can be said that increases in supply allow the diffusion of economic activity, i.e. its filtering down to lower urban ranks and in this sense its 'banalization'.

A **horizontally differentiated input** (say, industry-specific skills) in space, like the analogous good, behaves like a 'doubly' horizontally differentiated good: the wage is the same for all localized employers, which are homogeneous in purchasing power but not in technical needs, and labour chooses whether and where to supply itself according not only to its wage and to the cost of overcoming distance between itself and workplace location, but also to the employers' evaluation of the cost (i.e. lower productivity) of overcoming 'distance' between the required skill and the one under consideration. Consequently, the effect of the technical element is that labour does not necessarily supply itself to (and get employment from) the nearest employer. This two-dimensional variety produces overlapping labour markets, possibly defined by probabilistic rules, and fragmentation, although skill heterogeneity does encourage spatial agglomeration, in that wage competition among the workforce matters less relative to the employers' technical needs.

In an urban approach, this leads to the formation of dense metropolitan areas which attract labour from the whole territory. The location-skill trade-off is profitable to the employer and to the worker only within limits. Increases in labour demand lead to increases in the number of skills supplied: 'technical distance' falls hence employers' costs fall, but increased skill variety makes transport (commuting) cost less crucial to labour choice and therefore we may experience greater spatial agglomeration and correspondingly lower welfare. In the long run, however, we can foresee the spatial re-distribution of labour, because of the aim to reduce transport cost. The spatial distribution of skills, being no longer homogeneous, will constrain the location of economic activity, thus ossifying the production structure of cities. This problem can be smoothed if the number of potential skills is finite. In this case, in fact, at a given point in time the conditions of the previous model will re-appear for each skill, with market fragmentation among various competing employers located in different centres.

A **vertically differentiated input** (say, function-specific professions) in space, like the analogue good, behaves like a VD-cum-HD good: employers are homogeneous in technical needs but not in purchasing power and therefore in the wages they are prepared to offer, and labour chooses whether and where to supply itself according not only to transport cost, but also to its wage, which measures input quality and therefore is no longer the same for all workers. Consequently, the effect of the functional element is that labour does not necessarily supply itself to (and get employment from) the nearest employer: there is an incentive to labour to escape horizontal (spatial) competition in a vertical (functional) direction, through higher education; but the finiteness property defines a natural oligopoly, provided that education cost differentials do not reflect labour quality differentials fully. From a spatial viewpoint, markets do overlap and may even coincide.

In an urban approach, this leads again to the formation of dense metropolitan areas which attract labour from the whole territory. The location-quality trade-off is profitable to the employer and to the worker only within limits. However, in this case increases in labour demand lead to upward shifts of labour quality and therefore welfare increases. This may happen both with the appearance of the new function in a new centre and with the downward re-location of functions in the existing centre to make room for the new function in the largest centre. In any case, neither the number of production centres nor the shape of their locational distribution vary: thus the individual identities of production centres may change, but not the general form of the system. In the long run, however, here again we can foresee the spatial re-distribution of functions, no longer being homogeneous, will constrain the location of economic activity, thus ossifying the functional structure of cities. This problem cannot be smoothed, because the finiteness property hinders the fragmentation of functions over the territory.

Innovation in the Urban System

In the industrial economics literature, a number of firms (from 1 to n, but usually 2) compete on the market for one product, giving rise to the following issues: a) the product may be differentiated, horizontally or vertically, b) firms face the risk of entry into the market, c) firms can invest and innovate.

After covering the first point in the section above, the interconnected issues of entry/exit (including barriers and deterrence), investment, innovation and the evolution of industry/market structure receive special attention in this section. In an urban context, technological change is viewed here as the appearance of new activities in a specific city and in the urban system at large, hence technological diffusion is viewed as the adoption of an activity in different cities.

Entry and credible threat

As we know, the issue of entry is crucial to equilibrium results in competitive markets; therefore, it has attracted interest, especially in terms of the attempts of existing firms to deter potential competitors from entry. Among the first analyses, Bain (1956) and Sylos (1962) are well known. In a game theoretic framework, pre-emption strategies have been studied (Schmalensee, 1978), leading to the conclusions that **threats are effective only if they are credible**, and that **to be credible threats must involve an advance cost** to the existing firm(s) (Spence, 1977; Dixit, 1979).

The basic idea of this literature is that an irrevocable commitment of capacity-increasing investment alters the initial conditions of the post-entry game to the advantage of the established firm, without altering the rules of the game (Dixit, 1980). This may have potentially serious welfare effects, in that the welfare gain from larger quantities can be more than offset by the welfare loss from less diversity (Dixit and Stiglitz, 1977).

What is studied is **strategic interaction in investment decisions of firms** in the context of a new, growing market rather than a mature one (Spence, 1979). In such a market demand is greater than supply, i.e. there is no risk of excess capacity, therefore entry is possible and the finiteness property does not hold. The basic thrust of the argument is that constraints on growth and the timing of entry put firms in asymmetrical positions with respect to investment (Spence, 1979).

Costs of production are assumed to be of three kinds: fixed (setup) costs, constant average variable costs, and constant unit costs of capacity expansion (investment). Investment serves to build up capacity and, being a fixed cost, it creates scale economies that act as an entry barrier for potential competitors.

Being able to choose their capacity levels before their potential competitors make the entry decision, existing firms achieve one of the following results, according to the effect of their move on the entrant's expected profits:

- blockaded entry i.e. monopoly: the barrier effect is gained simply by pursuing one's own interest;
- effectively impeded entry i.e. Stackelberg equilibrium: the preferred point admits competition;

- intermediate case i.e. limit pricing: entry prevention is profitable but monopoly power is not unrestrained.

Excess capacity can enlarge the zone where entry is effectively impeded and restrict the ineffective zone. However, investment is constrained in that it is irreversible (credible commitment) and increasingly costly: advance commitments are not credible unless accompanied by positive investment, so firms are asymmetrical in entry times and in financial resources (Spence, 1979).

The optimal (NE) strategy for pre-emptive investments is to get to the target capital stock as soon as possible, and stop when the present value of net marginal profits of the capital stock is zero. From a static viewpoint it appears as over-investment, but this is because investment is irreversible: in the long run it is not over-investment, but optimal investment. Entry is harder when any of the following circumstances occurs: an absolute advantage in demand for the existing firm, a higher cross-price effect, higher fixed (setup) costs of the entrant, an increase in net absolute advantage (Dixit, 1979).

Innovation

There are three types of competition: price, product (differentiation) and R&D (missing e.g. in Arrow, 1962). An important ingredient in the performance of an industry over time is the extent to which the opportunities for cost reduction and product improvement are exploited. However, a precise microeconomic foundation for the rate and direction of technical progress is lacking, despite its link with market structure (Dasgupta and Stiglitz, 1980a). We consider here learning processes, the relationship between market structure, R&D investment and innovation, and asymmetry effects on market structure.

i) The learning curve

A model in the tradition of innovation as learning (Arrow, 1962) has been presented by Spence (1981). Competitive interaction and industry evolution occur in the presence of a 'learning curve' i.e. of a function relating unit costs to accumulated volume: with a learning curve the output decision is like an investment decision, in that it affects a relevant stock variable (accumulated output) which activates feedback effects. Thus, on an optimal output path, marginal profit is negative instead of zero, as with pre-emptive investment: at every time output should be the profit maximizing output, given the true marginal cost (as if the firm assumed learning has already occurred) both for the single firm and, with exogenous entry, for multi-firm competition.

The learning curve can be viewed as the result of economies of scale across market segments which are defined over time; like them, it can create substantial barriers to entry. These are smallest with a high learning rate, however excessive entry does not occur because negative profits appear well before it. If learning effects are not entirely firm-specific, spillovers reduce individual returns to investing in accumulated output, hence the pace of cost reduction. The result is a more efficient industry-wide cost reduction process.

ii) Market structure and innovative performance

In the basic model of stochastic invention, the choice variable is 'R&D investment' and an innovation production function is based on it (Reinganum, 1984). R&D investment features unit costs increasing with magnitude of investment and with efficiency achieved, but not with magnitude of output; the innovation production function shows returns to scale initially increasing, but eventually decreasing.

Interdependent firms compete for a flow of rewards over time, available only to the first innovator, and each firm bindingly commits itself to R&D investment, thereby purchasing the random variable 'uncertain success date', under a double condition of uncertainty: technological uncertainty, due to the stochastic relationship between R&D investment and innovation time, and market uncertainty, due to the innovation/success time of rivals.

With negative cross investment elasticities, as the equilibrium number of firms in an industry increases, the equilibrium level of firms' investment declines and yet the expected industry introduction date declines too (Loury, 1979). With no barriers to entry the number of firms is not given, so long as expected profits are positive. Thus, if the technology for innovation exhibits diminishing returns to scale, expected profits get to zero only in the limit: the outcome is atomistic competition with infinitesimal investment and greatest innovative activity. If instead there is an initial range of increasing returns (e.g. fixed costs), hence a barrier, there will be zero profits with a finite number of firms and excess capacity in the 'innovative technology' (special case of Chamberlin).

Loury's conclusions are sensitive to his specification of R&D costs: in fact he stresses fixed costs, assuming that the uncertain success date is purchased by paying R&D investment at time zero. To the contrary, Lee and Wilde (1980) stress variable costs, assuming that the purchase occurs by paying R&D fixed investment at time zero and R&D current investment at any period thereafter until the first innovation occurs. So Lee and Wilde's main results are opposite to Loury's.

iii) Market types and features of innovation

Dasgupta and Stiglitz (1980a and 1980b) argue that, for process innovation, on one hand demand growth, R&D technology and patent rights provide incentives to firms which affect innovation frequency and therefore performance; however, on the other hand, except in short run, both market structure and nature of inventive activity are endogenous and the innovation-concentration relationship is therefore not causal. They compare oligopoly with free entry (i.e. endogenous market structure), oligopoly with barriers to entry, and a socially managed industry.

Competition in the product reduces the level of innovation relative to monopoly, whereas competition in R&D increases the level of innovation, possibly beyond the socially optimal level; under certain conditions, it pays a monopoly to pre-empt potential competitors. When concentration is low, concentration and industry R&D move in the same direction. With free entry, both firm R&D (cost reduction) and monopoly power grow together with concentration; however, high concentration does not suggest lack of competition.

Optimal and firm R&D investment grow with market size, and they decrease with increasing costs of technology if demand elasticity is high, and vice versa. Even with free entry, cost reduction is likely to be less than optimal, despite industry R&D investment being more than optimal (with demand elasticity below unity): in fact, duplication of effort is likely, with low concentration as well.

iv) Asymmetry between firms and innovation

More realistic asymmetric models of innovation are of two kinds: 'inherited structure' or 'leader/follower' (Reinganum, 1984).

Inherited-structure, or monopoly-power, models have found:

- an inverse relationship between magnitude of innovation and the likelihood of invention by the current industry leader. When innovation is uncertain, as in the case of 'drastic' innovation in terms of effect on market shares, there is less drive for innovation from the highest profit makers (this is not true for minor innovations);
- lower R&D investment for the large firm and challengers, the greater the flow of current revenue to the large firm;
- a rate of individual investment on a particular innovation declining with the number of anticipated subsequent innovations: being current leader is not valuable when many innovations are still to appear.

Leader-follower, or innovation-timing, models have found:

- current leaders persisting as such through a pre-emptive patenting strategy with a non-stochastic innovation production process: post-innovation monopoly profits persist even after entry (purely deterministic innovation);
- R&D encouraged by licensing with similar production costs and discouraged when they are dissimilar: an ex-ante incentive to reduce R&D inefficiencies;
- the most efficient firm patenting, and then licensing to the leader, when ex-post licensing is permitted;
- increasing concentration of firms engaged in R&D, but not necessarily in the market, while developing an innovation.

Urban product innovation and the prevention of inter-urban competition

Referring to the previous section, in the two cases of homogeneous and horizontally differentiated goods, the long-run increase in consumers' purchasing power produces the **gradual diffusion of activities** to the whole urban system. Similarly, in the two cases of homogeneous and horizontally differentiated inputs, the diffusion of their use depends on the expansion of employers' demand. However, in the case of horizontally differentiated goods or inputs, this diffusion trend may be retarded by multiple location of activities in major centres; the lag will be greatest if the cities pursued a strategy of prevention of competition, both in a qualitative sense, through expansion of goods and/or input variety, and in a quantitative sense, through capacity expansion.

In the case of vertically differentiated goods or inputs, at the mature stage of natural oligopoly, innovation means a higher quality product or input, and the whole existing product/input structure shifts downwards and displaces the hitherto viable lower quality product/input. In this case, what happens is a 'functional filtering down' long run process, so that the hierarchical pattern of the urban system stays unaltered, although its contents change.

Recalling the results of Beath et al (1987), we can argue that large cities, with their economies of scope, may even decide not to produce some of the products that they would be able to supply and not to offer some of the inputs, if their dominant position is strong enough not to fear any loss of leadership. This result is especially true in periods of relative stability of the technological, economic and urban systems, whereas in 'crisis' periods dominant cities run the risk of losing their leadership to newer, more motivated competitors.

In fact, taking now into account the advances reviewed in this section, whereas innovation does find incentives in the spheres of demand, supply and institutions, in the long run the urban hierarchy as such and the nature of cost-reducing innovation are **both endogenous**, rather than causally linked.

On one hand, industrial innovation is encouraged by competition in strategic activities but not by competition in manufacturing: thus the many small cities may not have the ability to innovate, while the few large cities have it, but have also insufficient incentives.

On the other hand, as the competition for monopoly profits accruing through product differentiation is discouraged by the high costs of investment in physical and intangible urban infrastructure, at the lower and more populated urban ranks successful innovation occurs more often despite requiring lower expenses, thus giving rise to greater diversification and to the spatial expression of this horizontal division of labour: Dematteis' (1988) grid pattern (as shown in Pompili, 1989). Instead, at the higher ranks, quality (vertical) innovation is the dominating form, seldom occurring despite the higher levels of infrastructural investment.

The statement that threats aimed at deterring an increase in competition are effective only if they are credible, and that to be credible they must involve an advance cost to those who make them (in the context of an urban system, the cities that are already active), can be used to explain the extremely high levels of **social overhead capital** in metropolitan centres, in terms of 'hard' and 'soft' infrastructure geared to the productive needs of firms and to the re-productive needs of the workforce; besides, the statement can also explain the attempts of these centres to rejuvenate and expand this capital.

Social investment creates scale economies that act as an entry barrier for potentially competing cities. This produces no effects on the profit function, but reaction functions become discontinuous: the city that wanted to enter a new activity or function finds it better to give up in the face of the rival's preventive action if it expects not to recover, even in the long run, the increased fixed costs required for entry.

However, an alternative explanation does not need these competitive overtones, assuming that short run 'over-production', to which over-investment is related, is necessary in the learning process towards better efficiency and higher welfare. Large centres are already favoured by the existence of an absolute advantage in demand, although, in the case of old industrial cities, this advantage is offset by higher fixed costs, once the locational costs of congestion and of recurring redevelopment of extensive disused manufacturing areas are taken into account.

These asymmetry elements between urban centres, dependent on their history, may have potentially serious adverse effects on the global level of welfare in the system, especially in the long run. In fact, the welfare gain from larger quantities (scale economies) can be more than offset by the welfare loss from the spatial concentration of production and employment (especially high-quality employment) and from the consequences of this concentration, which raise important distributional issues.

Moreover, in the long run, as this concentration tends to occur in non-mature and therefore more dynamic activities, the gap between well-endowed centres and poorly-endowed centres may increase, with regard to activities, population and purchasing power. From this viewpoint, the lack of fiscal powers of local authorities (in Italy) on residents and local employment may act as a corrective mechanism, especially since central government may not be favourable to what in the short-run appear as blatant over-investments.

At the system level, these asymmetries may produce a lower propensity to innovate in largest metropoles and discourage innovative efforts by smaller cities, with the former in any case appropriating the new knowledge thanks to their financial power.

Supply Oriented Urban Dynamic Models

Recently, work has been carried on by our group in Milan under the label of the SOUDY (supply oriented urban dynamic) family of models (Camagni, Diappi and Leonardi, 1986, denoted as CDL from now on and Diappi and Pompili, (1988 and 1989, denoted as DP from now on), which are now presented. These models follow a theoretical approach whereby hierarchy is explained not in terms of output demand market areas, but rather in terms of localized external economies (location factors) and of input supply market areas (local labour markets). Although they take account of the evolutionary paradigm, it is possible to show their connections with advances in economic theory as they are interpreted in the previous sections.

A supply side Christaller-theoretic dynamic approach (SOUDY1)

While basically in line with the Christaller tradition, CDL have attempted some modifications rather than concentrating on equilibrium issues. Namely, the main innovations are to shift focus towards the supply-side determinants of urban growth and to expand the dynamic potentialities of the model (see also Camagni, 1986, and Pompili, 1986).

The traditional basic tenets are shared by CDL, except for the following changes. With time, input supply and output demand location and level vary, so distribution quickly becomes non-uniform in both cases. Thresholds are identified with exogenous **minimum effect production sizes** (technical entry barriers), whose corresponding demands translate into minimum urban sizes. Producers are price-setters (demand defines quantity only), so that they are able to earn excess profits with any production above the threshold. **Relocation costs are not zero**, as the process of gaining and losing activities occurs over time, and inputs are generally not ubiquitous. Increasing entry barriers and diminishing transport cost coefficients for higher-order commodities favour lower numbers of producers and higher profits; commodities can thus be ranked according to supply-side characteristics.

CDL also introduce new features. As for congestion in immobile input markets and negative social externalities, CDL introduce **locational costs** that, increasing with city size, reduce net profits, up to an urban size ceiling where they are zero. Between threshold and ceiling lies the **efficient city size range**, which is specific to each activity or hierarchical rank. As for agglomeration effects, they note that urban size is also a proxy for **diversification** of the input and output markets, hence for the availability of unequally distributed high-quality inputs, including the innovative spirit. As for **innovations**, technical innovation takes the form of reductions in the threshold. These favour supply of activities from centres originally too small for them, thereby increasing competition and reducing prices and profits so that larger centres may be forced to give up the activity. Urban innovation, instead, is displayed by centres when, growing towards a ceiling, they cross the threshold of a higher-order activity.

Although beneficial, **the decision to supply a new activity is not deterministic** but an object of choice, which is modelled stochastically. By the same token, the lack of innovativeness may drive a centre to decline. Simulation has shown that an urban hierarchy is created, or 'deepened', more quickly not only with low distance cost, but also with increasing returns to agglomeration and with slow technical progress.

A supply-side Loesch-theoretic dynamic approach (SOUDY2)

While partly building on the previous model, DP shift focus towards the Loeschian tradition. DP uphold the same traditional assumptions also shared by CDL. Recent developments in the theory of monopolistic competition constantly raise the problem of non-existence of price and/or location equilibrium (Gabszewicz et al., 1986). This problem of non-cooperative price-location NE can be solved by a sequential game, where the first stage determines location and the second stage determines price. 'Perfect equilibrium' (Selten, 1975) requires that, when producers choose location, they anticipate the effects of their choice on price competition. Under reasonable assumptions, the producers of a new commodity deliberately locate at the limits of the first producer's market area, until the whole territory is supplied and market area size corresponds to Christaller's thresholds.

On the production side, DP follow CDL in their new assumptions. In the theory of monopolistic competition costless relocation models seem to lose favour to sequential entry ones (Hay, 1976); entrants, by anticipating subsequent entry by future competitors, may choose an appropriate location, thereby gaining excess profits while deterring further entries. Entry is of course the first stage of a sequential game in three stages (entry, location, price). Thus, consequences on price and location are anticipated and then taken explicitly into account when making the entry decision in a perfect equilibrium context.

For any given level of demand, the equilibrium number of producers results from a **trade-off between transport costs and fixed costs**. Low entry barriers allow many producers and therefore stronger competition, while, with low transport costs, only a small number of producers can survive as horizontal differentiation is weaker. In both cases a costless relocation world implies the elimination of excess profits, however, given that relocation is not costless, excess profits exists even in equilibrium.

Consumer **demand is defined as a potential** over the whole territory (HPD assumption), with distance decay, and it is assumed to be related to purchasing power, thus increasing over time with the appearance of new production-supply units. This implies acknowledgement of Loesch's first assignment criterion, but also implies that inputs are likely not to be ubiquitous over time.

However, unlike CDL, DP are not explicitly concerned with size dynamics, expressed by them in terms of purchasing power rather than population, as DP view it as a result of innovation. Purchasing power derives from value added, increasing with functional rank, thus connecting production and consumption.

An additional requirement entering the locational decision is the local availability of funds for **investment** in fixed setup costs of a new supply point: these also come from purchasing power, so there is an issue of optimal **income distribution** between consumption, hence demand, and savings, hence investment; the income-investment link connects short and long run.

As for the **loss of activities**, these may occur because of regional saturation, i.e. the local demand requirements are no longer satisfied (e.g. population or demand distribution has changed), and/or because of local congestion, i.e. locational costs have become unbearable for some lower-order, less profit making activities, after the entry of new, higher-order activities.

Finally, inter-relations among activities must be taken into account, but supply does not depend strictly on the existence there of the activity of immediately lower rank and vice versa: this indeed closely corresponds to another of Loesch's assignment criteria (contra Christaller). The number of activities already existing in any centre is also an influence on the stochastic element and is taken as a proxy for diversification, as in CDL. However, all requirements structurally include the inter-activity synergic effect, so that explicit consideration is not necessary (contra CDL). Once the requirements are satisfied, **the decision to supply (or cease to supply) an activity from a centre is not automatic**, following CDL.

In sum, as in Loesch, the **main basic locational condition** for profit maximization is that supply units locate in order to minimize distance from consumer demand (with the constraint of a commodity-specific minimum threshold) and maximize distance from competitors (units supplying the same commodity); the latter, however, is not stringent, since there is horizontal differentiation and specialized areas may appear. A **further basic locational condition**, subject to the main one, is that the supply points of one activity will locate in order to minimize the distance from the supply points of other higher rank activities.

Within each iteration, Loesch is followed in assigning the lowest order activities first, however, a feedback from higher to lower order activities occurs over time. As in CDL, DP also want to explore whether a metropolis can emerge out of an initially uniform output demand and input supply distribution, without being exogenously picked out. Bearing in mind that commodities are still homogeneous, but that economic and geographical distance may not coincide and that supply is not deterministic, the process of appearance is sequential for each activity, so that a clear-cut functional rank does not emerge at any time, although the Loeschian features of intra-activity dispersion and inter-activity agglomeration are still a result. More precisely, a high level of development is favoured by fairly high accessibility, which helps higher-order functions to stimulate demand elsewhere through the income effect; poor accessibility and a high consumption share of income produce a rich metropolis versus poor periphery dualism.

Conclusion

Both Christaller's and Loesch's multi-sectoral spatial models were based on maximizing actors and produced a pattern of intra-activity dispersion and inter-activity agglomeration, although they differed in the derivation and structure of the resulting hierarchy of centres. Recent urban models have tended to follow newer, biological or other, paradigms, instead of the economic one.

However, monopolistic competition is now modelled in finer ways than at Christaller and Loesch's time. Notably, the urban hierarchy is viewed in terms not only of demand, but also of supply, including spatial externalities. In both cases, differentiation by consumer preferences (by product variety) applies beyond the spatial characteristic of the good and differentiation by consumer incomes (by product quality) is relevant too; in either case urban systems based on vast and dense metropolitan areas occur.

The economics of innovation also proves useful, notably in dynamizing spatial models. The gradual diffusion of activities in space is explained not just on the demand side by increasing purchasing power, but also on the supply side by technical progress reducing production and/or transport costs. This favours large cities, hence the perpetuation of the old hierarchies under a new guise, because of infrastructural investment costs, often to be made with still uncertain return prospects, and of the pre-emptive behaviour of these cities.

Thus, developments in the theory of imperfect competition provide a path to revitalization without discarding the insights that have been gained. This is already being done, e.g. in the SOUDY family of models, which incorporate the dynamic and stochastic characteristics of the new approaches within a strong theoretical economic framework, based on horizontally differentiated activities, technical and urban barriers to entry, local agglomeration economies and urban innovation.

There is in fact a growing consensus around the idea that urban development follows from choices in favour of 'innovative jumps', i.e. of the attraction of new activities or 'functions', where the strategic element is a supply one: the innovative potential of the locality. In the short run, innovative ability acts as an entry barrier, ensuring monopoly profits, or rents, to the innovative city; in the long run, incremental technical progress and learning effects allow other cities to 'catch up'. It follows that a metropolis may rest assured of maintaining its dominant role over time only in so far as it maintains a high quality and innovation rate.

What is needed next is a gradual, purposeful process of incorporation, which is especially necessary, and fruitful, for the relatively new approach of vertical differentiation. Differentiating economic actors by income, and indirectly by power, improves the understanding both of the effects of quality differences hidden behind nominally homogeneous activities and of the cumulative nature of hierarchical differences. This is already being done in the SOUDY2 model.

References

Allen, P.M. and Sanglier, M. (1979) A dynamic model of growth in a central place system, Geographical Analysis, 11, 256-272.

Allen, P.M. and Sanglier, M. (1981) Urban evolution, self-organization and decision-making, Environment and Planning A, 13, 167-182.

Arrow, K.J. (1962) The economic implications of learning by doing, Review of Economic Studies, 32, 155-173.

d'Aspremont, C., Gabszewicz, J.J. and Thisse, J.F. (1979) On Hotelling's stability in competition, Econometrica, 47, 1145-1150.

Bain, J.S. (1956) Barriers to New Competition, Harvard U.P., Cambridge.

Beath, J. Katsoulacos, Y. and Veph, D. (1987) Sequential product innovation and industry evolution, Economic Journal (Papers), 97, 32-43.

Beavon, K.S.O. (1977) Central Place Theory: a Re-interpretation, Longman, London.

Benson, B.L. (1984) Spatial competition with free entry, Chamberlinian tangencies and social efficiency, Journal of Urban Economics, 15, 270-286.

Berry, B.J.L. and Garrison, W.A. (1958) Functional bases of the central place hierarchy, Economic Geography, 34, 145-154.

Bowes, M. (1984) Profit maximizing versus optimal behaviour in a spatial setting, Southern Economic Journal, 50, 680-689.

Camagni, R. (1986) Innovation and the urban life cycle: production, location and income distribution aspects, in Nijkamp, P. (ed), Technological Change, Employment and Spatial Dynamics, Springer, Berlin, 382-400.

Camagni, R., Diappi, L. and Leonardi, G. (1986) Urban growth and decline in a hierarchical system: a supply-side dynamic approach, Regional Science and Urban Economics, 16, 145-160.

Capozza, D.R. and van Order, R. (1982) Product differentiation and the consistency of monopolistic competition: a spatial perspective, Journal of Industrial Economics, 31, 27-39.

Chamberlin, E.H. (1933) The Theory of Monopolistic Competition, Harvard U.P., Cambridge.

Christaller, W. (1933) Die Zentralen Orte in Sueddeutschland, Fischer, Jena.

Dasgupta, P.S. and Stiglitz, J.E. (1980a) Industrial structure and the nature of innovative activity, Economic Journal, 90, 266-293.

Dasgupta, P.S. and Stiglitz, J.E. (1980b) Uncertainty, industrial structure and the speed of R&D, Bell Journal of Economics, 11, 1-28.

Dematteis, G. (1985) Contro-urbanizzazione e strutture urbane reticolari in: Bianchi, G. and Magnani, I. (eds.), Sviluppo Multiregionale: Teorie, Metodi, Problemi, Angeli, Milano, 121-132.

Dematteis, G. (1988) La diffusione urbana: interpretazioni e valutazioni, in: Dematteis, G., Fubini, A. and Gambino, R. (eds.), Le Aree Metropolitane e le Reti Urbane, C.N.R. - P.F. Economia Italiana, Torino.

Dendrinos, D.S.and Mullally, H. (1985) Urban Evolution: Studies in the Mathematical Ecology of Cities, Oxford U.P, Oxford.

Diappi, L. and Pompili, T. (1988) Sistemi di citta come reti in interazione: un approccio dinamico di tipo Loeschiano, in Martellato, D. and Sforzi, F. (eds.), Studi sui Sistemi Urbani, Angeli, Milano (English version presented at the 28th European Congress of the Regional Science Association, Stockholm).

Diappi, L., Pompili, T. and Stabilinis (1989) Le citta come reti di funzioni interagenti: un approccio teorico e alcune simulazioni numeriche, in Diappi, L. and Mela, A. (eds.), I Sistemi Vzbani tra Concentrazione Metropolitana e Struttura Policentzica, Angel Milano (English version presented at the 29th European Congress of the Regional Science Association, Cambridge).

Dixit, A.K. (1979) A model of duopoly suggesting a theory of entry barriers, Bell Journal of Economics, 10, 20-32.

Dixit, A.K. (1980) The role of investment in entry deterrence, Economic Journal, 90, 95-106.

Dixit, A.K. and Stiglitz, J.E. (1977) Monopolistic competition and optimum product diversity, American Economic Review, 67, 297-308.

Gabszewicz, J.J. and Thisse, J.F. (1979) Price competition, quality and income disparities, Journal of Economic Theory, 20, 340-359.

Gabszewicz, J.J. and Thisse, J.F. (1980) Entry (and exit) in a differentiated industry, Journal of Economic Theory, 22, 327-338.

Gabszewicz, J.J. and Thisse, J.F. (1982) Product differentiation with income disparities: an illustrative model, Journal of Industrial Economics, 32, 115-129.

Gabszewicz, J.J., Thisse, J.F., Fujita, M. and Schweizer, U. (1986) Location Theory, Harwood, London.

Hay, D.A. (1976) Sequential entry and entry-deterring strategies, Oxford Economic Papers, 28, 240-257

Horstmann, I. and Slivinski, A. (1985) Location models as models of product choice, Journal of Economic Theory, 36, 367-386.

Hotelling, H. (1929) Stability in competition, Economic Journal, 39, 41-57.

Lee, T, and Wilde, L. (1980) Market structure and innovation: a reformulation, Quarterly Journal of Economics, 94, 429-436.

Lerner, A.P. and Singer, H.W. (1939) Some notes on duopoly and spatial competition, Journal of Political Economy, 45, 145-185.

Loesch, A. (1940) Die Raumliche Ordnung der Wirtschaft, Fischer, Jena.

Loury, G.C. (1979) Market structure and innovation, Quarterly Journal of Economics, 83, 395-410.

Menger, C. (1871) Grundsaetze der Volkwirtschaftslehre, Wien.

de Palma, A., Ginsburg, H., Papageorgiou, Y.Y. and Thisse, J.F. (1985) The principle of minimum differentiation holds under sufficient heterogeneity, Econometrica, 53, 767-781.

Pareto, V. (1896) Cours d'Economie Politique, Lausanne.

Pompili, T. (1986) Funzioni economiche, sviluppo urbano e tirannia della citta, in Camagni, R. and Malfi, L. (eds.), Innovazione e Sviluppo nelle Regioni Mature, Angeli, Milano, 163-187.

Pompili, T. (1989) Specializzazione e gerarchia nella regione economica milanese, Quaderni di Ricerca del P.F. Economia Italiana, C.N.R., Torino.

Prescott, E.C. and Visscher, M. (1977) Sequential location among firms with foresight, Bell Journal of Economics, 8, 378-393.

Reinganum, J.F. (1984) Practical implications of game-theoretic models of R&D, American Economic Review (Papers), 74, 61-66.

Rothschild, R. (1976) A Note on the effect of sequential entry on choice of location, Journal of Industrial Economics, 24, 313-320.

Schmalensee, R. (1978) Entry deterrence in the ready-to-eat breakfast cereal industry, Bell Journal of Economics, 9, 305-327.

Scholer, K. (1985) The welfare effects of spatial competition under sequential market entry, Southern Economic Journal, 52, 265-273.

Selten, R. (1975) Re-examination of the perfectness concept for equilibrium points in extensive games, International Journal of Game Theory, 4, 25-55.

de Serpa, A.C. (1986) Hotelling models: a general equilibrium approach, Southern Economic Journal, 52, 999-1009.

Shaked, A. and Sutton, J. (1982) Relaxing price competition through product differentiation, Review of Economic Studies, 49, 3-14.

Shaked, A. and Sutton, J. (1983) Natural oligopolies, Econometrica, 51, 1469-1484.

Shaked, A. and Sutton, J. (1987) Product differentiation and industrial structure, Journal of Industrial Economics, 36, 131-146.

Sonis, M. (1983) Competition and environment: a theory of temporal innovation diffusion, in Griffith, D. and Lea, A.C. (eds.), Evolving Geographical Structures, N.A.T.O. A.S.I. Series, Nijhoff, The Hague.

Spence, A.M. (1977) Entry, investment and oligopolistic pricing, Bell Journal of Economics, 8, 534-544.

Spence, A.M. (1979) Investment strategy and growth in a new market, Bell Journal of Economics, 10, 1-19.

Spence, M.A. (1981) The learning curve and competition, Bell Journal of Economics, 12, 49-70.

Sutton, J. (1986) Vertical product differentiation: some basic themes, American Economic Review (Papers), 76, 393-398.

Sylos Labini, P. (1962) Oligopoly and Technical Progress, Harvard U.P., Cambridge.

Walras, L. (1874) Elements d'Economie Politique Pure ou Theorie de la Richesse Sociale, Lausanne.

PART II - METHODOLOGICAL ISSUES

8. THE ARRIVAL OF NEW ACTIVITIES AND THE DUAL FORWARD-LOOKING EDITION OF THE (IMPROVED) LEONTIEF DYNAMIC INVERSE

Juan Sanchez-Garcia

Introduction

Growth is an unbalanced process of not only quantitative but also qualitative transformation. A crucial aspect of the qualitative transformation can be captured by the idea of new products and processes taking over from old products and processes. This transformational side of growth is captured by referring to the **diversifying economy**.

Let us postulate the concept of a **diversifying economy** as a way of embracing the growth-attrition process. We are not interested in growth as a 'g' rate but growth as diversification, which expands the growth rate 'g': this 'g' is the averaged statistical manifestation of diversification. The **diversifying economy** idea symbolizes the image of an umbrella under which a dynamic process is taking place. The aim of this section is to present the **diversifying economy** idea as a novel concept[1]. The aim of such a postulate is to capture a basic aspect of an economy as it is the modification of the 'basket of commodities' (the diversification), thus providing a more accurate concept in order to describe the economy's operation. The arrival of a new product (or service) is a problem of interest to a national as well as a regional economy.

Input-Output Analysis and the Diversifying Economy

Leontief (1965) touched on the subject of new products in the context of his inter-industry model when he delivered his dynamic input-output model to the economic profession. He sketched an analytical treatment of the above phenomenon although he did not pursue it further as far as we know. He criticised the way that conventional theoretical models were usually presented in terms of a system of differential or difference equations describing the development of a system of a **fixed** set of variables. Leontief wrote,

> '(the system) describes the position of an economy in the last year of, say, a period of twenty years in terms of the outputs and inputs of the same products and services which were employed to describe the original

position in the very first year of the twenty' (Leontief, 1965, p.130).

Leontief claimed that his dynamic formulation of the system of equations 'opens the path to the analytical treatment of... the introduction of totally new products'. (Leontief, 1965, p.130).

In Carter's words, Leontief 'describes successive input structures, with each year's outputs being produced from inputs produced in preceding years and without requiring that the inputs and outputs be identical or comparable over time'. However, she continues, 'while the approach avoids the unrealistic assumption that products remain qualitatively uniform, it is not helpful in pointing up regularities in how structures change' (Carter, 1970, p.9).

It is well known that the input coefficients of an industry in an input-output table are the weighted sum of the constituent coefficients of the detailed sub-industries. The weights take into consideration the relative size of each sub-industry's output. Hence a cluster of dynamically growing new activities whose outputs' sizes are not big enough can be absorbed into older and less dynamic categories. As Carter points out, 'whatever the new item [is, it has to wait] until it becomes large enough - old enough - to warrant new category space on its own' (Carter, 1970, p.12). However there are economists who see the need for an earlier consideration of new items in the interindustry grid if one wants to study the economy-wide impact of any technological innovation. Rosenberg says, 'a full accounting of the benefits of innovation must include an examination of interindustry relationships'. (Rosenberg, 1982, p.71).

Equally Metcalfe and Gibbon point out, 'it would appear essential to locate the new commodity within its network of interindustry input-output relationships before the full ramifications of the inducement mechanism can be explored'. (Metcalfe and Gibbon, 1983, p.21).

The outline of a mechanism which is able to attenuate some of the above deficiencies and, equally, is able to emphasize such a major economic issue as the introduction of new activities, is a worthwhile undertaking. Input-Output analysis is a good candidate to follow through the general equilibrium ramifications of any change. This is a strong argument for dealing with the introduction of new activities in an intertemporal dynamic input-output framework.

Leontief (1965) used a musical analogy for distinguishing between the polyphonic arrangement of a multisectoral analysis as opposed to the monolithic unisonity of the conventional aggregative theory of economic growth. In the context of the polyphonic arrangement, the arrival of a new product is like the introduction of a new instrument which changes the composition. The effect in the composition can be nice or bad, but what is clear is that it changes things. New instruments have continuously been introduced into the economy.

Hibernating Sectors : An Appropriate Analytical Dodge

The computable scheme for dealing with the introduction of new products is based on the analytical dodge of taking into account certain **hibernating sectors**. The anticipation of the fact that it is going to be a new activity at some stage over the time-span being considered will allow one to keep the matrices in the block of structural matrices of the same size.

Gossling has already pointed out the necessity of considering the economic phenomenon of diversification. He wrote: 'the arrival of a new industry... is simply handled by adding an extra column and row to main diagonal submatrices [of the time-phased structural matrix] for the industry's first and all subsequent years of operation'. (Gossling, 1975, p.43).

However we suggest that the use of the **hibernating analogy**[2] will provide a way of achieving insight which is not provided by the bordering approach, in particular in relation to the dual or price equation.

In some way the dormant mechanism bears a certain resemblance to the one mentioned by Lancaster in his 'new approach' to consumer theory article. Lancaster alludes - without naming - to the attempt of some economists in 'supposing the existence of a utility function for all possible goods, including those not yet invented'. (Lancaster, 1966, p.133). The problem with this assumption is that it needs to consider an infinite price for those non-existing goods in the map of a utility function for all possible goods. What we are proposing is a dynamic input-output setting with a matrix of structural coefficients taking into account those not yet invented commodities which will be in a dormant condition and their prices will be zero (as for 'free samples' of new items in shops, etc.) until the period when their capital capacity starts to be built up.

Let us take this analogy further and translate the standard input-output terminology into a biological analogy (Guyer and Lane, 1964). An input-output table may be seen as describing the combination of two metabolic processes, **katabolism** or destructive metabolism and **anabolism** or creative metabolism. The former is associated with the meaning of a row of an input-output table, that is to say with the process by which outputs are distributed and marketed as 'lots' of inputs to perform special functions elsewhere, while the latter depicts the role of a column, i.e. the process by which inputs are built up into outputs. This comparison allows us to frame the analogy of **hibernating sectors** into perspective by regarding them as sectors which have the property of staying in a state of diminished metabolic activity. The **hibernating sectors** will be depicted as those which present a self-anabolic mechanism and a non-katabolic one. The **hibernating sector** is founded upon a state of extreme sluggishness or stagnation of functions.

This paper focuses its attention on the wakening stage of the hibernating process, i.e. when a **rise in temperature** permits the sector to initiate the anabolic and katabolic processes. The arousal implies a **metabolic reorganization**. For the hibernation activity does imply meaningful activity before as well as after the dormant period; the former is thought about in terms of the technical feasibility, i.e. all the activities put into making the product technically feasible - R & D, prototypes, market research and so on - during the period called the new product development stage. Once the product is technically feasible it is placed in a sort of backlog condition - the 'backlog of structural improvements' as Carter (1970) called it - waiting for its wakening up to achieve the status of an economic, **suitable** product.

The question of what 'raises the temperature' is important. Is it changed demand conditions - external - or new technology in the sector concerned, finally reducing costs sufficiently to ensure profitable production? This is the old question (see Schmookler, 1966) of whether innovation is demand induced.

Now, before going further with the above analogy, let us describe concisely the inter-industry approach. The inter-industry approach is the result of some improvements made to the classical Leontief Dynamic Inverse model. This improved model has emerged as an unintended by-product of this author's main concern with new products (Sanchez-Garcia, 1987) this being the computable **forward** dual based on current and foreseeable future technological data, as well as information about the recent past. In this era of spreadsheets in personal computers it is not difficult to conceive an effective, unlimited accounting system which will embrace the structure of the past and the present and also give us a framework for projecting the future; all that we are doing is to put in a more explicit way things that accountants and growth modellers have already done.

A summary of the improved model is given next (see Sanchez-Garcia, 1987, for a detailed explanation). The primal or output equation can be written as MQ = E + R for short, where Q is the **column** of column n-vectors q_t, ..., q_{t+6} of industries' outputs, or **intensities of production**, postmultiplying the partitioned matrix M set of technological coefficient matrices for the same years; the result on the right hand side is the sum of two columns of vectors: the e_k for final consumption and the d_k

for capital replacements which are known at the beginning of year t. The normal structure can be seen below where G = I - A + K; H = I - A - K$\hat{z}$ wherein I = output matrix; A = input matrix; K = capital-output coefficient matrix; q_k = total gross output vector of industries in year k; d_k = predetermined (ie. known from industries' gross investment in years before year t) capital replacement vector in year k; $\hat{z}$ = diagonal matrix of (estimated) growth rates in final t+6th year; with k = t, t+1, ..., t+6.

On the other hand, the equation for the dual (or price equation) can be written as $P'M^{\triangle} = S'+W'$ where $M^{\triangle}$ represents the dual block of structural matrices in a partitioned form, which postmultiplied the **row** of row n-vectors p'_t, ..., p'_{t+6} of industries futures prices; S' is the row vector of 7 row n-vectors s'_t, ..., s'_{t+6} of estimated portions of industries' capital rental payments per unit of output and W' is the row vector of 7 row n-vectors w'_t, ..., w'_{t+6} of current (year t) and discounted future wages bills per unit of output for industries 1 to n. The $M^{\triangle}$ formal structure can be seen below where the main difference to be pointed out refers to the matrix T where T = I - A - $\hat{r}_{t-1}$K wherein $\hat{r}_{t-1}$ is a diagonal matrix of interest rates charged to the capital in year t-1 and to be paid in year t. The M structure of the primal must be read in a row-wise basis while the $M^{\triangle}$ structure of the dual must do it in a column-wise basis.

There exist two main differences between the Leontief model and our model:

i) Firstly, the existence in our model of an explicit initial condition in the dual or price equation, parallel to the Leontief-Petri's terminal condition of the primal (Leontief, 1970; Petri, 1972). The initial condition internalizes past information (the last year's rental, i.e. the cost of renting the capital) in the finite period under study.

ii) Secondly, the **forward** view of the economy that our dual or price equation formulation provides (compared with the Leontief's Dual backward view).

The forward-looking edition of the Dual of the (improved) Leontief Dynamic Inverse gives current discounted price levels for each line in a multi-industry system for the first year and a specified set of future years; hence the rates of return by line of business between adjacent years are also available.

Let us now come back to the above analogy and to portray it in the context of the output (Primal) and price (Dual) equations of the improved Dynamic Inverse.

Three different analytical levels are distinguished in depicting the output and price equations systems:

i) A first level which is related to the economic-commercial established world, i.e. those sectors already established whose products are commercially available. This first level stands for the working of the 'standard two-sided metabolic process', i.e. 'anabolism' and 'katabolism' of the hibernating analogy.

ii) A second analytical level is related to the world of backlog of new products; the book of blueprints. These are products which are not yet commodities at the beginning of the period under study but will in due course become available as commercial goods. This second level is outlined by the 'self-anabolic' stage of the hibernating analogy.

iii) Lastly, a third analytical level is presented as a bridge level between the second and first levels. This level is mainly associated with the investment set up needed to produce the new commodity. It is called the 'metabolic reorganization stage'.

The **hibernating sectors** in their dormant condition do not affect the solution of the system, in either the primal or the dual. The rows and columns of such sectors in the matrices of structural coefficients will be zeros. Only the identity matrix of the system of equations will show signs of the existence of new potential activities. The signs are the unity elements of the identity matrix which corresponding to the row/column of the **hibernating sectors**. The unity symbolizes the technical feasibility, referred to earlier as the self-anabolic mechanism. Those 1's secure the algebraical necessary consistence in order to compute the Dynamic Inverse. Some equations will be of the kind $1 \times 0 = 0$. The first symptom of the arousal is translated in the capital matrix by the shift of non-operative to non-negative elements in the set-up investment vector of the new activity.

Primal-Dual Interactions Employing Changing Technology Assumption

These relate to changes in processes and to the addition of new activities for new products and the fading away of old ones, in other words, to the inflationary cost pressures and industrial demand pressures associated with trends of rising capital/output coefficients, which have been reported for several Western countries from the mid 1950's to the early 1980's, can be added those pressures from the entry of new lines of business possibly relieved by the fading away of old ones.

New activities need capital to be set up. The new capacity will be built up by entrepreneurs. That will means expenses that later will generate a pay off. The business starts up with expenses, including payments to factors. There will be an initial build up of inputs before any output is produced.

The initial, or 'set up', investment vector of the new (n+1)th activity in year t is the new n+1 column of the K matrix times the output level in year t+1. The system of equations below, representing a time-span of two years, sets it out algebraically.

$$\begin{vmatrix} I_{n,n}-A_{n,n}+K_{n,n+1} & -K_{n,n+1} \\ 0 & (I-A+K)_{n+1,n+1} \end{vmatrix} \begin{vmatrix} (q_{n+1})_t \\ (q_{n+1})_{t+1} \end{vmatrix} = \begin{vmatrix} (e_{n+1})_t+(d_{n+1})_t \\ (e_{n+1})_{t+1}+(d_{n+1})_{t+1} \end{vmatrix} \quad (8.1)$$

The algebraic representation of the block matrices M and M will change slightly.

Assume a fictitious economy made up of 3 sectors. We are going to analyze it during a time period of 7 years. It is anticipated that there is going to be a new activity at some stage over the time-span being considered. All fixed capital shares a gestation period of 1 year and a lifetime of 3 years during which they operate at constant efficiency.

Let the new activity be introduced in the fourth year. In the first and second years the matrices of structural coefficients will contain a fourth column and row of zeros which describe the hibernating sector - see Figure 8.1 below. The fact of having a 7 year time-span and the fact that the new commodity becomes operational in the fourth year warrants the catching up of the capital replacement of the new capital set up for manufacturing the newly commodity, i.e. capital which was built up during the third year. This device allows us to keep matrices in the Dynamic Inverse of the same size. The new block of matrices, M, is shown in Figure 8.1.

The column and row vectors attached to the capital letters represent the technical evolution of the new activity.

The computational scheme needed to work out the output and prices equation will involve 28 equations for the primal and 28 equations for the dual - four of which will be of the style 1 x 0 = 0. In order to compute q, the vector of industries' intensities of operation, we need data for A, K, d and e. The output solution is sorted out on a once-and-for-all basis. M is inverted in one go.

Let us turn now to the dual formulation of the model using the hibernating analogy. The finite square block of matrices $M^{\blacktriangle}$ of the dual formulation is shown in Figure 8.2 below, when we are considering the introduction of a new activity in the

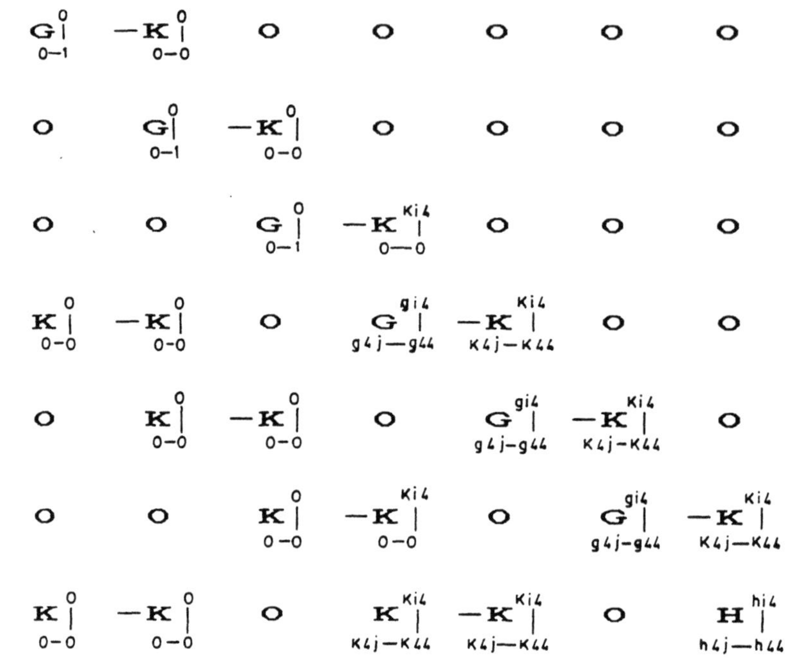

Figure 8.1 : M Block of the Primal Structural Matrices

fourth year of the finite period, i.e. t+3, and with its capital set up taking place a year earlier.

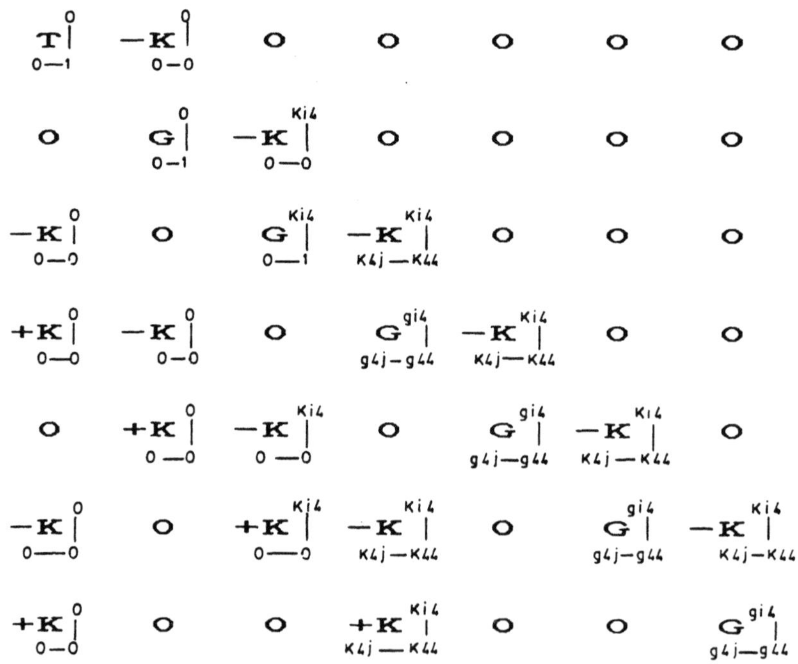

Figure 8.2 : M$^{\Delta}$ Block of the Dual Structural Matrices

As with the primal, the column and row vectors attached to the capital letters depict the technical incorporation of the new activity into the system, making use of the dormant sector provided by the hibernating analogy.

To get a clear picture of the working of the dual when a new activity is introduced, we have to work with the third column of the extended M^{Δ} matrix. The setting up of the capital for the new activity, in the primal, took place in year t+2 in order to be on stream in year t+3. In relation to the dual, in year t+2, the existing activities have a beginning and an ending capital stock, i.e. an extension of an existing capital stock; that is the reason why we have a difference of capital stock values involved. On the other hand, the new activity has only an initial capital stock.

From the numerical example presented in Sanchez-Garcia (1987) one interesting result is extracted: a positive price of a 'not-yet commodity' during the gestation time period of its capital set-up. There is a price attached to a not-yet commodity if we solve for outputs and assume that producers cover costs. The labour coefficient of the not-yet commodity is not operative; hence the fact that the labour is not paid means that it does not consume. That is consistent with the result obtained in the primal in relation to the output solution of the hibernating activity in period t+2, i.e. a nil output with a positive price.

An analytical explanation of this interesting outcome can be seen by obtaining the third element of the following dual equation, $P'M^{\Delta} = W' + S'$ in a partitioned manner. But before doing that, let us compare the output and price equation of the established economy in the first two years, when the hibernating sector is in its 'sleepy' condition.

The output equation for a year with sleeping sector is, in accordance with equation MQ = E + D and Figure 8.1.

$$Iq_2 - Aq_2 + Kq_2 - Kq_3 = e_2 + d_2 \tag{8.2}$$

or:

$$Iq_2 = Aq_2 + K(q_3 - q_2) + e_2 + d_2 \tag{8.3}$$

where $K(q_3 - q_2)$ covers enough for extension carried out during the second year, to be ready in the third year; it also takes care of the undetermined part of the capital replacement.

So, we have obtained an output; we sell it; the revenues should be enough for paying credits advanced to buy intermediate inputs (current costs); we know capital cost (cost of the capital used: value at the beginning of the second year, p'_1K, minus value at the end of the second year, p'_2K; labour cost; and future provisions cost ('depreciation') beyond the finite period; and future provisions cost, within the plan period. The price equation for the same year is, in accordance with equation $P'M^{\Delta} = W' + S'$ and Figure 8.2:

$$p'_2I = p'_2A + (p'_1 - p'_2)K + w'_2 + s'_2 + (p'_4 - p'_5)K \tag{8.4}$$

Let us see now the output and price equations in the third year, i.e. when the capital is set up for the new activity, in partitioned form; the second row and second column of the partitioned matrices depict the new activity.

Let one see first the output equation of the third year in accordance with $MQ = E+D$ and Figure 8.1, i.e. when capital for new activity is set up,

$$\begin{pmatrix} I_1 & 0 \\ 0 & I_2 \end{pmatrix} - \begin{pmatrix} A_1 & 0 \\ 0 & 0 \end{pmatrix} + \begin{pmatrix} K_1 & 0 \\ 0 & 0 \end{pmatrix} \begin{pmatrix} q_1 \\ q_2 \end{pmatrix}_3 + \begin{pmatrix} -K_1 & -K_2 \\ 0 & 0 \end{pmatrix} \begin{pmatrix} q_1 \\ q_2 \end{pmatrix}_4 = \begin{pmatrix} e_1 \\ e_2 \end{pmatrix}_3 + \begin{pmatrix} d_1 \\ d_2 \end{pmatrix}_3 \tag{8.5}$$

The output equation for the established economy in the third year will be:

$$I_1q_{1(3)} = A_1q_{1(3)} + K_1(q_{1(4)} - q_{1(3)}) + K_2q_{2(4)} + e_{1(3)} + d_{1(3)} \tag{8.6}$$

while the output equation for the hibernating activity is:

$$I_2q_{2(3)} = e_{2(3)} + d_{2(3)} = 0 \tag{8.7}$$

On the other hand, the price equation is, in accordance with $P'M^{\Delta} = W' + S'$ and Figure 8.2:

$$[p'_1\ p'_2]_2 \begin{pmatrix} -K_1 & -K_2 \\ 0 & 0 \end{pmatrix} + [p'_1\ p'_2]_3 \begin{pmatrix} I_1 & 0 \\ 0 & I_2 \end{pmatrix} - \begin{pmatrix} A_1 & 0 \\ 0 & 0 \end{pmatrix} + \begin{pmatrix} K_1 & K_2 \\ 0 & 0 \end{pmatrix} +$$

$$+ [p'_1\ p'_2]_5 \begin{pmatrix} -K_1 & -K_2 \\ 0 & 0 \end{pmatrix} + [p'_1\ p'_2]_6 \begin{pmatrix} K_1 & K_2 \\ 0 & 0 \end{pmatrix} = [w'_1\ 0] + [S'_1\ S'_2] \tag{8.8}$$

Price equation for the established economy in the third year is therefore:

$$p'_{1(3)}I = p'_{1(3)}A_1 + (p'_{1(2)} - p'_{1(3)})K_1 + (p'_{1(5)} - p'_{1(6)})K_1 + w'_{1(3)} + s'_{1(3)} \qquad (8.9)$$

The price equation for new activity in the third year, i.e. the year when its capital is set up, is therefore:

$$p'_{2(3)}I_2 = (p'_{1(2)} - p'_{1(3)})K_2 + (p'_{1(5)} - p'_{1(6)}K_2 + s'_{2(3)} > 0 \qquad (8.10)$$

If we want to balance the books, accountants may say, 'you can have an hypothetical price in commodity 4 - new activity - during the gestation period of its capital setting. The price that breaks-even includes a gestation cost, $(p'_{1(2)} - p'_{1(3)})K_2$; a future provision component during the finite period, $(p'_{1(5)} - p'_{1(6)})K_2$; and another future provision component that is provided for taking care of the capital used for setting the capital of the new activity beyond the finite period, $s'_{2(3)}$.

As with the replacement component of the primal, we have pre-determined future provisions $(s')_2$, and undetermined future provisions, e.g. $(p'_4 - p'_5)_2K$, in the price equation for the second year; it will be worked out with the solution of the whole price equation system.

Two different accounting levels, side by side, are provided in the input-output setting and they work; the first level refers to the existing activities, the accountancy of running an established business; while the second level refers to the cost of setting up a new line of business.

Conclusions

In Sanchez-Garcia (1987) the focus has primarily been on the introduction of new activity into an existing economy following on from Paolo Leon's (1967) seminal work in which the commercial 'suitability' of the outputs of such activities is most important in successful economic development. It is this focus, curiously, which has led to the creation of the forward Dual used in this paper: the economic process involved in the **setting up** of a new line of business, as well as its consequent operation, means that this business has very different 'prices' for two **consecutive** years; the first relates to unit set-up costs (at zero output) and the second to the usual unit costs of an established business. On their own, those results are nothing new, but in conjunction with the industrial interdependent system from which the new line of business has arisen, these price-results belong to two adjacent years in an intertemporal input-output system of costs and prices. Hence the forward Dual of this paper, previously presented in Sanchez-Garcia, arose as what is hoped is a 'suitable' tool in economic analysis.

This sort of technological change has one other result. Given that the new line of business has to be financed from some source or other, it may neatly be the fact that an old line of business is winding down and finds itself with cash to spare, but even so, the Primal and the Dual, when modified to take care of such technological change, have to be considered in conjunction with a system of banking and finance, money and credit. Nevertheless it is hoped that the results reported here are of sufficient interest to justify further investigation along these lines.

Notes

1. The **diversifying economy** is presented, using Good's terms (1962,p.1), as a 'partially baked idea' - p.b.i.for short. He defines a p.b.i. 'as a speculation, a question of some novelty, a suggestion for a novel experiment, a stimulating analogy, or (rarely) a classification. It has a bakedness "p" that is less than unity, or even negative. The bakedness of an idea should be judged by its potential value, the chance it can be completely baked, its originality, interest, stimulation, conciseness, lucidity and liveliness'. The intention of this research is to present the **diversifying economy** idea with a value $p \in (0,1)$.

2. In McCloskey's words: 'the metaphor is here an occasion for an instrument of thought, not a substitute' ... and... 'the more obvious metaphors in economics are those used to convey novel thoughts, one sort of novelty being to compare economic with non-economic matters'. (McCloskey, 1983, pp.507 and 503).

3. **Suitable** in the sense that the production decision is bound up with the expectation of a price level of the commodity under consideration that secures a spectrum of positive profit rates.

References

Carter, A.P. (1970) Structural Change in the American Economy, Harvard University Press, Cambridge Mass.

Good, I.J. (1962) The Scientist Speculates. An Anthology of Partly Baked Ideas, Heinemann, London.

Gossling, W.F. (1975) A Dynamic Model of Capital Replacement, in Gossling, W.F. (ed.) Capital Coefficients and Dynamic Input-Output Models, Input-Output Publication Company, London, 36-52.

Guyer, M. and Lane, C. (1964) Animal Biology, 5th Edition, Harper, New York.

Lancaster, K. (1966) New Approach to Consumer Theory, Journal of Political Economy, 74 (2), 132-157.

Leon, P. (1967) Structural Change and Growth in Capitalism. A Set of Hypotheses, Baltimore, The John Hopkins Press.

Leontief, W. (1965) Changements Technologiques, Consommation, Epargne et Emploi dans le Cadre de la Croissance Economique, Transactions of the Congress organized by the General Savings Bank and Pension Fund of Belgium on the Occasion of its Centenary, Brussels, pp.127-137, November 16-19.

Leontief, W. (1970) The Dynamic Inverse, in A.P. Carter and A. Brody (ed.) Applications of Input-Output Analysis, Vol.1, North-Holland, Amsterdam, 17-46.

McCloskey, D. (1983) The Rhetoric of Economics, Journal of Economic Literature, 21, 481-517.

Metcalfe, J.S. and Gibbon, M. (1983) Industrial Policy and the Evolution of Technology, paper presented to the Conference Technological Innovation and Production Structure: The Position of Italy, Milan, December.

Petri, P. (1972) Convergence and Temporal Structure in the Leontief Dynamic Model, in A. Brody and A.P. Carter (eds.) Input-Output Techniques, North Holland Publications Company, Amsterdam, 563-573.

Rosenberg, N. (1982) Inside the Black Box: Technology and Economics, University Press, Cambridge.

Sanchez-Garcia, J. (1987) The Diversifying Economy: An Interindustry Approach, Unpublished Ph.D., University of East Anglia (UK).

Schmookler, J. (1966) Invention and Economic Growth, Harvard University Press, Cambridge, Mass.

9. METHODOLOGICAL ISSUES IN THE DEVELOPMENT OF PREDICTIVE MODELS OF INNOVATION DIFFUSION

Neil Alderman

Introduction

Geographers and Regional Scientists are currently devoting a great deal of attention to the issue of technological change and its impacts and variation over space. Very little work appears to have addressed itself to the question of forecasting the uptake of new technologies in what is, after all, a period of great change and uncertainty, particularly within the sphere of manufacturing. Many agencies, both local and national, are concerned with the adoption and diffusion of new technologies and the implications this has for the development of appropriate policies to assist this process, such as the provision of suitable training facilities to enable companies to adopt and use the new technology efficiently. It is perhaps a not unreasonable supposition that if policy is to be responsive to the needs of technological change, the ability to satisfactorily forecast the patterns of diffusion of new technologies will be extremely important.

The development of predictive models that can be used to provide this forecasting facility for policy development at the local level is not, however, our primary concern here. The objective of this chapter is to draw attention to some of the methodological issues that arise in the context of forecasting the diffusion of new technologies within manufacturing industry at the local or regional level. Here we are concerned with models that represent the aggregate pattern of diffusion in a particular area, rather than models to predict innovation adoption by the individual firm or establishment.

These issues are illustrated by reference to a recent study that attempted to predict regional rates of diffusion of a number of key manufacturing technologies within the metalworking industries of Great Britain, through the application of Davies' (1979) model of process innovation diffusion. The paper explores the difficulties of measuring innovation diffusion through survey-based methodologies and discusses the implications of the dynamic context of technological change itself. The paper concludes with an evaluation of the usefulness of this model as a forecasting tool and identifies some of the requirements for establishing a viable method of technology diffusion monitoring.

Background

This work on diffusion forecasting was undertaken in the context of the concerns of policy-makers to be able to respond to changing patterns of innovation diffusion in different areas. The objectives of the project were to help lay the foundations for this by assessing the utility of Davies' model of process innovation diffusion (Davies, 1979) for forecasting the diffusion of new technology at the level of the manufacturing establishment, using data collected in CURDS on the adoption of a number of key manufacturing technologies (see Thwaites et al, 1982).

There is a considerable literature on technological forecasting and a wide range of methodologies (for example, Lanford, 1972). Most technological forecasting deals with long term forecasts of the direction of technology development (see Martino, 1972). Long term forecasts often give rise to spectacular errors. Little consideration appears to have been afforded the forecasting of the adoption of new technology by its users. Here the concern is with being able to forecast technology adoption over, say, three to five years. Three years was felt to be the appropriate time scale from the point of view of identifying a trend, devising policy to respond to that trend and implementing the policy. This sort of forecast is generally referred to as a medium term forecast (Lewis, 1982).

What is required is some form of analytical model that enables the pattern of diffusion of a technology through a population of users (in our case manufacturing establishments) to be identified. Empirical observation has demonstrated that the pattern of diffusion of an innovation through a population of firms or establishments is typically 'S-shaped' (for example, Mansfield, 1968; Metcalfe, 1970; Ray, 1984). Davies' model provides an attractive theoretical underpinning of a particular family of 'S-shaped curves'. The methodological issues raised by this work, however, should apply to any aggregate diffusion model based upon the 'S-shaped' curve.

Briefly, Davies' model identifies two types of innovation (termed types A and B), which are based on the cumulative lognormal and cumulative normal distributions respectively. The type of diffusion curve derives from assumptions made about the nature of the innovation and the learning process associated with its adoption. Type A diffusion patterns arise when the innovation is technologically simple, comparatively inexpensive and there is initially a rapid rate of learning about it on the part of potential adopters. Subsequent technological developments, however, mean that the innovation is soon superseded by alternatives and the rate of adoption declines (Figure 9.1).

Type B diffusion patterns arise when the innovation is technologically complex and expensive and there is, initially, a slow process of development as 'teething troubles' are overcome. During this stage adoption is slow, but subsequently a more rapid and sustained period of learning takes place and adoption proceeds swiftly, typically reaching a larger proportion of firms in the medium term than type A innovations (see Figure 9.1). In both cases it is the size of the firm that is the primary determinant of adoption with, on average, the larger firms adopting first (for further details see Davies, 1979; Alderman et al, 1988).

Operationalization of the Model

The two types of diffusion curve are represented by the following equations:

$$\Phi^{-1}(m_t/n_t) = a + b \log t \tag{9.1}$$

for the cumulative lognormal (type A) form; and

$$\Phi^{-1}(m_t/n_t) = a + b\,t \tag{9.2}$$

for the cumulative normal (type B) form. The state of diffusion is represented empirically by the proportion of establishments (strictly speaking potential adopters) that have adopted by time t, given by m_t/n_t in (9.1) and (9.2) above, where m_t is the number of adopters at time t, n_t the number of potential adopters at time t and Φ^{-1} is the probit transformation. The parameters a and b are to be estimated where $a = -\mu_D/\sigma_D$ and $b = 1/\sigma_D$, given $D_t = N(\log t \mid \mu_D, \sigma_D^2)$ for type A, and $D_t = N(t \mid \mu_D, \sigma_D^2)$ for type B.

The data used to calibrate and test the model were collected in 1981 for a study of new technology diffusion in manufacturing industry, undertaken in CURDS for the Department of Trade and Industry and the Regional Directorate of the EEC (Thwaites et al, 1982). Collection was by means of a postal questionnaire administered to all identified establishments within the following Minimum List Headings (HMSO, 1968):

MLH 331 Agricultural Machinery
MLH 332 Metalworking Machine Tools
MLH 333 Pumps, Valves and Compressors
MLH 336 Contractors' Plant and Machinery
MLH 337 Mechanical Handling Equipment
MLH 339 General Mechanical Engineering
MLH 341 Industrial Plant and Machinery
MLH 361 Electrical Machinery
MLH 390 Engineers' Small Tools and Gauges

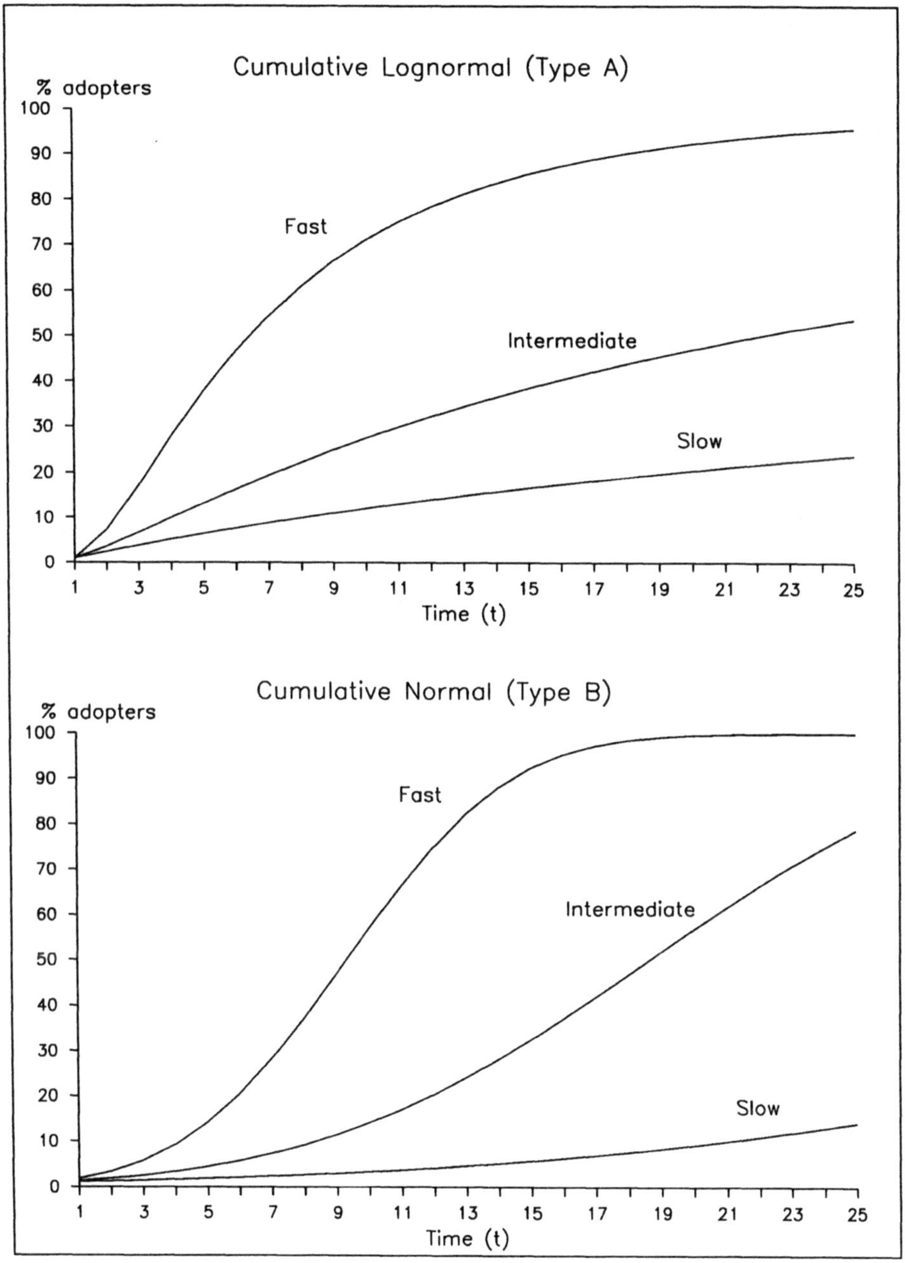

Source: Alderman et al (1988)

Figure 9.1: Typical Diffusion Curves.

This survey was followed up in 1986/87 in order to provide updated information on the adoption of selected technologies and to permit forecasts generated by the model to be tested against actual data. The use of extensive postal and telephone chase-up ensured an eventual response rate of over 95 per cent to this follow-up survey.

The innovations studied in this work fall into a number of groups. The first group are technologies for the control of metalworking machines (numerical control) (NC), computerized numerical control (CNC) and plug board control). The second are specific cutting technologies (electrical discharge machining (EDM) and coated carbide tools and inserts). The third refer to the general process of computerization (computers in commercial applications and microprocessors in manufacturing processes). The fourth group refer to product-related innovations (recirculating ball screws and microprocessors in products). Further details are provided in Alderman et al (1988).

Diffusion curves were estimated using a probit regression approach by means of the statistical package GLIM (Baker and Nelder, 1978). Various GLIM macros were programmed to allow forecasts of the diffusion of selected technologies within different regions of Great Britain to be generated and compared with the actual pattern as revealed by the follow-up survey. The results of these exercises have been presented elsewhere (Alderman, 1987; Alderman et al, 1988). Before this could be done, however, a number of methodological issues had to be tackled and it is these to which the chapter now turns.

Methodological Issues

The issues raised in the course of this research fall into two categories. The first concerns problems related to empirical measurement, and specifically those difficulties that arise through the use of a survey-based methodology. The second type of issue is substantive in that it concerns the nature of technological change and the way that the dynamics of this process affect attempts to model diffusion patterns.

a) Measurement problems

It needs to be stressed at the outset that the original survey described above was conducted in the context of a specific piece of research. At that time the idea of forecasting diffusion or of using the data to calibrate models for such a purpose had not been entertained. Many of the measurement problems described below are in part a consequence of the use of a particular data set for a different purpose to that for which it was originally generated.

i) Measuring time

Paradoxically, the definition of time in the context of an 'S-shaped' diffusion curve is a non-trivial problem. The first decision concerns the temporal units to use as the basic observations in the time series employed to estimate the shape of the curve. In practice this was dictated by the data available from the survey, which in this case was annual data. Dates of first introduction of the specified technology were provided by the respondent who filled in the questionnaire. Clearly, there were considerable differences between the lengths of time series available for estimation. Plug board control, for example, was observed to start diffusing in the early 1950s, whereas the diffusion of CNC is very much a phenomenon of the 1970s and 1980s. Thus, for some diffusion curves upwards of 25 observations were available, but for others it was less than ten. This inevitably reflected itself in the standard errors of the estimated diffusion parameters.

The apparent advantages of working with longer time series is offset by the need to rely on the corporate memory of respondents (which could involve documentary evidence or simply personal recall) in a cross-sectional survey carried out some time after the adoption event. Indeed, in many cases the particular respondent who completed a questionnaire may not have been with the company at the time the technology was introduced and the accuracy of responses could be considerably impaired. One effect of the inaccuracies in estimation of the date of adoption is the tendency for first adoption events to appear to cluster at five and ten year intervals. In practice this effectively precluded the analysis of a number of innovations that the survey originally investigated and the emphasis was therefore placed upon innovations diffusing during the 25 years prior to the 1981 survey.

Although not an option in the context of this particular project, one way of increasing the time series would be to use shorter units of time, say months or quarters. The disadvantages of this would be that respondent error could become magnified and forecasting periods would be correspondingly shorter. Moreover, the definition of adoption itself is fraught with difficulties, particularly if one is attempting to be very precise about the date adoption actually took place. In practice, the adoption of new technology can be a very long drawn out business. In recent research into the impacts of new technology, a time lapse of two years from the decision to adopt to the arrival of the equipment, was found to be not uncommon in the case of CNC (Alderman and Thwaites, 1987). This means that the accuracy afforded by shorter temporal units could well prove spurious.

This raises another measurement issue in terms of time: at which point in the process of adoption should the measurement be made? Is a technology adopted when the initial decision to do so has been made; does adoption occur when the equipment arrives at the factory; or is it when it has been finally commissioned and is beginning to contribute to productive output? It would be possible to ignore the first event given that decisions can be reversed if it is not left too late. However, there are technologies, particularly those in their infancy, that have to undergo many trials and on-site development before they can be said to be part of the production process. Early experimentation with computer-aided-design (CAD) systems or advanced cutting tool materials might come into this category. Pragmatically, adoption is usually taken to refer to the arrival of the technology in the factory, but the distinctions mentioned above can lead to distortions in the data.

Accepting the fact that there is bound to be measurement error in survey-based research designs of this nature, there is a more fundamental difficulty that relates to the length of the time series used for estimation purposes. In the case of the CURDS data the diffusion curves could be estimated on time series data up to 1980, the last full year covered by the original survey. The first date in the time series, however, is a non-trivial issue, since its choice directly affects the estimates of the parameters of the cumulative lognormal curve (Davies, 1979).

The solution chosen was to make the assumption that the diffusion process could be considered to have started in the same year regardless of location. This implies that all establishments were aware of the innovation and that the learning process started at the same time in each region. Whilst there may be very good technological reasons why the diffusion process might start at different times within different industries, at the regional level this is a less tenable argument given that each region has a share of each industry and of those large establishments that it is expected will be among the earliest adopters. In practice, this assumption had also to be applied to analyses for individual industries in the absence of any indication to the contrary.

The choice of starting date proved to be less than straightforward as well because, whilst secondary sources and other knowledge of the innovations would suggest one particular date, the empirical data sometimes suggested a slightly different one. Particular difficulties arose where there were clear recall problems on the part of respondents, as referred to above. This necessitated the removal or adjustment of obvious outliers in the data where the claimed adoption dates were inconsistent with other sources, such as trade journals or information provided by suppliers.

The assumption of a uniform or global starting date for the diffusion process across all regions also meant that for some regions no cases of adoption were observed in the early years of the time series. This is consistent with expected sampling variation, because in the early years the survey is sampling comparatively rare events. Nevertheless it was found to bias the goodness of fit and the curves therefore had to be estimated using only those years in which adoption had been observed, but with time still measured from the global start date.

ii) Entrants and Exits

The survey-based methodology used to provide data for estimating diffusion curves in this research means that observations are only available from a cross-sectional sample of establishments that have been in existence for all or part of the time series and remain so at the time of the survey. It is patently obvious, however, that industrial populations are not stable over time; there are both new entrants and establishments that exit for one reason or another (for example, closure, change of product or change of function, or, in the case of a regional time series, long distance relocation). The use of a sample survey allows entrants to be identified (in the CURDS survey information was elicited as to the date of start-up of the establishment), but clearly, not the exiters. Figure 9.2 shows that the effect of this can be to cause the observed proportion of adopters to fall between successive years if the rate of adoption is slow.

The implication is that using the proportion of adopters m_t/n_t to estimate the level of diffusion attained by time t will tend to lead to an over-estimation of the proportionate level of diffusion in the early years, declining to zero at time T (the last year in the time series), if n_t is constant. The effect of entrants (increasing n_t) will typically be to positively skew the diffusion curve. The precise magnitude of the error in estimating the level of diffusion in this way will depend upon the adoption characteristics of the exiters.

The critical factor is thus the behaviour of the exiters, about whom there is no information. The solution adopted in the context of this project was to generate estimates of the sample exit rates that should have been observed on the basis of published secondary data in the UK Census of Production. This provides figures for the number of establishments in the assumed population for each industry. From the census data the net effect of entrants and exits within the population can be identified. By assuming the observed sample entry rate to apply to the population as a whole, the population exit rate can be inferred. This exit rate can then be applied to the sample and the required correction factor is given in equation (9.3):

$$S_t = \frac{m_t P_T n_T}{P_t} \qquad (9.3)$$

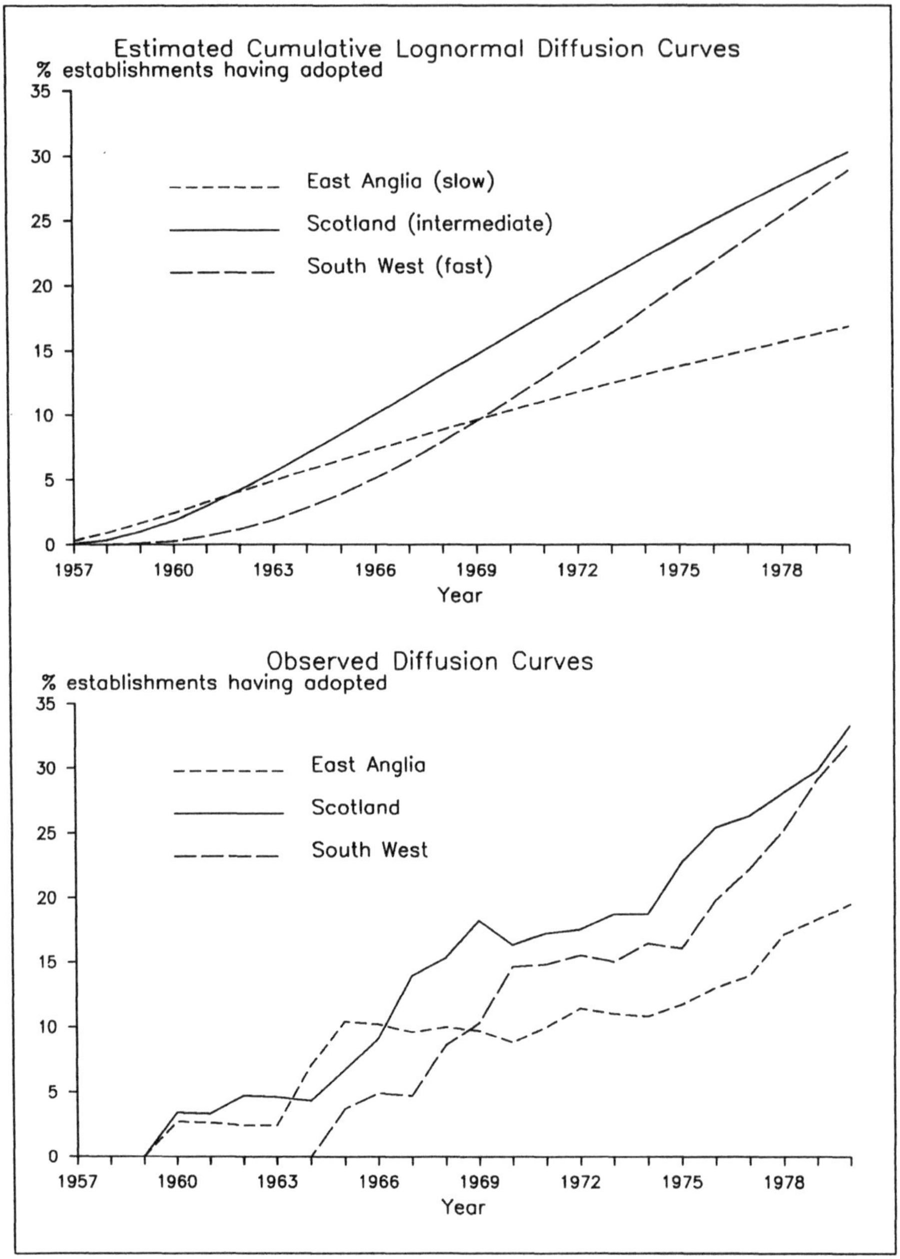

Source: Alderman et al (1988)

Figure 9.2: The Diffusion of Numerical Control.

where m_t = adopters in the sample;
n_t = total establishments in the sample;
P_t = total establishments in the population;
and S_t = corrected proportion of adopters in the sample
for $t = 1 \ldots\ldots T$

Equation (9.3) assumes that exiters will not have adopted. Whilst this is intuitively reasonable, for long time series it is perhaps more likely that at least some of the exiters will have adopted by the time they disappear. Indeed, when exit is attributable to corporate rationalization or takeover and merger activity, it will often be more than a possibility. The effect of this may be estimated by:

$$S_t = \frac{m_t}{n_t} \; \frac{n_t P_T n_T (1 - \theta) + \theta}{P_t} \qquad (9.4)$$

where θ measures the propensity of exiters to adopt relative to the survivors. It can be seen that if $\theta = 0$ equation (9.4) reduces to equation (9.3). If $\theta = 1$, that is exiters mirror the population as a whole, then $S_t = m_t/n_t$ and the sample estimate is that given by the data as it stands. Unfortunately there is no way of estimating θ. The approach adopted was to estimate diffusion curves with values of 0.0, 0.5, and 1.0 for θ. The results were evaluated simply in terms of goodness-of-fit of the respective curves to the data (Alderman et al, 1988). This was relatively straightforward in the case of industry diffusion curves, but secondary data on the numbers of establishments by region had to be approximated with information from other sources, which introduced additional assumptions. Evaluation was therefore restricted to the industry diffusion curves in this particular exercise.

The results indicated that the curves with the best fit to the data arose under the assumption that $\theta = 1.0$; in other words exiters mirrored the survivors in terms of adoption behaviour. Although this might appear counter-intuitive, it could reflect the fact that, rather than having similar adoption characteristics, the exiters are more akin to the entrants over the time period, since, typically, new entrants were not observed to adopt immediately and the inclusion of them could be compensating for the effects of exit. Table 9.1 shows that in virtually all cases new entrants had a lower level of adoption in 1980 compared to those establishments in existence at the time the innovation was assumed to have started to diffuse. There are good reasons why this should be the case. On average, new entrants will be smaller than the existing stock of establishments and ceteris paribus less likely to adopt. Moreover, high levels of investment in complex equipment would not be expected immediately, except perhaps in the case of green-field branch plants.

The conclusions reached in this part of the analysis were that the exclusion of exiters from the sample did not bias the estimated annual proportions of adopters and

Table 9.1

Proportions of Adopters in 1980

Innovation	Amongst establishments existing at start date	New entrants since start date
NC	0.24	0.31
CNC	0.27	0.14
Plug board	0.30	0.15
Coated Carbides	0.32	0.21
Computers	0.44	0.28
Microprocessors in processes	0.09	0.08
Microprocessors in products	0.20	0.14

it was acceptable to use the data as it was derived from the survey returns; a fortunate outcome given the difficulties of obtaining accurate figures for regional populations of establishments within each industry, at least in the British case.

iii) Saturation level

The saturation level is the proportion of establishments that ultimately adopt the innovation. The implicit assumption in the estimation of these types of diffusion curves is that the saturation level is 100 per cent, in other words all establishments will eventually adopt. It is quite possible, of course, that the saturation level is predicted to be attained so far into the future that it effectively becomes meaningless and the fact that not all establishments do adopt is not critical. This situation is particularly likely in the case of type A innovations, because they will tend to be superseded by alternative technologies.

The problem involved here is principally one of identifying the true set of potential adopters. If some members of the sample are not in fact potential adopters then the true saturation level will be less than 100 per cent of the full sample. To some extent, judicious use of supplementary survey information might allow some non-potential adopters to be identified. In the context of the CURDS survey, information concerning the range of metalworking operations was used to identify those establishments that would not have been expected to have adopted specific technologies. For example, establishments not engaged in metal cutting operations were excluded from the set of potential adopters of plug board controlled metal cutting equipment or coated carbide cutting tools.

In the case of some technologies, such as computers, it is reasonable to assume that all establishments could potentially use at least a microcomputer for commercial applications and so there is no a priori justification for excluding any of the sample. The identification of the set of potential adopters in relation to product innovations is much more problematic. In the case of microprocessor applications in products it might, to some extent, be possible to identify those establishments manufacturing products that seem suitable for the incorporation of microprocessors, but this would have proved time consuming and beyond the expertise of the research team in this instance.

In previous studies the problem of the saturation level has been circumvented by concentrating on small populations and large firms (e.g. Mansfield, 1963). In the studies reported by Nabseth and Ray (1974) ideas of a technologically feasible maximum or a technological ceiling are applied to define the set of potential adopters. In Davies' chapter on the tunnel kiln in brick-making the set of potential adopters is restricted to those firms using particular types of clay (Davies et al, 1974). These approaches are clearly not viable when studying a very large population of establishments or firms and innovations that have a wide range of possible applications.

An alternative solution, within the survey design, would be to include some kind of filter that sought to identify those establishments that would not adopt, because the technology was inappropriate to their operations. This method was tried in the follow-up survey, whereby respondents were provided with the opportunity to indicate whether or not they felt specific technologies were applicable to their establishment. Perhaps not surprisingly, considerable numbers indicated this to be the case. Analyses of these responses are not yet available, but there are good reasons for treating such information with caution. For a start, responses represent the perceptions of the person responding to the questionnaire at the point in time they fill it in and these perceptions are quite likely to alter as circumstances change and more is learned about the innovation in question. There is also a danger that a response of 'not applicable' might be a way of covering up for a lack of understanding or poor managerial performance where technology is concerned. Nevertheless, this aspect of the survey design did at least help to avoid the problem of non-response when an innovation was not considered applicable.

Even when the set of potential adopters has been defined as accurately as the data will allow, it is still possible that adoption may not proceed to 100 per cent saturation. Experiments were undertaken with a different software package that treated the saturation level itself as a parameter to be estimated. At best, however, only minor improvements in the fit of the curve were obtained and when the innovation had not diffused very far by the end of the observed time series the estimated saturation level was often extremely low; so low, in fact, for it to be unrealistic in the light of other knowledge about the innovation, such as market expectations or post-survey adoption rates.

The least unsatisfactory assumption about the saturation level thus appears to be that it will indeed be 100 per cent. This would also seem preferable from a theoretical point of view, since, even if at some point in time it is in reality less than that, it is unlikely that it would remain constant: the effect of both entry and exit, as discussed above, and the evolution of the innovation could be expected to affect the saturation level. In other words it would assume constant conditions which the model itself assumes to be changing.

Notwithstanding the problems of measurement described here and the difficulties arising from the use of sample survey data to provide an indication of the state of diffusion of new technologies, the overall conclusions reached were that, in the absence of demonstrable bias, the use of this kind of data provided models with good fits to the data and forecasts that were subsequently shown to be highly accurate over the medium term.

(b) The dynamic context to technological change

Having already referred to the dynamic nature of industrial populations and the issue of changing saturation levels it is pertinent to consider the dynamics of technological change itself, which will obviously have an impact upon the pattern of diffusion.

i) Changes in the innovation

One of the major criticisms of models of diffusion, such as that proposed by Davies, is that innovations do not retain the same characteristics with the passage of time. They may undergo quite substantial change and it is argued that this means that the set of potential adopters is constantly changing (see for example, Rosenberg, 1976; Gold, 1981; McArthur, 1987). To some extent this is true. Davies' model is not able to account for fundamental changes in an innovation, which might, for instance, cause an innovation to change from a type B to a type A. It might be postulated, for example, that the advent of the microcomputer could constitute just such a shift, with the result that the diffusion of computers would follow a different trajectory from the beginning of the 1980s.

It is important to note that Davies' model explicitly accounts for changes to the innovation that may be considered incremental and evolutionary. In fact it is the evolution of the innovation that allows it to be adopted by more firms, as improvements lead to greater expectations of profitability (faster pay-back) and lower risk. It should also be recalled that Davies' model was developed as a model of process innovation diffusion. It has already been shown how product-related innovations are more problematic and consequently likely to be less satisfactorily forecast than the diffusion of process innovations.

To some extent the problems experienced here stem from the way technologies are treated in generic rather than specific forms. Thus, CNC is regarded as a homogeneous technology with a clearly definable set of potential adopters, whereas one of the features of the last decade has been the expanding range of applications of CNC. Whilst it is likely that most of the early adopters of CNC were introducing CNC lathes (the most common application identified in the 1981 survey), today CNC is applied to all kinds of metal cutting machinery and increasingly to other forms of metalworking operation, of which a good example is the CNC power press for metal forming. If the model is to forecast well it may be necessary to anticipate such developments or to define technologies in much more specific terms (as Davies did in his original work).

The case of CNC illustrates another aspect of technological change, which is the convergence of different technologies. As a control technology it is lending itself to more and more forms of metalworking operation. One of these is EDM. Some of the most sophisticated metal cutting machines now being marketed are CNC spark erosion (EDM) machines (Gough, 1979), so that the diffusion trajectories of these technologies now overlap.

ii) The speed of change

As the development and diffusion of new technologies proceeds at an ever faster rate the ability to calibrate aggregate models based on some form of 'S-shaped' curve will become impaired. The time scales required in order to produce time series of sufficient length may not be available. In the experiments undertaken by Alderman et al (1988) it was found that more than ten years had to have elapsed before reasonable forecasts could be achieved. Furthermore, increasing product differentiation as a part of competitive strategy (Parker, 1985) and as a response to increasingly specialized customer requirements (Mensch, 1979) will exacerbate the problems of defining innovations as identified above.

iii) Integration

The CURDS survey work carried out so far has restricted itself to identifying the date of first introduction of a technology. Whilst it may be argued that this event is the most important one on the learning curve of the individual establishment, recent research suggests that, as the penetration of particular technologies proceeds within the establishment, so the problems faced become more difficult to resolve and the need for policy towards provision of advice and training is perhaps all the greater (Alderman and Thwaites, 1987).

The use of dates of introduction as a measure of diffusion fails to identify how the technology is applied within the establishment and the extent to which it is used. One of the principal current trajectories in manufacturing is the move towards integration through the computerization of the entire manufacturing process. In this way, many of the innovations studied are converging, in the sense that they form individual components of an integrated system. The rate and extent to which integration takes place within the manufacturing sphere can be expected to influence the diffusion of some of these components.

The Usefulness of Davies' Model as a Forecasting Tool

The forecasting ability of Davies' model was tested by generating predictions of the level of diffusion beyond the end (1980) of the time series used to estimate the diffusion curve. These were adjusted such that the estimated rate of change was applied to the observed level of adoption in 1980 rather than that estimated from the fitted diffusion curve, in order to avoid introducing any discrepancy due to the residual in the estimating equation. Table 9.2 shows some of the best results for forecasts to 1986 of the diffusion of coated carbide tools and inserts and computers in commercial applications. The data here have been aggregated and as such tend to cancel out forecasting errors that were apparent at the regional level.

Table 9.2
Aggregate Ex Ante Forecasts

Innovation	Year	Adjusted Forecast	% of sample	Actual	% of sample	Difference
Coated Carbides	1981	113.5	27.1	111	26.4	- 0.7
	1982	124.2	29.6	128	30.5	0.8
	1983	135.5	32.3	138	32.9	0.6
	1984	147.1	35.1	146	34.8	- 0.3
	1985	159.1	38.0	157	37.4	- 0.6
	1986	171.4	40.9	170	40.5	- 0.4
Computers	1981	168.3	43.6	177	45.7	2.1
	1982	184.5	47.8	189	48.8	1.0
	1983	201.0	52.1	204	52.7	0.6
	1984	217.5	56.3	223	57.6	1.2
	1985	233.9	60.6	244	63.0	2.4
	1986	250.1	64.8	263	68.0	3.2

It is clear from these forecasts that even over six years the predicted levels of diffusion match those actually obtained to a remarkable degree and, over the three year forecast period that is probably appropriate from the point of view of policy response, the errors here are marginal. Certainly, they are superior to naive forecasts or linear extrapolations of the data. Not all the technologies were forecast as accurately as this. Part of the explanation for this lies with some of the difficulties discussed earlier, in particular the shortness of some of the available time series and, in the case of CNC, the possibility that the type of diffusion curve itself had not been correctly identified (Alderman et al, 1988). Nevertheless, considering the turbulent period of recession and subsequent partial recovery covered by both the end of the estimation period and the forecasting period, a priori expectations as to the performance of these forecasts was not high. The fact that some of them have proved to be remarkably accurate provides some grounds for confidence in the ability of the model to forecast during a more stable economic period.

Davies' model has a number of advantages as a forecasting tool. It has a simple structure and can be readily estimated using proprietary software. The generation and adjustment of forecasts is straightforward and the input of new data is easy. It is most suitable for forecasting change over the middle part of a diffusion curve, when it is least sensitive to data error or error in the assumptions concerning the saturation level, and when the rate of adoption is most rapid. This is when the requirement for policy responses from relevant agencies is likely to be greatest.

The fundamental disadvantage with the model as a forecasting tool lies with its data requirements. The CURDS survey provided a unique opportunity for testing the model with a large quantity of data at the establishment level. This kind of survey work is usually only undertaken as a one-off exercise. In order to become a practical forecasting tool, this type of aggregate diffusion model would require some form of continual (e.g. annual) monitoring with a sufficiently large sample. An arbitrary requirement of at least 50 observations (establishments) for each region or industry was applied in our case, but even this was probably too low when the rate of diffusion was slow.

There is a need for clear definition of the innovations to be monitored; they should obviously be as homogeneous as possible. It is also necessary to be able to identify in advance those technologies whose diffusion it will be important to forecast (for example, Liff, 1986). Given sufficient knowledge of the characteristics of the innovation, one of the strengths of the model lies in the ability to identify the most likely type of diffusion pattern.

This technology monitoring exercise not only needs to identify the changing nature of the hardware itself, but also how it is being applied by the early adopters so that the opportunities for wider diffusion can be understood and anticipated. One of the problems encountered in the testing of the model lay with the long early tail to the distribution of adopters, particularly for the type B innovations, that corresponds to the period of 'teething troubles' referred to before. Successful forecasting would seem to depend upon the correct identification of the point of 'take-off' of the diffusion process.

It will also be necessary to monitor the population of manufacturing establishments. The forecasting period for our tests was characterized by an unprecedented decline in the number of establishments, followed by some degree of recovery. This meant that in the follow-up survey more than one fifth of the original respondents had disappeared for one reason or another. It might be hoped that the coming years will see less of a hiatus in the industrial population and a more stable environment for diffusion forecasting.

Conclusion

This chapter has reported some of the methodological issues arising from attempts to use Davies' model of process innovation diffusion as a medium term forecasting tool. Notwithstanding all the difficulties encountered and the instability in the environment over the forecast period, the results proved remarkably encouraging. Some of the requirements for a suitable monitoring system that would enable the model to be used as a practical tool have been suggested.

As the pace of technological change increases and the competitive position of firms and regions become increasingly bound up with it, so the need to be able to anticipate the problems and difficulties engendered by rapid technological change at a local level becomes more crucial. Many agencies, particularly those concerned with the provision of technological advice and training assistance, need to be able to identify potential bottlenecks in the technology adoption process in order to be able to respond at the right time. Private sector consultancies are also becoming more involved with the identification and installation of new technology in manufacturing firms so that there is a commercial imperative for good forecasting tools as well.

Successful diffusion forecasting will lead to a better anticipation of the requirements of firms and improve the timing and appropriateness of policy responses from both the public and private sectors. What this particular research has shown is that, in conjunction with suitable mechanisms for monitoring the adoption of new technologies, there are simple methodologies with the potential to provide useful forecasting tools given careful application.

There is also, however, a need to extend our understanding of the diffusion process itself. More needs to be known about the actual learning process in the diffusion of technology and how acquired knowledge is transmitted to manufacturers further along the diffusion path. Davies' model is essentially a demand-side model, dealing as it does with the actions of the adopters. It is, of course, recognised that it does not incorporate all the influences upon diffusion that undoubtedly exist. The actions of suppliers also affect rates of diffusion (Brown, 1981; Stoneman, 1983) and the role of Government policy has also been shown to have an influence in some areas (Gibbs, 1984).

Most importantly, though, there is a need to improve our understanding of the process of technological change itself so that our forecasting models can capture more of its dynamic nature. If it is the integration of manufacturing technologies that will be the key feature of technology diffusion in the 1990s, it raises the question of whether it is possible to measure integration in any meaningful way in order to be able to forecast its trajectory. Current forms of predictive diffusion model may prove inadequate for this task.

Acknowledgements

The research for this chapter was undertaken jointly with Alfred Thwaites of CURDS and Steve Davies of the University of East Anglia, with the financial support of the Training Agency. The assistance of Kieran McInerney and Trisha Gillespie with the follow-up survey is gratefully acknowledged. Any views or opinions expressed or errors and omissions are the responsibility of the author alone.

References

Alderman, N. (1987), Modelling regional innovation diffusion: a probit regression approach, paper presented at the Fifth European Colloquium on Quantitative and Theoretical Geography, Bardonecchia, Italy.

Alderman, N. and Thwaites, A.T. (1987), New technology and its implications for training in the engineering industry. Final Report to the Engineering Industry Training Board, CURDS, University of Newcastle upon Tyne.

Alderman. N., Davies, S. and Thwaites, A.T. (1988), Patterns of Innovation Diffusion, Technical Report, CURDS, University of Newcastle upon Tyne.

Baker, R.J. and Nelder, J.A. (1978), The GLIM System: release 3, Numerical Algorithms Group, Oxford.

Brown, L.A. (1981), Innovation diffusion: a new perspective, Methuen, London.

Davies, S. (1979), The diffusion of process innovation, Cambridge University Press, Cambridge.

Davies, S., Smith, R. and Lacci, L. (1974), Tunnel kilns in brickmaking, in L. Nabseth and G.F. Ray (eds.), The diffusion of new industrial processes, Cambridge University Press, Cambridge.

Gibbs, D.C. (1984), The influence of investment incentives on technological change, Area, 16, 115-120.

Gold, B. (1981), Technological diffusion in industry: research needs and shortcomings, Journal of Industrial Economics, 29, 247-269.

Gould, P.J.C. (1979), Advances in E.D.M technology, Metallurgia, 46, 315-320.

Lanford, H.W. (1972), Technological forecasting methodologies: a synthesis, American Management Association.

Lewis, C.D. (1982), Industrial and business forecasting methods, Butterworths, London.

Liff, S. (1986), The Technology Monitor 1983-84, Manpower Services Commission, Sheffield.

McArthur, R. (1987), Some aspects of technical innovation, diffusion and technical change: the diffusion of the colour scanner in the United Kingdom, in K. Chapman and G. Humphrys (eds.), Technical Change and Industrial Policy, Blackwell, Oxford.

Mansfield, E. (1963), Intra-firm rates of diffusion of an innovation, Review of Economics and Statistics, 30, 348-359.

Mansfield, E. (1968), Industrial research and technological innovation, Longman, London.

Martino, J.P. (1972), Technological forecasting for decisionmaking, Elsevier, New York.

Metcalfe, J.S. (1970), The diffusion of innovations in the Lancashire textile industry, Manchester School of Economics and Social Studies, 2, 145-162.

Mensch, G. (1979), Stalemate in technology: innovations overcome depression, Ballinger, New York.

Nabseth, L. and Ray, G.F. (eds.), (1974) The diffusion of new industrial processes, Cambridge University Press, Cambridge.

Parker, R.C. (1985), Going for growth: technological innovation in manufacturing industries, Wiley, Chichester.

Ray, G.F. (1984), The diffusion of mature technologies, N.I.E.S.R., Occasional Paper, 36.

Rosenberg, N. (1976), Perspectives on technology, Cambridge University Press, Cambridge.

Stoneman, P. (1983), The economic analysis of technological change, Oxford University Press, Oxford.

Thwaites, A.T., Edwards, A. and Gibbs, D.C. (1982), Interregional diffusion of production innovations in Great Britain. Final report to the Department of Trade and Industry and the Commission of the European Communities, CURDS, University of Newcastle upon Tyne.

10. A LONGITUDINAL ANALYSIS OF THE GROWTH OF FIRMS

Bert van der Knaap and Marina van Geenhuizen

Introduction

It is surprising to note how little empirical evidence there is to support the product life-cycle theory as an explanatory tool for regional and industrial change (cf. also Walker, 1979). Usually the level of analysis chosen is for classes or subclasses of industry, but not at the level of individual firms. Therefore there exists very little comprehensive knowledge about the spatial behaviour of individual firms and their responses to spatial structural change and renewal. More recently also other authors have made a plea for more micro research at the level of the firm (e.g. Thomas, 1987) to improve our knowledge and understanding of the current process of industrial change. Especially important in this context is the role of product-market strategies over time, their spatial implications and the spatial conditions required for the different strategies.

This paper will report the results of a pilot study carried out in the Netherlands. The study is based on a review of the literature and seven case studies. The focus of this pilot study was to examine the temporal relations between product-market strategies and the locational behaviour of firms within the context of a dynamic market. First of all, the growth of firms will be briefly discussed and it will be demonstrated how this led to different organizational responses. In addition to organizational change, the spatial pattern of urban-industrial growth has also changed. Different locational patterns seem to emerge for different types of industry, which in turn are related to different types of cities. Second, the relation between organizational structure, strategy and markets is discussed with respect to the nature and direction of change and changes in innovative capacity within firms. Third, the relation between internal adjustments (within the firm) and external changes will be discussed. Finally, in the conclusions some directions for further research are formulated.

Structural Change

The growth of firms and organizational change

The rapid economic growth during the 1960s has been of considerable importance in determining the structure and organization of firms. This change also has had its consequences for the locational requirements of firms. In addition to this - and this is a more recent phenomenon due to rapid technological change - the structure and organization of the production process itself has changed. Historically the industrial firm is seen as having grown from a one plant-one product unit with one location to a multi-plant, multi-product organization with a large number of locations for each unit (cf. Lasuèn, 1971; Watts, 1981). This organizational growth had a considerable impact on the spatial structure of production.

At least two major effects of such changes can be observed:

(a) The organizational structure of a firm has become an important factor in the operation of the firm and creates separate spatial conditions for the operation of the multi-plant firm. This development has been documented and discussed by G. Törnqvist (1970 and 1978), who used a simple conceptual model to demonstrate the nature of the complex relations between spatial flows between the different units of one production organization (internal flows) and flows between units of other production organizations (external flows). Using the concept of the production column Törnqvist pointed out the spatial separation of the various corporate activities. As a consequence of the process of change the emphasis shifted, from primary production processes, with a dominance of physical flows, to secondary and tertiary processes, with an emphasis on organization and management, in which information flows become increasingly important.

(b) The external environment is no longer homogeneous for the firm, but has become a variable relative to specific parts of the production organization. This suggests that the external conditions suitable for the location of different production units vary, which will give rise to a spatial segmentation of possible locations. They become more specific for one or two production functions and less suitable for the whole array of activities encompassed within the whole organization.

During the 1970s another development occurred which changed again the organizational structure of the firm. In this period there was a strong tendency towards mergers, diversification and increases in the scale of production within one firm. This process seems to have come to an end in this specific organizational format. Firms tended to reduce their direct influence over a wide range of functions and focused on their main tasks. Concomitant with this development was the increased independence of the different production units. A major reason for this type of change was the change in economic conditions (saturated markets) and the beginning of the economic recession. There was therefore a shift in firm strategy towards an emphasis on risk minimalization by risk avoiding and risk sharing behaviour (Krumme and Hayter, 1975; Fredriksson and Lindmark, 1979).

The first strategy contributes to the increase in importance of subcontracting and also the growth of the number of small firms (see Keeble and Wever, 1986). This leads to a further increase in the primary and secondary producers' markets. The large firm operates on the primary market which is a rather stable market and dominates from here its production relations with the small producer. In this context it is important for the small firm operating on the secondary market - which is unstable - to develop linkages with more than one large firm. This risk sharing behaviour from both the large and the small firm increases the role of material and non-material networks. Not only will the number of networks increase, but also the movements on the network. This changing organizational structure of the firm is reflected in the growing demand for producer services. The latter development is an important generator of demand for non-material services and therefore fosters the growth and strengthening of non-material networks.

From some of the structural changes, such as the mergers and concomitant diversification within firms, the subsequent focussing on core-activities and the associated increase in subcontracting relationships, it may be suggested that the product life-cycle has lost its importance in the growth strategies of individual firms during the 1970s and 1980s. The concept of the product life-cycle will be examined in detail in one of the following sections.

Urban-industrial change

Industrial development and urban growth have generally been considered to be synonymous during periods of rapid industrialization. However, in the industrialized countries since the 1950's this relationship has changed, all be it gradually in the beginning. These changes in the spatial pattern of industry have become evident in the Netherlands in two ways. First, a downward shift in the location in the urban hierarchy can be noted. Second, concomitant with this downward shift, there is also a noticeable shift away from the urban core towards the more peripheral regions. This shift is by no means observable at the same level of intensity and in the same way for all industrial sectors. To understand these differences a distinction has to be made between traditional and modern sectors (cf. Van der Knaap and Louter, 1986; Atzema et al., 1986). The traditional (industrial) sector contains the 'older' industrial activities, such as textiles, ship-building and heavy metal industries. This division also relates to services, which can be separated into tertiary and quaternary services. The latter part is the modern segment sometimes referred to as commercial or producer services.

For the urban system as a whole one can note that the larger and more centrally located the city, the larger the share of the service sector in its economic composition. Within industry and services generally there is also a shift towards the more modern segments, namely modern industry and commercial services. This development is illustrated in Table 10.1. The shift in the traditional industry from large to medium sized cities and smaller settlements can be clearly noted. A number of developments, however, do not fit the general pattern described above. It is surprising to note that there is a relative under-representation of commercial services in the medium-sized cities in the core region (the Randstad). This is in contrast to their presence in the medium-sized cities in the intermediate adjacent zone and the more peripheral regions of the country (see also Van Dinteren, 1986). It is also interesting to note the spatial variation in the intermediate zone. Moving away from the core towards the East there is a re-orientation between traditional industries and services. Moving towards the South one can observe a continuing specialization in traditional industry and for certain larger medium-sized cities a new orientation towards modern industrial activities. From these observations it may be concluded that the relation between industrial development and urban structure has become rather more complex in recent years. How do these observations relate to technological change at the micro level?

Recent studies by Kok (1985) and Kleinknecht and Mouwen (1985) have attempted to measure the degree of innovativeness of individual firms. Several indicators were used to define the degree of innovativeness, including on the input side, investments in research and development (R & D) and on the output side, the number of new patents registered and the number of incorporated innovations. From these studies it appears that as an explanatory factor the (relative) location or orientation of the firm with respect to the urban core region has diminished in importance. Firms located in large urban areas do not exhibit a larger degree of innovativeness than firms located elsewhere, although firms located in the intermediate adjacent zone have a somewhat higher innovative performance.

This changing pattern of urban-industrial activity is not dominated by a high degree of industrial mobility. Movement only occurred to a limited extent in the 1950s and early 1960s (Reinink, 1970; Wever, 1985). Much more important seems to have been the internal structural change of the firm in terms of its adaptation towards growth and decline, responses in innovative behaviour and changing composition of products for intermediate and finite markets. Relocation behaviour

Table 10.1

Location quotients[a] of traditional industry, modern industry and quaternary services in different cities (large (L), medium-sized cities (M), smaller settlements (S)).

Traditional industry

	Urban hierarchy			Centrality Medium-sized cities		
	L	M	S	Core	Interm.	Perif.
1950	68	105	119	83	102	118
1963	49	100	137	78	98	128
1978	35	88	140	81	91	138
1984	36	85	135	81	95	134

Modern Industry

	Urban hierarchy			Centrality Medium-sized cities		
	L	M	S	Core	Interm.	Perif.
1950	91	114	95	120	105	63
1963	74	123	99	113	109	73
1978	70	123	99	104	101	93
1984	64	127	98	99	104	95

Quaternary Services

	Urban hierarchy			Centrality Medium-sized cities		
	L	M	S	Core	Interm.	Perif.
1963	198	81	47	91	98	113
1978	179	91	67	82	120	95
1984	170	89	79	87	110	103

a. the Netherlands = 100 (labour)

Source: Van der Knaap and Louter 1986, pp. 38, 46.

can only be observed in a limited number of cases and for innovative firms these tend to be short distance moves (De Jong, 1987).

This raises questions about the long term behaviour of firms in relation to their product-market strategies and their information requirements in order to operate and continue to operate successfully in different urban environments (Davelaar and Nijkamp, 1987). In the next section the relation between markets, strategy and the behaviour of firms is examined within a theoretical context.

Organizational Structure, Strategy and Markets

The increased attention paid to cyclical patterns of growth has stimulated a large number of authors to describe the growth path of a firm within this overall economic context. This has led to a number of related theories such as the product life-cycle theory (Vernon, 1966), the innovation cycle theory (Rothwell and Zegveld, 1985) and the dynamic market theory (De Jong, 1985). Vernon's formulation of the product life-cycle theory can be considered as one of the first attempts to model the dynamics of growth. Usually three or four stages in the growth path are identified, by which the development of a product is described relative to its design, production mode and associated skill levels, market development and location of production. The relation between these factors is summarized in Table 10.2.

In the innovation cycle theory technology is the driving force. The role of technology varies throughout the product life-cycle and associated with this, the nature of the innovations change. The authors while accepting this interrelation between them, make a distinction between process and product innovations. During the life cycle the emphasis changes from key product innovations to process innovations, accompanied by relatively unimportant product improvements. In the second stage of the cycle, when the technology has been developed and mass production is being achieved, there is little room left for product innovations and product change and the emphasis is upon efficiency of production.

In the dynamic market theory the third stage of the life cycle is an important theme where the required level of technological development is associated with market saturation. In the first stage of the product development infant firms, which usually suffer from a shortage of capital, are vulnerable to co-operative agreements, mergers or take-overs. This process of relative concentration can lead to an agglomeration of similar producers (for example, through horizontal merger) or with other producers in the same industry (for example, a vertical merger). When moving from the expansionary stage to maturity another cycle introducing organizational change can occur. The takeover of a growth firm by (usually) a

Table 10.2

The Product Life Cycle : Spatial Aspects of the Various Stages

Stage	Sub-Stage	Result	Location	Skill Level	Urban/ Rural	Market
RESEARCH AND DEVELOPMENT (introduction)	- fundamental research - applied research - development prototype	discovery invention prototype	university company lab. company lab.	high	large urban areas	- - test
PILOT PRODUCTION	- development and testing production process - small scale test production	production process product	´near´ HQ and R&C lab			introduction
STANDARDIZED PRODUCTION (maturity)	- automated, standardized mass production	homogeneous products	´detached´ production plants	low	peripheral rural areas	growth maturity saturation

Source: adapted from Kamann (1986, p.139).

larger one secures growth for the larger firm without taking risks and without losing profits. In this was it is able to continue its performance through replacement of old activities.

Although the concept of the product life-cycle can be considered as a descriptive tool, it has generated a wide range of criticisms that have gone beyond the concept itself (cf. Van der Knaap and Linge, 1989) and dealt with some of the underlying principles of industrial production and organization:

(a) **The structure of production.** The theory is geared towards large multi-product, multi-plant firms with a sharp distinction between large and small firms. It is implicitly assumed that small firms are always dominated by large firms (Taylor, 1986). In this way large firms can determine which location is the most advantageous with respect to labour cost, tax advantages, access to markets, etc. It should be noted however that small firms operate differently and even within large firms it is possible to follow individual strategies (cf. Harrington, 1985).

(b) **Sectoral and technological determinism.** First of all it is surprising to note that the only evidence for the existence of a product life-cycle is within industry and not within other sectors. This is partly explained by the fact that innovations and technological change play such an explicit role in the process of change. The assumption is that a given technology can be associated with a specific type of production in a specific stage in the life-cycle. Daly (1985) pointed out that in biotechnology, for example, both new firms in their pioneering stage and older firms in a mature stage can be using the same technology while their strategies may and usually will vary considerably.

(c) **Product differentiation and level of aggregation.** The theory is unsatisfactory as it simplifies commercial reality. Enterprises do not compete with homogeneous products, but with distinct qualities or prices aimed at specific market segments. Each of these specific products has its own cycle, on a rather different level than the sectoral cycle. Moreover, enterprises usually have a combination of several products and these may vary in age or stage in the life cycle. Given these observations, the question then arises as to what level of aggregation is appropriate for testing the theory.

(d) **Unicausality and timing.** A basic assumption of the theory is that one technology is replaced by the next in order to move forward along the cycle. In some sectors, such as the textile machinery industry, there is a strong emphasis on continuous innovation as a means of survival (cf. Walker, 1979). The techniques of production become the key factors in such cases. A related point is the time it takes to develop a full cycle. In 'high-tech' industries this can be rather short; in some cases less than eight years (Markusen et al, 1986), while in other instances it may take 30 years or more (Oakey, 1985). The necessity for a firm to engage in a full cycle is not directly apparent and is certainly not always the case for small firms, which may remain in the first or second stage and leave the ultimate mass production to other firms who have a better knowledge of the consumer market.

(e) **The role of the environment.** A commonly accepted hypothesis is the comparative advantage of the metropolitan environment for technological and economic progress. There is however little empirical evidence for the incubator theory (Davelaar and Nijkamp, 1986). On the contrary, Gudgin and Fothergill (1984) concluded that there was little difference in the availability of key factors such as skilled workers between the urban and the rural environments. Innovation activity may even suffer from dis-economies (bottleneck factors) in the urban environment.

(f) **Organizational dominance.** The product life-cycle theory is based upon the principles of scientific management, (viz. Taylorist principles) which results in a specific approach to the organization of production. Other approaches based on the principles of specialization are also possible based on a different organization of work. In this second approach the emphasis is on a decentralization of (operational) decision making in production, marketing and sales, while at the same time strategic decision-making power on matters such as investment and financial management are being centralized. This process may lead to the elimination of the intermediate layers of decision taking which will have severe consequences for the future distribution of skill and jobs.

The above criticisms of the product life-cycle theory, as well as the current extensions of this theory, have highlighted the interplay of three important intervening factors, viz. the organization of the firm, the strategy of the firm and its markets. The market is no longer a static element but a dynamic factor which adjusts itself and through such adjustment creates room for further alterations to meet changing social and economic conditions.

It is hypothesized here that the nature of the interrelations between the determining factors will vary with different spatial environments. The environment is not only operating as a constraint, but also possesses dynamic qualities leading to the creation of new conditions, which in turn will lead towards structural change and in some instances to renewal. From the above discussion it appears that within firms the renewal strategy formulated within the product life-cycle concept is only one type of strategy available from a whole range of alternatives. The different strategies will vary according to variations in specialization or diversification, increase or reduction in the scale of production, introduction of new products or the opening of new markets, or orientation towards specific markets and the relative importance of innovation in the production process.

The role of the market is also seen as an important element in the analysis of the locational dynamic of the firm. This locational dynamic refers both to relocation behaviour and structural adjustments at the same location. For firms with specialized products there is a considerable difference in product strategy for stable and unstable markets. On the one hand, firms with a diversified product range may have a stronger strategic orientation on both growth and non-growth markets. This could explain the behaviour of older and more diversified firms. They often have a smooth endogenous growth path with a constant renewal of products and little locational dynamics. While on the other hand a young and specialized firm may experience leaps in growth resulting from the production of a new product in an unstable market. Such a firm can have a strong locational dynamic. Given these considerations, the longitudinal empirical analysis outlined below focused on the following questions:

(1) What is the direction and nature of strategic change in individual firms, concerning their products, markets and innovative behaviour? What is the influence of market characteristics and organizational characteristics on this evolving strategy? What is the relevance of the concept of the product life-cycle in relation to strategic change?

(2) In what respects does this strategic change and its determining factors vary with different spatial environment?

(3) How are different strategies related to different means of growth (internal, external) and different locational adjustments?

Patterns and Causes of Strategic Change

Aim and method of the case studies

In the pilot phase of the research project seven case studies of industrial firms were carried out, to test the viability of a longitudinal analysis from 1950 up to the present, to discover answers to the questions raised above. The sample of firms was based on a selection of previously mentioned firm characteristics, aimed at capturing as much variety as possible, viz. different sizes, one-product and multi-product, single-plant and multi-plant, traditional and modern sectors, and the incidence of strategic 'jumps' in recent years (Appendix 10.1). Within the overall conceptual framework described above the following selected themes have been explored:

(a) Nature and direction of strategic change (products and markets), with special reference to the product life-cycle.

(b) Innovation strategy.

(c) Method of growth (internal or external), and related locational dynamics.

(d) Background to strategic change (position of firm, market dynamics).

Longitudinal studies are not novel and some company-history analysis has been published, for example, Siemens (Krumme, 1970), Boeing (Krumme and Hayter, 1975), firms in biotechnology (Daly, 1985), and Philips (Muntendam, 1987; Froling and Zoon, 1988). The study of Siemens provides an insight into the spatial organization of the firm (multi-plant, multi-regional) and its ability to adjust locations and activities to external changes (labour market). The study of Boeing focuses on the spatial implications for employment of corporate strategies, in particular in relation to the life paths of products. The corporate histories in biotechnology provide the opportunity to compare different strategies within the same industry, but this analysis has no spatial perspective. The studies of Philips have a strong accent on strategic adjustment in a changing economic environment, both in broad terms and in detail for recent years. Unfortunately these studies are very limited in the context of the present study as they differ in spatial perspective and focus on strategy. Studies dealing empirically with the evolution of firms from the perspective of restructuring and innovative behaviour are very rare indeed.

As the empirical analysis is retrospective, annual reports and in-depth interviews have been used as data sources.[1] Based on the experience with the seven cases, it can be concluded that the development path of firms and the most important influences upon it can be reconstructed satisfactorily by this method. However, there are some restrictions in the interpretation of collected data due to a lack of detail from the 1950s and 1960s, for example concerning R & D. In addition to this, only a few of the firms were found to publish annual reports.

In the interviews a structured schedule was used and in most cases corporate personnel were interviewed who had been with the firm for at least 25 years (or 10 years in cases of establishment after 1950). Nevertheless, the quality of the data was affected to a certain degree by response error (memory gap) or bias. In cases of doubt additional data sources have been used, to supplement the interview material, for example, economic journals. At the same time, some particular themes in the interview were devoted to the collection of hard data, in order to guarantee comparability of the firms. This type of data referred to the different growth strategies, different types of innovations and to the timing of the changes (viz. the decision for change or the implementation).

Within the framework of the case studies no general access problems with respect to the company were experienced, but some problems may be expected in the case of product development if it is surrounded by strict confidentiality requirements (production for military purposes or competition in top class research fields). In the next section the results from the case studies are discussed.

Nature of change

An enterprise may be in a relatively static or dynamic economic environment. The manner in which the firm responds to its economic environment depends on the nature of environmental change and the structure of the firm, for instance the ability to sense changes. The case studies revealed the following three basic patterns of strategic adjustment to market dynamics:

(a) The market changes gradually and the firm's strategy develops incrementally with it. This development path is characterized by many gradual adjustments, such as increases or reductions in production, penetration of new geographic markets and introduction of new production methods, such as 'just-in-time'. In this type of situation two or three changes towards new products and/or markets can occur within a period of approximately forty years. However, large diversified or conglomerate firms experience a much more complex development path than that described here.

(b) The market change is sudden or at a much greater rate than usually experienced and this requires fundamental strategic readjustment. A major change in market demand, for example due to a sudden drop in oil prices, makes a major shift towards new products and markets or a complete reorientation of the business necessary.

(c) The market changes gradually, but the firm fails to develop with it, until a fundamental readjustment is inevitable. This type of situation seems particularly prevalent during the stage of market success, when the management may fail to sense changes that result in competition from more advanced products or cheaper imitations.

One reason for the last mentioned pattern is associated with the structure of many modern large enterprises, which may hinder strategic awareness (Johnson and Scholes, 1984). These enterprises are likely to be multinational, split into divisions and structurally specialised. Such a firm was illustrated within the case studies where it did not perceive the influence of a particular cheap technology in time. The resultant drop in sales coincided with high integration costs of an acquisition. In order to survive, the vertically integrated production firm experienced a dramatic transformation into a marketing and service organization, leaving R & D and production to others.

Direction of change

The product life-cycle presents a unidirectional change process. Although this can be true for sectors of industry as a whole (Webbink, 1985), as previously stated this is rather unrealistic in the context of individual firms. The case studies show that within firms product life-cycles are often truncated and this can happen in different stages of the cycle. Sudden change in market perspectives is one reason, the focusing of attention on main activities is another. Firms can even develop in the opposite direction to the sectoral cycle, for example, to exploit and flourish in a small market niche within saturated markets. The relation between firm and product life-cycle is therefore rather complicated for several reasons. Multi-product firms, diversified into several sectors, have cycles that do not run simultaneously. Some firms have continuous fundamental R & D in certain product areas, while their current products are in several later stages of the cycle as well. Other firms may be more or less specialized in one (sub)stage of the cycle, for example in the production of prototypes. In addition, the theory of the product life-cycle is not specific about certain important business strategies. For instance, in the late growth stage of the cycle investments can be made in machinery (cost saving) or in R & D, staff and market research. The difference between an in-depth or more broadly oriented investment pattern seems to be a key factor in the performance of the firm (De Nationale Investeringsbank, 1987).

The previous examples illustrate that enterprises develop according to several principles of which the product life-cycle is only one. The case studies revealed other important market dynamics such as specific instability in markets, general economic fluctuations, removal of trade barriers (Common Market, 1962 and nowadays) and changing prices of raw material (Common Market), etc.

Changes in innovative performance

Technological innovation can be an important strategy for survival in a competitive environment. Nevertheless, there is a large variation in the degree of importance attached to innovative capacity between individual firms. Much attention has been paid to different innovation strategies and related typologies of firms (e.g. Freeman, 1982; Porter, 1983; Morphet, 1987). The case studies indicate that innovation differs both between firms and within firms over time. Young firms appear to increase their innovative capacity after some years of existence, due to better know-how and better financial circumstances, which enables them to bear more risks. Competition can stimulate this development. Other young firms increase their R & D efforts after a merger with a large firm that gives the necessary financial support. The amount of innovation may also increase after a switch from production for stock towards production to specific contract.

Firms also change the character of their innovations, for example from normal problem solving (technical progress) towards radical product innovation (completely new products). This can be illustrated by one firm that established an R & D department (electronics) in 1974 and increased its R & D staff to 11 per cent of total staff and its R & D expenditure to 8 per cent of turnover (1986). Since 1974 the firm has continuously developed and introduced new products (for example, cattle management systems, hands free access control systems) and new generations of existing products. The decision of this firm to become leader in certain product areas is obviously related to the rapid changing of technologies within its products.

In the context of empirical enquiry it is necessary to pay attention to the measurement problem associated with innovations. Identifying and counting numbers of product and process innovations is an inaccurate measure due to the different degrees of novelty and the interrelations between innovations (Van Geenhuizen, 1988). In many firms innovation is a continuous activity achieving both small and more substantial results. This also raises questions about the reliability of innovation indicators, like R & D departments and R & D budgets. Small firms without an R & D department and budget may have continuous innovation incorporated in their production, based on learning from mistakes and on product specification by customers.

Method of Growth and Locational Dynamics

The case of Gist-brocades

Gist-brocades is a large company (approximately 6300 employees) mainly producing food ingredients, pharmaceutical intermediate products and pharmaceutical consumer products. The historical development of the company has been characterized by frequent diversification, which has meant that traditional products (yeast and alcohol) have declined in importance in favour of antibiotics, steroids and enzymes. The development of these new products was realised by means of a continuously high level of innovative performance on the part of the firm (since 1970 the R & D budget has amounted to 10 per cent of turnover).

A close examination of the three major product groups reveals different strategy paths within the company (Table 10.3). The production of yeast reflects the influence of a mostly saturated market, in which the company increased its share by means of cost reduction, quality improvement and acquisition. Diversification into the related area of bakery ingredients has been another adjustment to market contraction. Most recently Gist-brocades has expressed the intention of buying a large company in the USA, which will make it world market leader in yeast and will open up to the firm major market networks for bakery ingredients in the USA. On the other hand, the strategy for raw penicillin and intermediate products generally reflects a growth market, only interrupted by short-term market cycles. Since 1965, growth of production volume has been based on two major innovations concerned with the scaling up of new production processes. The development path of enzymes reflects a combination of a growth market and (subsequent) stagnation. At first, production growth was based on a new product used in detergents, and subsequently on a broad product and market differentiation (food industry). Recently the market has become saturated and the company is adjusting in two ways. First, R & D is being directed towards new products, mainly fine chemicals produced by modern biotechnology, and secondly a new division is being established in which enzymes for the food industry will be integrated with yeast and bakery ingredients.

Gist-brocades has realised this growth and diversification mostly through acquisition: six take-overs in the stagnating yeast market and four take-overs in the related market of bakery ingredients. Growth in the enzymes market has first been achieved through internal development and later through acquisition (two take-overs). Antibiotics seems to be the only product group with an exclusive internal growth path. On the whole, Gist-brocades has been extremely active in acquisition, due to the growth characteristics of its markets (partly saturated) and high level of its R & D expenditure which makes fast growth necessary to support it. Recently external development has become preferred for research as well, with technology in the field of biofine chemicals being developed in a joint venture with another large company (Shell). An inventory of joint production or research (Appendix 10.2) shows that Gist-brocades has been operating in an international network since the 1960s and possibly earlier. This contrasts with some other cases in which firms have only become part of networks in the 1970s or more recently.

Almost all take-overs of Gist-brocades were aimed at increasing market share outside the Netherlands. This strategy resulted in a further internationalization of production within approximately ten additional plants abroad. Concomitant with this

Table 10.3

Gist-brocades : Strategic Performance

Product	Strategy
Yeast (bread)	
1950 - 1970	. Continuous rationalization (cost reduction)
	. Continuous growth of production volume, market expansion (third world)
1970	. Introduction of product innovation (better quality)
1970 - +1975	. Spatial concentration of production (cost reduction)
	. Diversification (bakery ingredients)
1984 - now	. Market expansion (bakery ingredients) (Italy, France, Spain)
1987	. Reduction of production (yeast)
Antibiotics[a]	
± 1946	. Introduction of penicillin
from 1950	. Growth of production volume (export)
from 1950 - now	. Continuous process innovation (increase of productivity by strain improvement and optimization of fermentation)
1965, 1972	. Introduction of new processes for production of intermediate semisynthetic antibiotics
1965 - now	. Continuous growth of production volume based on these new processes
from 1986	. Market expansion (China, India and other third world countries)
Enzymes	
1950 - 1965	. No important strategic change
1965	. Introduction of new product (used in detergent)
from 1965	. Growth of production volume, based on this new product and subsequent product improvement
from 1975	. Product - and market differentiation into the area of food industry
	. Market expansion (USA)
from 1985	. Diversification (biofine chemicals)
	. Market expansion (food ingredients)

[a] Raw penicillin and intermediate products

Source: Annual Report.

activity was an increase in Gist-brocades' jobs outside the Netherlands (25 per cent of all jobs in 1970 against 40 per cent in 1986). The international external growth contrasts sharply with a second strategy implemented within the company since the early 1980s, namely focusing on its main activities, which resulted in the sale of approximately five plants within the Netherlands. This product-related adjustment had no specific spatial (urban/regional) orientation.

Experience in three of the remaining case studies

Locational dynamics are related to external as well as internal development. The locational dynamics of two firms, UCO and Sensor, that concentrate on internal development are summarized in Table 10.4, by means of simple event-history description. An event is the transition of a firm from one state to another state, for example, relocation, or discontinuity in growth pattern.

Table 10.4

Locational dynamics[a]

Firm	Observation period	I	II	III
UCO (contact lenses)	1968-1987	0,5	-	1,0
Sensor (geophones)	1967-1987	1,0[b]	1,5	2,0

I - relocation
II - change in number of plant locations
III - discontinuity in number of jobs

[a] Number of events per 10 year
[b] Except relocation within the same town.

Short-term stability of market demand was obviously a differentiating factor between the two firms. UCO, which produces in a stable market, relocated only once after its establishment and initial growth. But Sensor, which is producing in the unstable oil-market, has relocated several times in the same time span and established an additional office and production plant abroad. A difference in spatial orientation of relocation parallels this experience, namely relocation into a less urbanized region (UCO) and relocation within a highly urbanized area (Sensor). In addition, Sensor showed steep fluctuations in the numbers of jobs it provided (a decrease of 40 to 50 per cent in a few years). Figure 10.1 displays the contrasts between the two firms in terms of strategy as well as locational adjustments. This figure depicts the number of changes (events) and the period of time between successive events (episodes) for four different items. The bars represent the episodes and the numbers the time-span of an episode in years. The dislocation between the end of one bar and the beginning of the next represents an event. For example, Sensor changed the number of its operating locations three times. After ten years of operation the firm opened a sales office (1977), but this was subsequently closed one year later. In 1982 the firm opened an additional production plant which is still operational.

Given the two examples, one could hypothesize that small specialized firms in unstable markets will continue to be located in urban core regions, because of their permanent information requirements and sudden increases in labour demand. The same may hold true for small firms that are specialized in the first (sub)stages of the product life-cycle. On the other hand, one could hypothesize that small firms, specialized in products that follow the full path of the life-cycle will be established in urban core regions in the first stage and relocate into less urbanized regions during the stage of expansion.

The case studies also present a firm whose locational dynamics are not clearly associated with characteristics of the market or individual characteristics like specialization. This firm (AVEBE) is producing in a rather stable market, but it shows heavy locational dynamics. The main determining factor seems to be the role of the government in various respects, namely the introduction of laws concerning the treatment of waste water, the pressure to merge with another firm (to guarantee employment in the region) and the pressure to reorganise production, as a condition for financial help. However this firm may be exceptional.

A first tentative typology

From the preliminary results obtained so far initial types of firms may be suggested, based upon size of firm, level of specialization, market dynamics, strategy and locational adjustment. One additional type is characterized by the mismatching of strategy to market change. At this stage of the analysis there is insufficient empirical data to include organizational characteristics, like position in production networks or corporate status. Bearing in mind the limitations of the pilot study, five preliminary types can be distinguished:

(1) Small, specialized firms experiencing gradual growth and operating in stable markets; this incremental pattern is associated with modest locational change.

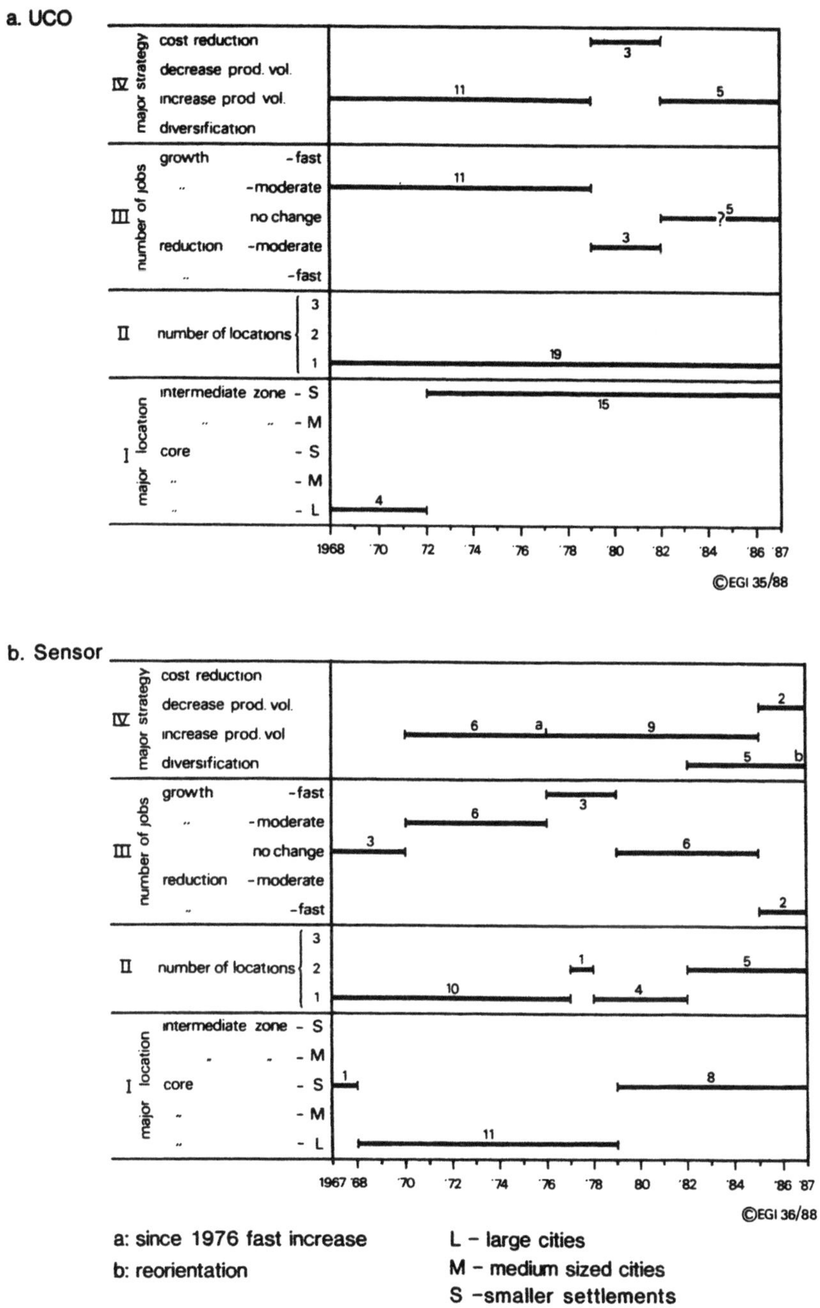

Figure 10.1: Strategic performance and locational adjustment

(2) Small, specialized firms experiencing substantial growth or decline and operating in unstable markets; the fundamental readjustment in strategy is associated with strong locational dynamics.

(3) Diversified firms experiencing gradual growth in expanding markets and entering new expanding product-market areas; this incremental pattern is associated with modest locational change.

(4) Diversifying firms moving into different sectors and experiencing continuous growth in stagnating markets and at the same time shifting into new expanding product-market areas. They tend to have recently focused on main activities; this complex pattern is associated with major locational change.

(5) Specialized firms, operating in gradually contracting markets, but without changing their strategy until a fundamental readjustment is inevitable; this pattern of experience is related to strong locational dynamics. If the strategy is gradually adjusted, by means of diversification into expanding product-market areas, there is resemblance with type 3 described above.

As mentioned earlier, the seven cases have been selected to depict as much variety as possible. This explains the relatively high number of types presented here. In order to provide a sustainable base for a typology, of course, more theoretical and empirical results are needed.

Conclusions

The changing patterns of urban-industrial activities raise questions about the long-term growth strategies and related locational adjustments of individual firms. In this context the first aim of the study was to test the viability of a longitudinal analysis of individual firms, on the basis of several case studies. The methodology shows that the development path of firms can be reconstructed satisfactorily as far back as 1950. However, there are some restrictions in the interpretation of material and events due to lack of data and response bias.

The second aim was to identify different strategies and determining factors of change, especially with reference to the product life-cycle. A strong and influential factor was found to be the dynamics of the market, both as a result of product life-cycles and other market influences. Equally important is the influence of interrelated individual characteristics of the firm, like size, level of specialization and position in the production network. With reference to the product life-cycle it could be concluded that the growth path of firms is by no means unidirectional with most firms often truncating their cycles. Moreover, multi-product firms have a number of different cycles running at the same time and other firms are permanently specialized in one (sub)stage of the cycle.

The method of growth (internal or external) and related locational dynamics have also been investigated within the framework of different markets and individual characteristics. Market saturation and high level of R & D costs seemed to be the main reason behind fast external growth (acquisition) in the case of large diversified firms. In addition to this, short-term market instability seemed to be the main reason for strong locational dynamics in the case of small specialized firms.

Two directions of further research can be indicated. First, a typology of firms could be elaborated, in which organizational characteristics, market dynamics and strategy are related to locational adjustment. In addition to this, hypotheses could be formulated concerning the relation between different urban environments and different types of firm. In this context event-history design (Tuma and Hannan, 1984) can be used to analyse development paths and to define relationships between event probabilities and explanatory variables. Second, an explanatory model of product-market development could be constructed, in which special attention is given to the innovative behaviour of the firm. Interrelations between the market, organization of the firm and strategy may vary with different urban environment and within different industry and service sectors. Hence, in the next stage of the research at Erasmus, firms will be investigated in spatial environments that contrast both in levels of urbanization and in degree of concentration of industry or sector.

Note

1 Writing corporate histories is rather time-consuming. As far as older firms are concerned, the analysis of annual reports from 1950 took approximately four days with one outlier of twice as long. The (additional) interview took approximately two hours, but in the case of complex histories two such interviews were held, with different corporate managers. Travelling time included, these latter interview took one or two days.

References

Atzema, O.A.L.G., Grit, S. and Lensink, E. (1986) Regionale verschuivingen in de stuwende werkgelegenheid in Nederland (1963-1978), in J. Buursink and E. Wever (eds.), Regio en ontwikkeling, Geografisch en Planologisch Instituut, Katholieke Universiteit, Nijmegen.

Daly, P. (1985) The Biotechnology Business: A strategic analysis, Frances Pinter, London.

Davelaar, E.J. and Nijkamp, P. (1986) De stad als broedplats van nieuwe activiteiten, Stedebouw en Volkshuisvesting, 2, 61-66.

Davelaar, E.J. and Nijkamp, P. (1987) Het grootstedelijk milieu als informatie - en incubatiepool voor innovaties, Planning, 30, 15-23.

Dinteren, J.H.J. van (1986) Commerciele kantoren en Nederlandse Steden, Geografisch Tijdschrift, 77 (3), 263-272.

Fredriksson, C.G. and Lindmark, L.G. (1979) From firms to systems of firms: a study of interregional dependence in a dynamic society, in F.E.I. Hamilton and G.J.R. Linge (eds.), Spatial analysis, industry and the industrial environment, Vol.1 : industrial systems, Wiley, New York, 155-86.

Freeman, C. (1982) The Economics of Industrial Innovation, Frances Pinter, London.

Froling, A. and Zoon, K. (1988) Ruimtelijke gevolgen van veranderingen in het strategisch management van Philips. Geografisch Instituut, Rijksuniversiteit Utrecht, Utrecht.

Geenhuizen, M.S. van (1988) Vernieuwend bedrijfsleven op lange strek; verslag van een vooronderzoek, Economisch-Geografisch Instituut, Erasmus Universiteit, Rotterdam.

Gudgin, G. and Fothergill, S. (1984) Geographical variation in the rate of formation of new manufacturing firms, Regional Studies, 18, 203-206.

Harrington, J.W. (1985) Corporate Strategy, Business Strategy and Activity Location, Geoforum, 16 (4), 349-356.

Johnson, G. and Scholes, K. (1984) Exploring Corporate Strategy, Prentice-Hall, London.

Jong, M.W. de (1985) Dynamische markttheorie, Stenfert Kroese, Leiden.

Jong, M.W. de (1987) New economic activities and regional dynamics, Economisch-Geografisch Instituut, Universiteit van Amsterdam, Amsterdam.

Kamann, D.J.F. (1986) Industrial Organization, Innovation and Employment, in P. Nijkamp (ed.) Technological Change, Employment and Spatial Dynamics, Springer, Berlin.

Keeble, D. and Wever, E. (1986) New firms and regional development in Europe, Croom Helm, London.

Kleinknecht, A. and Mouwen, A. (1985) Regionale Innovatie (R en D): Verschuiving naar de Halfwegzone? in W.T.M. Molle (ed.) Innovatie en region. Onderzoekprogramma Technologie en Economie, 's, Staatsuitgeverij, Gravenhage.

Knaap, G.A. van der and Louter, P. (1986) De middelgrote steden, Economisch-Geografisch Instituut, Erasmus Universiteit, Rotterdam.

Knaap, G.A. van der and Linge, G.J.R. (1989) Labour environment and industrial change, Routledge, London.

Kok, J.A.A.M., Offerman, G.J.D. and Pellenbarg, P.H. (1985) Innovatieve bedrijven in het grootstedelijk milieu, Geografisch Instituut, Rijksuniversiteit, Groningen.

Krumme, G. (1970) The interregional corporation and the region, Tijdschrift voor Economische en Sociale Geografie, 61, 318-333.

Krumme, G. and Hayter, R. (1975) Implications of corporate strategies and product cycle adjustments for regional employment changes, in L Collins and D.F. Walker (eds.), Locational dynamics of manufacturing activity, Wiley, London.

Lasuen, J.R. (1971) Multi-regional economic development : an open-system approach, in T. Hagerstrand and A. Kuklinski (eds.), Information systems for regional development - a seminar, Cleerup, Lund, 167-229.

Morphet, C.S. (1987) Research, development and innovation in the segmented economy: spatial implications, in G.A. van der Knaap and E. Wever (eds.), New Technology and Regional Development, Croom Helm, London.

Markusen, A., Hall, P. and Glasmeier, A. (1986) High tech America: the what, how, where and why of the sunrise industries, Allen & Unwin, Winchester MA.

Muntendam, J. (1987) Philips in the world, a view of a multinational on resource allocation, in G.A. van der Knaap and E. Wever (eds.), New Technology and Regional Development, Croom Helm, London.

De Nationale Investeringsbank (1987) Koplopers en achterblijvers in de bedrijvenwereld, 's Gravenhage.

Oakey, R. (1985) High-technology industries and agglomeration economies, in P Hall and A. Markusen (eds.) Silicon Landscapes, Allen & Unwin, Boston.

Porter, M.E. (1983) The technological dimension of competitive strategy, Research on technological innovation, management and policy, 1, 1-33.

Reinink, G.J. (1970) Industriale bedrijfsmigratie in Nederland in de jaren 1950-1962: een onderzoek naar verplaatsingsfactoren, SISWO, Amsterdam.

Rothwell, R. and Zegveld, W. (1985) Reindustrialization and Technology, Longman, Harlow.

Taylor, M. (1986) The product-cycle model: a critique, Environment and Planning A, 18, 751-761.

Thomas, M.D. (1987) The innovation factor in the process of micro-economic industrial change: conceptual explorations, in G.A. van der Knaap and E. Wever (eds.) New Technology and Regional Development, Croom Helm, London.

Törnqvist, G. (1970) Contact systems and regional development, Cleerup, Lund.

Törnqvist, G. (1978) Swedish industry as a spatial system, in F.E.I. Hamilton (ed.), Contemporary Industrialization, Longman, London, 86-110.

Tuma, N.B. and Hannan, M.T. (1984) Social Dynamics, Models and Methods, Academic Press, Orlando.

Vernon, R. (1966) International Investment and International Trade in the Product Cycle, Quarterly Journal of Economics, 80, 190-207.

Walker, W.B. (1979) Industrial Innovation and International Trading Performance, Tai Press, Greenwich, Connecticut.

Watts, H.D. (1981) The branch plant economy, Longman, London.

Webbink, A.H. (1985) Groot en klein in de industrie, Economisch Instituut voor het Midden- en Kleinbedrijf, Zoetermeer.

Wever, E. (1985) Regionaal-ekonomisch perspektief, Geografisch en Planologisch Instituut, Katholieke Universiteit, Nijmegen.

Appendix 10.1
Main Characteristics of industry firm[ab]

	1	2	3	4	5	6	7
Age	117	58	20	19	68[c]	152	89
Size (nr.of workers)	6,300	330	60	20	2,610	190	3,170
Status	independent	independent	subsidiary	independent	independent (cooperation)	independent	independent
Products	yeast food ingredients pharmaceutics[d] industrial enzymes	control systems[e] electr. specials	geophones related products	contact lenses liquids	starch derivatives	carpets synthetics	weighing mach. cash desks refrigerators services
Multiplant	X		-	-	X	-	X
Market	intermediate finite	intermediate finite	intermediate	finite	intermediate	intermediate finite	intermediate finite
Major strategy change (recent)	-	-	X[f]	-	-	X	X

a. 1. Gist-brocades N.V. (Delft) 4. UCO (Wieringerwerf) 7. Mij. Van Berkels Patent N.V. (Rotterdam)
2. Nedap N.V. (Groenpo) 5. AVEBE ba (Veendam)
3. Sensor Nederland B.V. (Voorschoten) 6. Van Besouw N.V. (Goirpe)

b. One additional firm is in transports (NedLloyd N.V. in Rotterdam)

c. Major change in corporate identity after 1950 (due to merger or otherwise)

d. Intermediate (bulk) products and consumer products

e. Cattle management systems, access and security systems

f. Reorientation (products, markets)

Appendix 10.2
Gist-brocades: Joint research and production[a]

Year	Partner	Country	P/R[b]	Field
1957	Organon	the Netherlands	P	Pharmaceutical products
±1960	Organon	the Netherlands	R	Pharmaceutical products
1961	Orgachemie	the Netherlands	P	Enzymes
1961	Fabr. de Levure et d'alcool (j.v.)	France	P	Yeast
1961	Comp. Uniao Fabril (j.v.)	Portugal	P	Pharmaceutical products
1965	TNO & Produktschap voor veevoeders	the Netherlands	R	Additives (cattle feed)
1971	Farbenfabr. Bayer	FRG	P	Pharmaceutical products
1971	Syntex Corporation	USA	R	Pharmaceutical products
1975	Nord Westdeutsche Hefe - Union (j.v.)	FRG	P	Yeast
1976	Universities	the Netherlands and abroad	R	Pharmaceutical products
1981	University of Nottingham	Great Britain	R	Pharmaceutical products
1981	Shell Research	the Netherlands	R	Biotechnology
±1983	University of Dusseldorf and Unilever	FRG the Netherlands	R	Enzymes
1984	C.L.B.[c]	the Netherlands	R	Coagulation (blood)
1984	Chiron Research	USA	R	Molecular biology of yeasts
1986	Univ. Groningen and Plant Genetic Systems	the Netherlands Belgium	R	Protein Engineering
1986	G.R. Spinks (j.V.)	Great Britain	P	Bakery ingredients
1986	Produktschap voor veevoeders	the Netherlands	R	Cattle feed (phosphate project)
1987	Shell (j.v.)	the Netherlands Great Britain	R+P	Biofines

a Not exhaustive (due to characteristics of data source)
b P = production, R = research
c Central Laboratory for Bloodtransfusion Service

Source: Annual Report

PART III - REGIONAL ISSUES: SOME EMPIRICAL EVIDENCE

11. A STUDY OF THE PERFORMANCE AND TECHNOLOGY LEVEL OF MANUFACTURING INDUSTRY IN DANISH REGIONS 1972-1981

Ulf Christiansen

Introduction

The improvement of the performance of manufacturing industry is basic to the solution of the structural problems of the Danish economy. The share of industrial employment in Denmark is among the lowest in the Community (16-18 per cent of total employment in 1985). During the last 30 years the geographical redistribution of industry has increased the level and nature of technology in the country's peripheral areas. However, very little high-tech industry had moved or been created there by the beginning of the 1980s, but medium-tech and low-tech as well as food processing industries experienced growth in such regions, which supports the development of local economies through industrial dispersal and technological change. One question remaining to be answered about industry's performance is whether the productivity of **existing** industrial plant is determined by both regional conditions and the general technological level of the industry in question, or whether 'technology' is alone responsible. This question is addressed in the paper by two simple regression models. One model tests for differences between industry groups (groups defined by their technology level) in the country as a whole. Another tests for differences between regions within industry groups.

The analysis showed that productivity varies among industries assumed to be at different levels of technological development and R & D as well as at different levels of exposure to international competition. It also showed that there are no (or only very slight) productivity differences between regions within each industry group. A possible explanation is the ease with which technology has been transferred during the 1960s and 1970s. The detailed results of the analysis have been published in Danish in Christiansen (1988).

Denmark's Industrial Geography: Definition of Regions

Denmark lies in northern Europe as a stepping stone between the Scandinavian peninsula and Germany. Geographically it is a small country - 43,000 sq.km. It is connected by land to the rest of Europe only at the southern border of Jylland; the rest of the country consists of about ninety inhabited islands of which eleven have more than 10,000 inhabitants. The largest islands, population wise, are Sjaelland, Fyn, Amager, Bornholm, Lolland and Falster.

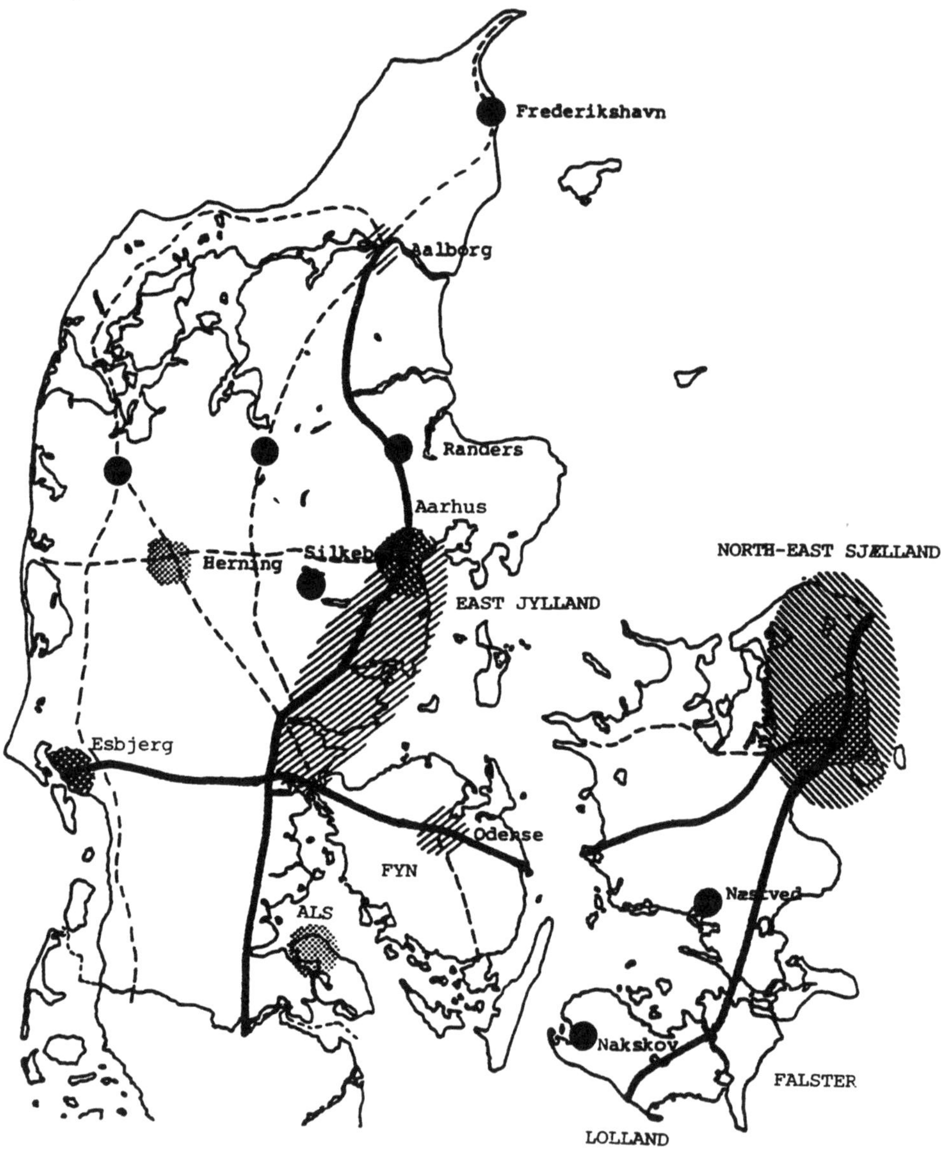

Figure 11.1 : Industrial areas (hatched), larger towns and cities and main national road connections in Denmark

During the last fifty years, the Danish population has increased by about 1.6 million people (from 3.55 million in 1930 to 5.12 million in 1981). In 1982, the population figure went down for the first time in more than 150 years but since 1983 it has stabilized around 5.12 million again.

The country has a reputation as a producer of agricultural products, but it is actually quite urbanized and industrial (with some qualifications to which we shall return at the end of the chapter). The largest urban concentrations are, in the first place, Copenhagen with a metropolitan built-up area of about 1.3 million inhabitants, then Arhus on the eastern coast of Jylland (250,000), Odense on the island of Fyn (170,000), Alborg in northern Jylland (150,000) and Esbjerg on the western coast (80,000). Several towns in Jylland and a few on Sjaelland have between 50,000 and 60,000 inhabitants (1987).

The communications systems in Denmark are well developed. Roads, domestic airlines and telecommunications function well even though the Great Belt (Storebeaelt) acts as a brake on exchanges between eastern and western Denmark. This will change after 1993, when the tunnel and the bridge, which are to connect Fyn with Sjaelland, have been built. This shorthand sketch is illustrated on Figure 11.1.

Denmark's first period of industrialization led to concentration in and around the major cities. The country's second industrialization began in the late 1950s. At that time the geographical distribution of industry might be characterized in the following way:

- An industrial region in north-east Sjaelland comprising 50 per cent of the industrial workforce. It had a dominant position due to the national functions of the Copenhagen agglomeration. In this chapter called: Metropolitan region.

- Three medium-sized industrial areas: east Jylland in and around Arhus and the cities of Odense and Alborg. They all have a rather large industrial workforce and some national service functions.

- Two smaller industrial areas: the town of Herning and the island of Als. In the period covered, these areas have had a significant industrial growth. They have no national service functions.

- Five industrial towns: Naestved, Silkeborg, Randers, Frederikshavn and Nakskov, which are seen as industrialized towns with a limited range of services (Antonsen, 1964).

Concerning the balance between the major sectors of the economy, the statistics show that in 1960 for the first time employment in industry was higher than in agriculture with the growth rate of the economy reaching 4.7 per cent per annum between 1958 and 1973.

Industrial production increased after the first oil-crisis in 1973-74, with the exception of the year 1975. During 1976-1980 the growth was smaller than from 1971-73 and at the beginning of the eighties there were grave problems with very low profitability, an extremely low volume of investments and low capacity utilization. The decline in industrial employment for the country as a whole in the period 1971-81, after the 1960's boom, was nevertheless only about nine per cent, whereas it was 31 per cent in the metropolitan region. Outside the metropolitan region, the number of employed persons in manufacturing industry actually increased by 11,000 persons. As regards location, generally speaking manufacturing establishments moved west, i.e. away from the metropolitan region, during the 1960s and 1970s. At the end of 1981 365,000 persons were employed in manufacturing industry proper (establishments with more than five employees), 275,000 of whom were employed in establishments with 20 or more employees. This is the total upon which our 1981 investigation is based (Christiansen, 1983).

In the **regional subdivision** of the present study we thought it important to reflect the history of the industrial development, i.e. the situation of 1960 compared to 1972 and 1981. At the same time county borders had to be used as regional boundaries because of the way the database for manufacturing was constructed in 1972. It was also necessary to form regions large enough to provide statistical reliability. This led us to construct five regions as follows:

No.I - the metropolitan region, formed from the municipalities of Copenhagen and Frederiksberg, Copenhagen, Frederiksborg and Roskilde counties;

No.II - the remaining eastern island region, formed from the counties of Vestsjaelland and Storstrom. It covers the rest of the island Sjaelland and the total area of the islands Lolland, Falster and Mon. (The island of Bornholm far to the east in the Baltic sea was not included in the study, as its manufacturing sector is minimal).

No.III - the region of Fyn, with its main industrial centre, Odense.

No.IV - the region of northern and western Jylland, formed by the counties of Nordjylland, Viborg and Ringkjobing. In this region manufacturing has been growing in the 1970s.

No.V - the region of eastern and southern Jylland, formed by the counties of Arhus, Vejle, Sonderjylland and Ribe. Together with the metropolitan region, this region is both the most industrialized and the most modern one in the country.

The regionalization is shown in Figure 11.2.

Figure 11.2: Regions in Denmark as defined in this study.

A Classification of Manufacturing Industries According to Technological Level and Exposure to International Competition

The analysis is based on the national and regional production figures in 1972 and 1981 of 149 sectors of manufacturing industry. Industry has been classified into three main categories. Two of these contain branches which are heavily exposed to international competition. The third category contains branches which are less exposed on account either of various protective measures, or of the nature of the product (language barriers for printing and publishing, high transportation costs for beverages, for example). The two competitive categories have been subdivided further according to their technological level and the nature of the product; the less competitive (protected) category only according to nature of product.

All groups were formed using the standard international classification of industry (ISIC). The 149 industrial branches were classified into 11 groups, which were then allotted to the three main categories. This modified industrial classification is shown in Table 11.1.

The general idea behind the classificatory scheme is that competitiveness and technological capacity is higher in production units that are staffed with many engineers and other professionals, high level technicians and well educated and experienced workers, than would be the case in units that employ a large proportion of less skilled or unskilled workers.

The categorization is based on the work of a Swedish economist, Lennart Ohlsson (see Ohlsson, 1980). In a study of Swedish industry using data from the period 1959-1975, he defined a **protected** industrial sector, a **competitive raw materials**-based sector and a **competitive footloose** sector. They were further subdivided in a way similar to the one used in this study.

The competitive footloose sector corresponds to the two competitive categories, we have defined and the protected one to our third category. The raw materials-based sector relates to mining and forestry, two industries that either do not exist, or are of very minor importance in Denmark.

The rationale behind this division of manufacturing industry is the close correlation between the long-term market growth of an industry in the period 1960-1975 (as well as the development of selected factor intensities) and the relative number of engineers, technicians and business school graduates employed (Ohlsson, 1976).

Similar comprehensive studies do not exist for Danish industry, but a few investigations with different aims corroborate Ohlsson's statistics on the number of high-level technical personnel in the various branches of industry. Thus, a regional study of industry in Nordjylland by Ibsen and associates (Ibsen et al, 1980) indicated that the relative number of well-educated technicians is high in the electronics and mechanical industries and low in the textile and clothing industry and in the food industry.

In a study about the impacts of high technology (information technology, bio-technology) on industry, trade and financial services in the Oresund region

(Oresundgruppen, 1981) it was also found that the number of engineers and science graduates was very high in pharmaceuticals, electronics and instrument firms; certainly not a surprising result, but a corroboration of the Swedish results. Finally, a national study (Moller, 1985) on research and development intensity and export performance in Danish industry in the 1970s confirms that the industries in our category 1 also had R & D intensities well above the national average, as well as the highest growth in R & D. Group 1E, however, formed an exception.

It therefore seems well warranted to use the percentage of engineers, technicians, etc., as an indicator of technological and competitive capacity, if we follow Ohlsson and Moller. The above mentioned Danish regional studies and, again, Ohlsson (1980) further justify the detailed categorization itself.

We shall now take a closer look at the modified classification used in the present study (see Table 11.1).

Industry category 1 (i.e. group 1) contains industrial branches with a relatively high proportion of engineers and other technicians in their personnel (HT) and branches with an average or medium proportion (MT). They all belong to the chemical industry or to the mechanical, electronic (including instruments) or metals-'treating' industries. Group 2 contains competitive industries with a low proportion of technicians. Group 3 contains the 'protected' or partly protected industries. The industries directly influenced by the pricing and quota policies of the EEC for agricultural products and fisheries are included in this group. Meat packing and slaughtering, the dairy industry, sugar refineries and industrial bakeries are included, among others, in group 3A. The part of the food and beverages industry which is only weakly or not at all regulated by the EEC have been put in group 3B. Producers of building materials are to be found in group 3C and printing, etc., in group 3D.

The distributions of production and employment of the industrial groups so defined vary greatly over the regions. Group 1A, pharmaceuticals and paints, is practically only found in the metropolitan region, whereas group 1B, mechanical industry, has a regionally much more balanced distribution. This should, however, not lead anyone to the conclusion that it has had a balanced employment growth in the period under consideration. The metropolitan region, I, lost employment and Jylland gained very considerably, especially region IV, which contains the largest area eligible for aid to industry. On the other hand the electronics sector, 1C, kept its employment in the metropolitan region and developed, relatively speaking, very much in region IV. But the bulk of the employment in this group is still located east of the Great Belt. For the rest of the chemical industry, group 1D, the metropolitan region (I) and eastern and southern Jylland (V) are dominant. Group

1E is fairly evenly spaced all over the country.

Group 2A encompasses a lot of older industrial enterprises, some of them dating from the 18th century, such as glassware and chinaware plants, paper- and

Table 11.1
Classification of Industrial Sectors

Group	Code	Characterized by	Industrial Sectors
	1A	HT: high proportion of engineers and scientists	Pharmaceuticals, colours and paints.
	1B	HT: high proportion of engineers and technicians	Mechanical industry
1	1C	HT: high proportion of engineers and technicians	Electronics and instruments, electrical machinery
	1D	MT: average proportion of engineers and technicians	Chemical industry except group 1 and without plastics and oil refining
	1E	MT: average proportion of engineers and technicians	Cables, household appliances tools, vehicles (ship-building was excluded from the investigation).
2	2A	LT: low proportion of engineers and technicians	Plastics, glass, china, paper and paper products, certain metal goods
	2B	VLT: very low proportion of engineers and technicians	Textiles, clothing apparel, leather and shoes, wooden furniture, musical instruments
	3A	Regulated or strongly influenced by the agricultural policies of the	Dairies, meat packing, canned fish and vegetables, sugar, bread, etc.
	3B	Weak foreign competition on the home market	Breweries, metal foundries, cans
3	3C	Relatively weak foreign competition on domestic market	Products for use in building and construction
	3D	Relatively weak foreign competition on domestic market	Printing and publishing, book-binding (daily newspapers and publishing of books not included)

saw-mills. Factories for plastics goods are also in the group. The former are located in the capital and in the old provincial towns, the latter in all regions. Jylland, and more especially the area round Herning in region IV, have most of the clothing and apparel industry from group 2B; furniture production is also more evenly distributed, but in small towns and villages within each region.

The subgroups in group 3 are also regionally diversified. Slaughtering, meat-packing and fish-canning in particular are located in Jylland (group 3A). 85 per cent of the printing industry (3D) is located in region I and in region V. These regions are the most urbanized regions of the country. The regional and national distribution of employment in 1981 within the defined industries is given in Table 11.2.

Table 11.2
Regional distribution of employment in industry, 1981
(1000 employees, establishments with 20 or more employees)

Region Industry Group	I	II	III	IV	V	Country
1A HT: Pharmaceuticals etc.	8.1	0.5	0	0.2	0.4	9.2
1B HT: Mechanical industry	8.4	4.3	3.4	7.0	17.6	40.6
1C HT: Electronics, etc.	15.6	1.4	1.5	4.5	3.2	26.1
1D MT: Chemical industry, etc.	5.1	1.1	0.6	1.4	3.9	12.2
1E MT: Cables, household	5.3	2.0	2.5	4.7	7.6	22.0
2A LT: Plastics, glass, china, etc.	10.7	5.8	5.5	3.4	9.3	34.8
2B VLT: Textiles, clothing, apparel, etc.	2.1	1.6	3.0	13.9	12.6	33.1
3A Dairies, meat packing. etc.	7.9	6.7	5.2	12.8	17.5	50.0
3B Breweries, etc.	9.3	1.8	2.9	2.1	3.8	20.0
3C Products for use in building, etc.	4.4	0.9	0.7	4.9	3.6	14.3
3D Printing, etc.	5.3	0.3	0.4	0.6	2.2	8.8
Total region/country	82.2	26.4	25.7	55.5	81.7	271.7

Productivity and Value Added: The Data

Productivity is usually defined as a quotient between a measure of production and one or more measures of the production factors. Most productivity indexes are partial. The two indexes of labour and capital productivity are most commonly used. But the concept of total factor productivity is a better measure than either of these, as it simultaneously takes both labour and capital effects on production into account (see e.g. Bosworth, 1976).

The partial indexes are defined:

$$P_L = Y/L \quad \text{(labour productivity)} \qquad (11.1)$$

and

$$P_K = Y/K \quad \text{(capital productivity)} \qquad (11.2)$$

and total factor productivity as

$$P_{Tot} = Y/(aL + bK) \qquad (11.3)$$

In the three expressions above Y, L and K denote a production index, a labour index and a capital index respectively; a and b are weights.

In our study an index of labour productivity is used. Labour productivity measurement has a long tradition so that at least a superficial comparison with other authors' results may be possible. The difficulties involved in getting reliable data for capital use has also played a role in this decision.

The index that we construct is defined as value added measured in DKK divided by number of employees. The statistical unit on which these variables have been measured is the establishment. For reasons of confidentiality the data have been aggregated to three establishments as a new unit. The new units were formed in such a way that all three component establishments have the same or nearly the same productivity. The labour productivity in establishment j in the year t may be written as:

$$P_j(t) = Y_j(t)/L_j(t) \qquad (11.4)$$

cf.(11.1) above.

Value added is as usual defined as total sales Sj (t) with materials input Mj (t) subtracted. Materials input consist of all inputs which are not labour and capital.

Obviously the productivity measure as defined in (11.4) has some drawbacks. Instead of using the number of employees in the denominator it would have been better to use number of hours worked per year because the comparison between years then becomes more precise. The length of the working week may shorten over time (as is actually the case) and part-time work exists and need not form the same proportion in the years compared, etc. The numerator in (11.4) also stands for more than just the result of the production effort. Price and quantity changes over which the firm has no control (such changes that the firm in principle cannot predict) cannot rightly be ascribed to the firm, etc.

If we integrate P_j (t) over all establishments in an industrial sector in a region or nationally, we get a first index of productivity as

$$\mathrm{Sum}(j)\ P_j\ (t) \tag{11.5}$$

This is a simple summary index, which is used in the description of production and productivity change by region in the next two sections of this chapter. The sectoral productivity is determined by the same factors as the productivity of the single establishment, but may of course change over time due to births and deaths of establishments, if their productivity differs from the average of the sectoral productivity. Regional differences in sectoral productivity may obviously also be caused by heterogeneity in the regional sectoral composition of plants.

Production Change from 1972 to 1981

For each industry group a price index for the change from 1972 to 1981 was constructed. Price changes were measured as the changes in the quotient of gross factor income (GFI) in current prices versus GFI in 1975 prices. This required a correspondence between the industrial sectors defined in the national accounting system and the SIC-code, which cannot be totally obtained, however, but the correspondence was found to be reasonably good for our purpose.

Table 11.3 shows the change in the production from 1972 to 1981 in the various industrial groups in the country as a whole. It is clear that the technologically most developed groups have had the highest growth (+41 per cent), while the group mostly producing for the domestic market and/or regulated by the EEC had the weakest performance. It is the decline in the production of building materials (-24 per cent) and in the printing industry (-22 per cent) which make themselves felt.

Tables corresponding to Table 11.3 have been constructed for all five regions. From these tables it may be inferred that the metropolitan region, I, still has the largest part of total industrial production (33 per cent approximately). East, south and a part of western Jylland, V, follow directly with 30 per cent. The growth of production was highest in north-west Jylland, IV, with V following immediately behind, see Table 11.4.

Table 11.3

Value Added by Industry 1972 and 1981, Country, 1981 prices

Region Industrial Sector	Value added, DKK 1000m, 1981 prices 1981	 1972	 Change 1972-81	% Change 1972-81
1A HT: Pharmaceuticals etc.	2.83	1.28	1.54	120.6
1B HT: Mechanical industry	7.46	6.15	1.31	21.3
1C HT: Electronics, etc.	4.62	2.43	2.19	90.1
1D MT: Chemical industry, etc.	3.90	2.69	1.21	45.2
1E MT: Cables, household appliances, etc.	4.01	3.62	0.39	10.9
SUM Group 1	22.82 (40.1%)	16.17	6.65	41.1
2A LT: Plastics, glass, china, etc.	6.67	-	-	-
2B VLT: Textiles, clothing apparel, etc.	5.61	4.66	0.95	20.5
SUM Group 2	12.28	-	-	-
3A Dairies, meat packing, etc,	11.72	9.11	2.61	28.7
3B Breweries, etc,	4.73	4.20	0.53	12.6
3C Products for use in building, etc.	3.47	4.54	-1.07	-23.6
3D Printing etc.	1.93	2.47	-0.54	-21.6
SUM Group 3	21.85	20.32	1.54	7.6
TOTAL	56.95	-	-	-

Table 11.4
Percentage Change of Production 1972-81

Region / Industrial Sector	I	II	III	IV	V
1A HT: Pharmaceuticals etc.	115.1	216.4	8.1	85.0	189.7
1B HT: Mechanical industry	-21.2	29.6	71.7	87.7	30.0
1C HT: Electronics, etc.	68.1	141.4	(-)	106.3	133.0
1D MT: Chemical industry, etc.	22.5	0.6	61.3	78.1	94.8
1E MT: Cables, household appliances, etc.	-33.4	0.6	46.4	109.5	34.8
SUM Group 1	23.4	34.6	(-)	94.5	47.8
2B VLT: Textiles, clothing, apparel, etc.	-50.4	-39.1	7.7	65.4	29.3
3A Dairies, meat packing etc.	- 0.4	29.0	14.8	59.6	33.8
3B Breweries, etc,	- 4.7	50.1	4.7	32.7	51.2
3C Products for use in building, etc.	-31.0	16.4	-46.5	-17.4	-16.2
3D Printing etc.	-31.5	6.3	-29.8	43.7	0.1
SUM Group 3	-15.3	25.5	1.0	26.1	23.7

Note: (-) : not calculated.

The table shows among other things a very characteristic decline in the production of group 3 in the metropolitan region.

The Growth in Productivity

The productivity for industry branch 'i' in 1972 measured in 1981 prices is defined as the product of its productivity in 1972 measured in 1972 prices and an index of price increase V(i). Results are given in Table 11.5 for country and regions.

Table 11.5

Percentage Growth of Productivity from 1972 to 1981

Region / Industry Group	I	II	III	IV	V	Country
1A HT: Pharmaceuticals etc.	107	224	148*	106*	124*	113
1B HT: Mechanical industry	23	33	31	37	25	26
1C HT: Electronics, etc.	78	140	(942)	102	83	95
1D MT: Chemical industry, etc.	60	29	5	35	78	56
1E MT: Cables, household	23	31	33	8	34	23
2B VLT: Textiles, clothing, apparel, etc.	93	55	102	89	97	92
3A Dairies, meat packing. etc.	47	21	20	36	33	33
3B Breweries, etc.	29	57	39	53	67	41
3C Products for use in building, etc.	9	25	24	10	8	11
3D Printing, etc.	19	20*	27*	17	20	19

Note: * less than 500 employed in 1981.

() Value added very low in 1972 in region III.

It clearly appears that pharmaceuticals and paints (1A) have had the highest productivity growth in all regions. Then follow the electronics industry (1C) and the technically less advanced industries in group 2B (textiles, clothing apparel, shoes, furniture). The other advanced chemical sector (1D) follows. The mechanical industry (1B) and the metal goods industries (1E) both show low productivity increases. The food industry (3A) has not experienced any large productivity increase. The building materials industries and the printing industry have had the lowest productivity increases of all.

In all groups there seem to be clear differences in productivity growth. Industry group 1A had a productivity increase twice as big in Sjaelland region outside the metropolitan region as in the metropolitan region itself. This is also true for the electronics group, 1C, which also had a high productivity increase in northern and western Jylland. Groups 1B and 1E in the iron and metal industry also show a regional variation. For both groups the metropolitan region had an increase lower or equal to the national average. Group 2B had a high increase in productivity in all regions with the exception of Sjaelland outside the metropolitan region. For the rest of the regional variation the reader is referred to the table.

New technology development is often divided into process and production equipment innovations and the development of new products. Both types of development happen all the time in all sectors of industry, but with different emphases on each of them. In the seventies some sectors were clearly stressing the rationalization of their production processes - mainly in order to cut costs - whereas others were able to make significant changes in their products and product range.

The clothing industry, 3B, and the brewing industry, 3B, have rationalised and shed labour nationally. In these industries the rationalization explanation is corroborated by high growth in productivity coupled with low growth in production (Tables 11.4 and 11.5). Regionally the clothing and textiles branch managed to hold its own only in north-western Jylland (IV), in which region, incidentally, the town with the most modern production plants, together with a century-old tradition in the industry, is also located (Herning).

The high-tech industries present another picture. Among these both pharmaceuticals, 1A, and electronics, 1C, have had the best performance in productivity and production. This concords with independent evidence of a high innovation rate regarding both processes and products for the pharmaceutical industry and a high product innovation rate in the electronics industry. Important changes in the production processes for this industry were, however, expected for the 1980s (Oresundsgruppen, 1981). Looking at the regions, the most important development was the growth of the electronics industry in northern Jylland, region IV. This took place at the same time as the build-up of the new Alborg University centre. Nevertheless, Sjaelland (region I and II) was still preponderant.

The pharmaceuticals industry was still nearly exclusively tied to the metropolitan region even though branch plants existed in other parts of the country. The intimate collaboration with university research and teaching hospitals that firms in this sector must have is no doubt responsible, together with the ability of the region to attract foreign-born researchers.

A Model of National Production Functions for Each Sector of Industry

A traditional aggregate production function may be specified as:

$$Y_{kt} = f(K_{kt}, L_{kt}, H_{kt}) \quad (11.6)$$

where

Y_{kt} = value added in industry group k in the year t, K_{kt} = use of capital equipment in industry group k in the year t, L_{kt} = number of employees in industry group k in the year t, H_{kt} = time use in industry group k in the year t.

In order to specify (11.6) the data only allow us to use one explanatory variable. This is variable L as neither K nor H have been observed.

Value added in industry group k may then be written (subscript t having been dropped):

$$Y_k = \alpha_{ok} L_k^{\beta_{1k}} \quad (11.7)$$

i.e. productivity is

$$Y_k/L_k = \alpha_{ok} L^{\beta_{1k}-1} \quad (11.8)$$

If β_{1k} = 1, then α_{ok} is defined as the partial productivity of labour in sector k. Taking logarithms, we get :

$$\log Y_k = \log \alpha_{ok} + \beta_{1k} \log L_k \quad (11.9)$$

In case that β_{1k} = 1, the relation between production and employment (Y and L) will be linear. Phrased differently:

$$Y_k/L_k = \alpha_{ok} = \text{constant for industrial sector k.}$$

The linearity may give rise to different interpretations. The most interesting interpretation is that there are neither economies, nor diseconomies, of scale in production for the industrial branch in question.

Country Results 1972 and 1981

For the country as a whole it was tested whether β_{1k} = 1. Results are given in Table 11.6.

The form of the curve may also be influenced by the position of the industry in the business cycle. If the industry is in recession large firms may retain labour at lower capacity utilization when waiting for 'Better times', at least in the short run, whereas small firms are perhaps not in the same position for financial reasons. Price rigidities also play a role; then Y will slope gradually downwards for increasing values of L and $\beta_1<1$.

Table 11.6

Estimates of β_1, 1972 and 1981

Region / Industry Group	Number of Observations	β_1	Number of Observations	β_1
1A HT: Pharmaceuticals etc.	34	1.16*	33	1.44**
1B HT: Mechanical industry	121	0.99	120	1.05
1C HT: Electronics, etc.	62	0.97	70	1.01
1D MT: Chemical industry, etc.	64	0.85*	55	0.85
1E MT: Cables, household	90	0.98	86	1.00
2A LT: Plastics, glass	186	1.06	145	1.07
2B VLT: Textiles, clothing, apparel, etc.	299	1.08*	204	1.44**
3A Dairies, meat packing. etc.	181	0.96	149	0.98
3B Breweries, etc.	55	0.91	52	0.89+
3C Products for use in building, etc.	101	1.00	88	0.93+
3D Printing, etc.	77	1.01	57	1.00

\+ $p < 0.1$
* $p < 0.05$
** $p < 0.01$

The stochastic specification of the model (11.9) is:

$$Y_{ik} = \beta_{ok} + \beta_{1k}\, l_{ik} + e_{ik};\ e_{ik} = N(O,\sigma^2_k) \qquad (11.10)$$

where

$$e_{ik},\ i = 1, \ldots, nk; \quad k = 1, \ldots, K$$

are independent and normally distributed, K = 11 and Y_i is the logarithm of value added for the ith data set in industry sector k.

l_{ik} is the logarithm of the number of employees in data set i in industry sector k. β_{ok} is defined as $\log \alpha_{ok}$.

Table 11.6 shows that the industrial branches 1A (advanced chemistry) and 2B (clothing, shoes, furniture, etc.) had values of β_1 larger than 1 in **1972**. The productivity grows with employment due to heterogeneity or to some of the more specific reasons mentioned in the last section.

For industry branch 1D β_1 was smaller than 1 in 1972. All other branches had values of $\beta = 1$. In **1981** five of the industrial branches had values of $\beta_1 = 1$; the heterogeneity is, so to say, increasing. This could be expected as the economy had entered a period of recession in 1979-1980 at the time of the second oil crisis. In such a situation economic results between branches are likely to diverge more than in periods of high growth (which was the situation in 1972). Branches 1A, 2A and 2B had β_1 values significantly larger than 1, whereas branches 3B and 3C had values smaller than 1. That 3C (building materials) had a smaller value may be explained by the slow-down in construction which began in the late 1970s. The low value for the breweries in 3B may very well have been caused by their keeping too large a labour force up to and beyond 1981. Certainly, considerable reductions of their personnel have taken place after 1981.

The industry branches 1B, 1C, 1E and 3A and 3D have shown stability over time and allow us to estimate β_{oK} in both years. β_o is found by solving equation (11.10) with respect to β_{ok} with $\beta_{1k} = 1$:

$$\beta_{oK} = Y_{\cdot k} - l_{\cdot k}. \qquad (11.11)$$

In the expression (11.11) the points (.) indicate mean value estimates. In this case they are simple arithmetic averages.

It is now possible to test hypotheses about the productivity parameter β_{ok}, when its standard error has been estimated. The groups for which the estimation of productivities in **both** 1972 and 1981 are meaningful on a country basis numbered only five, three in the high technology category one and two in the 'protected' or regulated category three (see Table 11.7).

Table 11.7

Productivity in Industry Groups, 1981 and 1972 (1000 DKK*)

Group	Industrial Sector		1981		1972	
			β_{ok}	β_{ok}	β_{ok}	β_{ok}
	1B	HT: Mechanical Industry	2.2408	174.1	1.7958	62.5
1	1C	HT: Electronics etc.	2.2168	164.7	1.7248	53.1
	1E	MT: Cables, household appliances etc.	2.2355	172.0	1.7947	62.3
	3A	Dairies, meat packing etc.	2.3355	218.5	1.8273	68.8
3	3D	Printing etc.	2.3022	200.6	1.7863	61.1

*β_1 = 1 in both 1972 and 1980.

The question now arises as to whether the productivities within each group actually differ from one another (intra-group differences) and whether there is a significant difference between groups. Is productivity in the mechanical industry (1B) higher than productivity in the electronics industry (1C), as the bare figures indicate, or must they in fact be considered equal, when the statistical uncertainty of the β_{ok} estimates is taken into account? Furthermore, is the productivity higher in the partly regulated food industry (3A) than in the industries in group 1, which, at least in principle, should be high-tech?

The answer to the first question about group 1 is negative. The differences in productivity between the three subgroups is not statistically significant[1]. However, the electronics group, 1C, may lie systematically below 1B and 1E, as its β_o is lower in both 1972 and 1981. In order to confirm this one would need data for the years between 1972 and 1981.

The other question (about the differences between the technology intensive industries and the food industry) must on the other hand be answered clearly in the affirmative. The food industry has a statistically significant higher productivity in both years, and the difference is larger in 1981 than in 1972. This is no doubt a reflection of the good performance of this industry after Denmark's entry into the Common Market in 1973. The advantages of the common agricultural policy were still operating in 1981 in spite of the recession.

Regional Results 1981

By adding subscript j, which stands for the category 'region', to expression (11.10) the regression model also takes into account the possibility of introducing the regional parameters β_{1jk} and β_{ojk}. 'j' takes values between one and five as the number of regions is five. The number of sectors was reduced to nine, as the sectors 1A and 3D were excluded. In this model the parameters and the variance of the residuals vary with region and industrial sector.

The regression analysis is performed as follows: first we estimate the parameters for each combination of region and industry sector. Then we test, whether the variance of the residuals σ_{jk}^2 may be considered the same in all five regions for each sector (Bartlett's test). As this is the case, we go on to test whether the slope-parameters β_1 are equal to 1, i.e. the hypothesis:

$$H_{o1} : \beta_{1jk} = 1, \ j = 1, \ \ldots, \ J \text{ is tested } (p = 0.025)^3$$

The hypothesis was accepted for all sectors except sector 2A, which comprises old, low-technology industries (Table 11.1). Region V contributes about two-thirds to this result for 2A. If we exclude the data for this sector in region V, then the hypothesis is also accepted here.

Finally the hypothesis of equal productivities within sectors among regions is tested:

$$H_{o2} : \beta_{ojk} = \beta_{ok}, \ j = 1, \ \ldots, \ 5$$

H_{o2} is accepted for all nine sectors and we obtain the estimates in Table 11.8.

An Interpretation of the Results

In order to interpret the equality of productivity between regions in the same sector we must imagine ourselves to be back around the 1950s and 60s. Until then industrial production had been concentrated in the Copenhagen region and in some

Table 11.8
Productivity in industry groups

Industrial Sector			Productivity (1.000 DKK) β ok	β ok
1B	HT:	Mechanical industry	2.2437	175.27
1C	HT:	Electronics etc.	2.2299	169.79
1D	MT:	Chemical industry etc.	2.4797	301.79
1E	MT:	Cables, household appliances etc.	2.2411	174.22
2A	LT:	Plastics, glass, china etc.	2.2458	176.12
2B	VLT:	Textiles, clothing apparel etc.	2.1703	148.01
3A		Dairies, meat packing etc.	2.3385	218.02
3B		Breweries etc.	2.3743	236.76
3C		Products for use in building etc.	2.3352	216.37

larger cities and a few other industrial areas. The relocation of industry towards more peripheral areas, especially in Jylland, which, incidentally, was also characterized by local firm creation and growth, began in the sixties and gathered momentum at the end of the decade.

At the beginning of the period, productivity was undoubtedly highest in the older industrial regions. It has since grown considerably in the western parts of the country, by the employment of the surplus labour from agriculture. The data support this hypothesis to some extent, as the growth in productivity was high in Jutland for such groups as machinery, electronics and agro-industry.

According to several authors (quoted in Wichmann Mathiessen, 1985), the spread of industry in the sixties either concerned large, integrated firms or mass producing sub-contractors. Differences in regional infrastructure tended to disappear, both for the training and education of the labour force and for transportation of goods and persons. These changes as well as the expansion of other segments of the public sector in the economy made it possible for all types of industrial firms to establish themselves in peripheral regions during the 1970s.

Notes

1. The variance of log y - log x is found as:

$$V(\log y - \log x) = V(\log y) + V(\log x) - 2\,\mathrm{Cov}(\log x, \log y)$$

We calculate estimates of V (log y - log x) for industries 1B, 1C and 1E and test for equality:

$$F(1C, 1E) = 1.36$$

the degrees of freedom being $(f_1, f_2) = (67,84)$, is not significant at the 5 per cent level.

A joint value of s^2 for 1B and 1C may then be calculated, since the differences between their variances is even smaller than the one found between 1C and 1E. We get:

$$s^2(1C, 1E) = 0.0196$$

Equality between the logarithms of the productivities p (1B) and p (1C) is then tested by student's:

$$t = 1.14$$

which is not significant at the 10 per cent level. The productivities for 1B, 1C and 1E must then be considered as equal.

The joint weighted estimate of their logarithm is:

$$\log p = 2.2220 \text{ and thus}$$

$$p = 170.0$$

A joint estimate of V (log p) is = 0.0186.

2. Here we test whether the difference between the productivities in group 1 (1A and 1D being omitted) and group 3A is statistically significant. We find a very clear difference, as both the variance and the mean value in 3A lie very significantly higher than the corresponding joint values in group 1.

$$f(1., 3A) = 1.924$$

the degrees of freedom being:

$$(f_1, f_2) = (148,273)$$

which is significant at the 5 per cent level.

Testing the difference between the means, gives:

$$U = 5.84$$

which is highly significant. There is thus not the slightest doubt that the food industry group 3A has a higher economic productivity than the mechanical, electronics and metals goods industries of group 1.

3. When looking for regional differences we first test:

$$H_{o1}: \beta_{1jk} = 1,\ j = 1, \ldots, J$$

$$T = F(5, \Sigma(n_{jk} - 2))$$

For the industry sectors where Ho1 is accepted, we test the hypothesis:

$$H_{o2} : \beta_{ojk} = \beta_{ok}, \; j = 1, \ldots, 5$$

in a one-way analysis of variance with:

$$V = F\,(4, \Sigma\,(n_{jk} - 1))$$

The test results are shown in the following table:

	H_{01}		H_{02}			
Industrial Sector	T	P	β_o	v	f	p
1B	0.86	0.51	2.2437	1.01	(4,114)	0.41
1C	0.23	0.95	2.2299	0.74	(4,64)	0.57
1D	2.60	0.04	2.4797	0.73	(4,48)	0.58
1E	0.79	0.56	2.2411	0.21	(4,79)	0.93
2A	5.42	0.0001	2.2458	0.54	(4,140)	0.71
2B	1.07	0.38	2.1703	0.91	(4,197)	0.46
3A	2.60	0.03	2.3385	0.08	(4,142)	0.99
3B	1.00	0.43	2.3743	0.71	(4,44)	0.59
3C	1.61	0.17	2.3352	0.16	(4,82)	0.96

Acknowledgements

The study has received support from the Danish Social Science Research Council (grant 14-3236) which is gratefully acknowledged. The statistical analysis has been discussed with Mr N Kousgaard, at the Statistical Institute of the University of Copenhagen. The testing of regional differences is largely due to his efforts. I thank him for a very fruitful co-operation. For any errors in data, model specification and interpretation of results the author is wholly responsible.

References

Antonsen, K. (1964) Placeringen of Danmarks Industri 1938 til 1960, Landsplanudvalgets Sekretariat, Copenhagen.

Bosworth, D.L. (1976) Production Functions, Saxon House, Lexington, Mass.

Christiansen, U. (1983) Dansk industris mobilitet efter 1960, SBI-byplanlaegning 46. Statens Byggeforskningsinstitut, Hörsholm.

Christiansen, U. (1988) Industriudvikling og produktivitet. En analyse of regioner og byom råder i Danmark, SBI-meddelelse 69, Statens Byggeforskningsinstitut, Hörsholm.

Ibsen, F., Nielsen, K. and Nielsen, K. (1980) Jern- og metalindustrien i Nordjylland, Amtskommunernes og kommunernes Forskningsinstitut, Copenhagen.

Moller, K. (1985) En teoretisk og empirisk analyse of danske virksomheders forskning og produktudvikling og Danmarks udenrigshandelsmønster, The Copenhagen Business School, Copenhagen.

Ohlsson, L. (1976) Svensk verkstadsindustris internationella specialisering, Industriens utredningsinstitut. Stockholm.

Ohlsson, L. (1980) Världshandelsförändringar och Nordens framtida industrisysselsattning, Nordiska ministerradet, Oslo.

Wickmann Matthiessen, C. (1985) Danske Byers Vaekst. Det kgl. danske geografiske selskab. I kommission hos C.A. Reitzel, Copenhagen.

Oresundsgruppen/v Nerp (1981) Öresundsregionen og teknologien i 80'erne, Nordisk Embedsmandsgruppe for regionalpolitik, Copenhagen.

12. INTERACTION OF INCOME GENERATION CAPACITY AND TECHNOLOGY IN THE REGIONS OF HUNGARY

Peter Klekner

Introduction

The goal of regional development policy in Hungary is two-fold: firstly to utilize more effectively the physical and human resources of the economy in the regions, which will enhance national economic performance, and secondly to bring about regional convergence in terms of the levels of employment, the supply of basic services, standards of living and so on.

In order to achieve this second goal, at the end of the sixties regional policy in Hungary was designed to reduce the regional differences in the level of employment and to improve the industrial structure of backward regions. Therefore, in the first half of the 1970s, creating new workplaces became the major aim in the depressed areas of the country, which at that time possessed a surplus of labour. This programme could be viewed essentially as an **extension** of the economy. Another, and complementary, aim was to reduce the disproportionate dominance of Budapest in the economic and intellectual life of the country, which was to be arrived at through spatial rebalancing of key elements of the economy (R & D, industrial management and higher educational activities etc.) within a context of growth at the national level. In the event, however, from the second half of the 1970s and particularly in the 1980s, a period of scarcity of financial resources and labour led to a shift away from greater extension of the economy towards **intensification** and **efficiency**, particularly efficiency in labour utilization, accompanied by rising productivity.

Over the past 15 years, therefore, there have been significant changes in the regional structure of the economy in Hungary and, due to the impact of regional policy, a partial decentralization of productive forces has taken place. At the same time concentration within larger industrial units has also taken place.

One consequence of these changes has been that regional differences in the level and sectoral structure of production have decreased, but production is now concentrated in larger local establishments as well as within national enterprises. However, regional differences in income and in the distribution of higher order functions, such as management and R & D, have hardly changed at all, especially when changes are measured at the level of smaller spatial areas.

Recently, qualitative changes and new trends in socio-economic conditions have identified the weaknesses of the old system resulting in a call for the reformulation of policy and the identification of a new strategy for regional and local area development. The new strategy, among other things, pays close attention to development at the micro-scale and the role of technical change in enhancing local income generation capacity.

In this chapter, attention will be focused first on the analysis of R & D activities at the national and sub-national levels in Hungary. Secondly, the chapter investigates the county and local area patterns of income generation and productivity gain, before going on to explore the relationship between the level of technological activity and income generating capacity in the regions.

Research and Development: the National Picture

Taking together research and development in industry, higher education and the public sector, Hungary has expended over two per cent per annum of the Gross Domestic Product (GDP) on this activity over the past two decades (Table 12.1).

Table 12.1

Gross domestic expenditure on R & D (GERD) at the national level in Hungary (current expenditure*)

	Million Ft					
	by enterprise	government and non-profit organizations	higher education	sums from abroad	total	Percentage of GDP
1970	5,491	1,457	543	-	7,491	2.0
1980	13,865	2,944	2,109	111	19,029	2.5
1984	17,472	3,035	2,736	89	23,332	2.4

Source: Tudomanyos kutatases fejlesztes, KSH, Budapest, 1972, 1982, 1985.

* figures in constant prices are not available.

The figures also show that approximately two thirds of research is undertaken by industry. The figures in Table 12.1 suggest that the highest growth in R & D expenditure may have been within higher education and the slowest growth in the public sector. The table also suggests that the greatest expansion of R & D effort came in the 1970s, but that it has not retained its share in GDP through the 1980s.

These suggested trends are to some degree confirmed through examination of the manpower engaged in R & D (Table 12.2) which shows that 64,000 researchers were at work in 1970, rising to 85,000 in 1980 but falling back to 78,000 as the effects of the world and other recessionary pressures affected the Hungarian economy. These figures suggest that expenditure and employment on R & D in Hungary is on a par with all but the most research intensive economies in the Western world (e.g. see OECD, 1988). Similarly, in the recession, the falls in R & D employment have tended to be relatively greater in the technical and administrative grades than in the scientific grades (for a comparison of performance in the UK, see CSO, 1988).

Table 12.2
Manpower directly employed on R & D tasks at the national level in Hungary (number of persons)

	Scientists	Technicians	Administrators	Total	Percentage of working population
1970	23,721	29,239	11,459	64,419	1,29
1980	38,705	30,849	15,802	85,356	1,68
1984	36,834	27,769	13,793	78,396	1,59

Source: Tudomanyos kutatases fejlesztes, KSH, Budapest, 1972, 1982, 1985.

County Differences in Technological Capacity

The territory of Hungary is divided into 19 counties, plus the capital city, all of which function as administrative units (Figure 12.1. See also Appendix 12.1 for the relationship between counties and sub-counties in local areas). The boundaries for the counties were determined some 25 years ago on an historical basis and are now seen as increasingly problematic areas for analysis, given the emergence of new development processes and changes over time in the basic economy. Despite these shortcomings, this is the level of spatial analysis that has been traditionally used in Hungary and remains the only spatial areas for which sub-national data are collected and made available.

In Hungary, due to the previous political and economic situations (central direction of the economic and social system; strong concentration and centralization of economic activity), the weight of the capital has become disproportionately large in the economic and intellectual life (encapsulated in higher order labour functions) of the country. This concentration of activity in and around Budapest has hardly

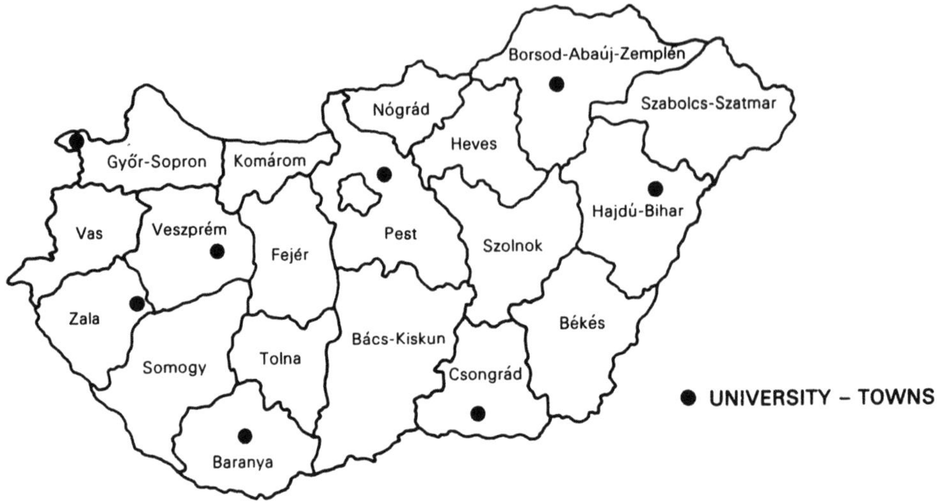

Figure 12.1: Hungary - counties.

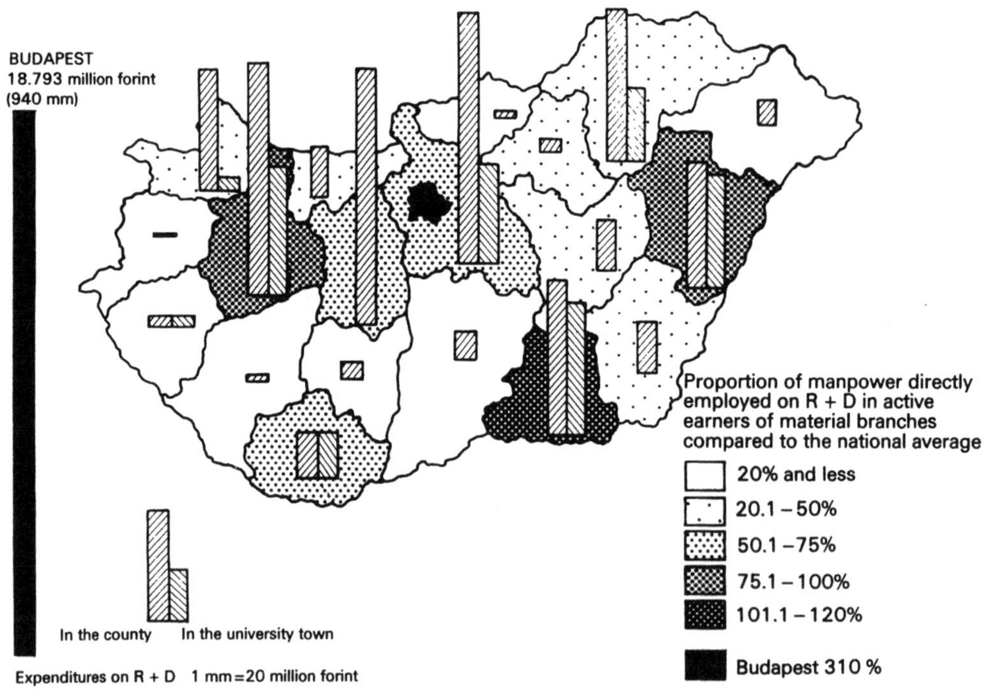

Figure 12.2: Hungary: R & D activities by counties and University towns in 1986.

changed over time in spite of the efforts to bring about change through strong policy, applicable in those areas outside the city and its immediate surroundings. Data made available for the first time in 1986 at the county and university town levels show that today the level of concentration of research and development activities is still very high, with more than 70 per cent of R & D expenditure, two-thirds of all research workers and half of the number of research units and institutions located in and around Budapest (see Table 12.3 and Figure 12.2). This dominance is across the whole spectrum of research from governmental and university to that undertaken in enterprises (Table 12.4).

Table 12.3

R & D Organization, Manpower and Expenditure by Counties and University Towns, 1986

Counties and the University Towns	Number of R & D organizations	Manpower directly employed on R & D	R & D Manpower of which Scientists	Expenditure Million FT
Budapest	669	50,839	25,347	18,793
Baranya	81	1,944	1,208	223
of which: Pecs	79	1,910	1,200	220
Bacs-Kiskun	17	678	315	133
Bekes	11	756	186	245
Borsod-Abauj-Zemplen	56	2,322	993	720
of which: Miskolc	43	1,731	772	363
Csongrad	137	3,656	1,931	747
of which: Szeged	127	3,178	1,812	645
Fejer	24	2,455	1,074	1,250
Gyor-Sopron	51	1,449	716	579
of which: Sopron	23	326	184	52
Hajdu-Bihar	77	3,225	1,745	631
of which: Debrecen	75	3,089	1,679	567
Heves	22	570	322	73
Komarom	8	624	205	233
Nograd	4	241	89	37
Pest	51	4,050	1,363	1,226
of which: Godollo	40	2,187	758	497
Somogy	9	220	129	44
Szabolcs-Szatmar	23	745	325	120
Szolnok	16	711	212	260
Tolna	2	247	53	88
Vas	6	162	80	19
Veszprem	31	2,414	855	1,150
of which: Veszprem	23	1,675	647	645
Zala	16	355	158	65
of which: Keszthely	15	326	147	60
Total	1,311	77,663	37,306	26,636

Source: Tudomanyos kutatas es fejlesztes, KSH, Budapest, 1988.

Table 12.4

Regional proportion of R & D capacity (per cent)
(on the basis of staff numbers)

Settlements	Total	R & D capacities of enterprises	institutes of Academies of Science & of Central Government	of universities
Budapest	66.5	69.5	78.5	48.3
University towns	18.2	5.9	14.1	40.1
Other towns & villages	15.3	24.6	7.4	11.6
Country	100.0	100.0	100.0	100.0

Source: Bartke and Nemes Nagy (1988).

In terms of total R & D capacity the capital is followed in order of importance by the university towns: Szeged, Debrecen, Godollo, Pecs, Miskolc, Veszprem, Gyor-Sopron and Keszthely which play important research roles in the life of the nation (see Figures 12.1 & 12.2). In these towns there are 18 per cent of total research staff and 34 per cent of the national research units or institutions. Outside of the capital and the university towns, the only counties with reasonable concentrations of research effort would seem to be in Fejer (see Figure 12.2 and Table 12.3).

Table 12.3 also illustrates the high concentration of scientific staff in and around the capital. Perhaps more problematic from a spatially distributed industrial development point of view is the distribution of regional research institutions concerned with the technical sciences: 64 per cent of all such research units are also concentrated in Budapest (Table 12.5). In the traditionally industrialized zones of the country, in North-Hungary: and North-Transdanubia only 12 per cent and 11 per cent of technical R & D units respectively are to be found. The remainder of the country has very little science-based research indeed.

The less developed regions - (i.e. Hungarian Great Plain as well as South- Transdanubia) - have only about 10 per cent of technical research units, although 37 per cent of industrial workers are employed and 28 per cent of national

Table 12.5

Regional proportion of R & D units 1984 (per cent)

Regional Districts	All units	Technical sciences	Agrarian sciences	Social sciences
Central part	54.8	67.0	41.4	57.3
Budapest	50.2	63.6	25.3	55.4
North-Hungary	6.2	11.7	1.8	4.1
North-Transdanubia	9.0	11.4	17.3	5.9
South-Transdanubia	8.0	2.9	10.5	9.5
North-Great Plain	9.4	3.2	15.4	7.0
South-Great Plain	12.6	3.8	13.6	16.2
Country	100.0	100.0	100.0	100.0

Source: Bartke and Nemes Nagy (1988).

income is produced in these regions. The extent of the dominance of Budapest in research terms is, however, perhaps best illustrated by the fact that the capital also dominates the agricultural research field, although there is representation in other regions outside the industrialized zones of the North.

The regional proportions of R & D units within industrial enterprises show a similar situation with about 63 per cent of all industrial R & D units situated in Budapest in 1986 (see Figure 12.3). In 1975 more than 65 per cent of these units were to be found in Budapest so the weight of the capital has hardly decreased over the past decade despite policy-based attempts at dispersal. However, in the traditionally industrialized zones of the country the number of R & D organizations has increased slightly and so their proportional contribution to national R & D activity has grown marginally: in North-Hungary from eight per cent to nine per cent and in North-Transdanubia from 13 per cent to 17 per cent of the national total. In the remainder of the country the R & D capacity within industrial enterprises appears to have decreased to some degree.

In addition to pure research workers, Scientific and Technical workers in industry can also make a significant contribution to technological change and local economic advance. If we take into consideration the proportion of Scientific and Technical workers employed in industry and commerce we observe that these types of workers are also very much concentrated in the Budapest area where 60 per cent of the national resources are employed. The share of higher order occupations in industry is also high in North-Transdanubia (in counties Gyor-Sopron and

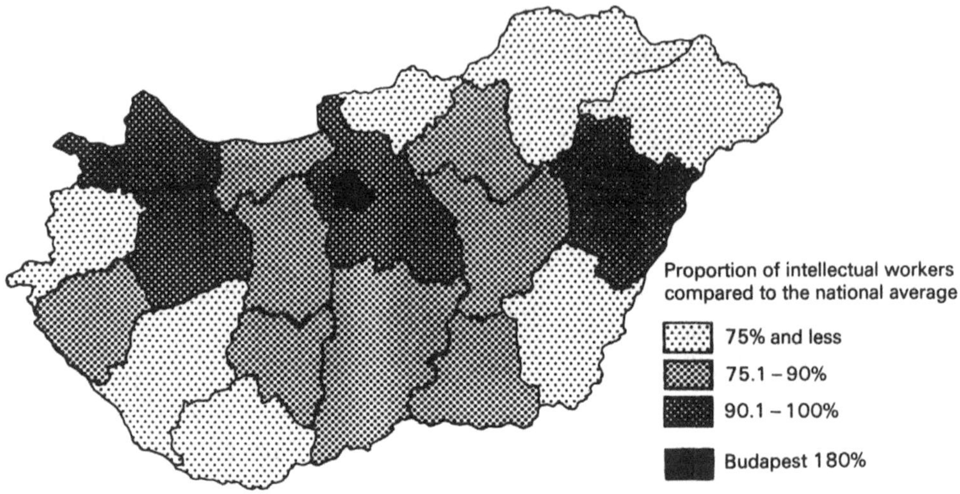

Figure 12.3: Hungary: Regional pattern of industrial R & D activities in 1986.

Veszprem), in county Pest and Hajdu-Bihar. At the same time it is very low in North-Hungary (the home of the industrialized counties) and in South-Transdanubia, where it was less than two-thirds of the national average in 1986 (see Figure 12.3).

The Spatial Pattern of Income Generation in Industry

Having established spatial variations in research activity in Hungary, what is of interest is how the pattern relates to the pattern of output and income within the economy and whether any changes in research capacity have been accompanied by changes in income, or if they are independent. As mentioned before, over the past 15 years a process of concentration in industry has taken place in Hungary. As a result of this process large enterprises have been established, with 20-30 factory units scattered about the country. Unfortunately, performance and output data of the enterprise is not disaggregated to establishment level, but is handled collectively in the centre. Thus, the data which might indicate regional differences in the effectiveness of the economy are not directly available.

For this reason regional levels of income can only be inferred through indicators constructed from secondary data on local wages and investment. The estimating method used assumes that local income potential is embodied in local labour and 'dead labour' (capital investment). So our starting point for estimating local income is the quantity of wages paid locally and the contribution to income derived from the value of fixed assets available at each industrial site throughout the country. These data were collected on an area basis through regular surveys of approximately 8000 business establishments which make up the Hungarian manufacturing sector. The total income generated by each industry is also known at the national level.

On the assumption that a unit of wages or net capital contributes equally to total industrial income, then national income can be disaggregated to the counties and Budapest using the following formula:

$$I_a = W_a \times 1.5 + (I_i - W_i \times 1.5) \times C_{as} \qquad (12.1)$$

where I_a = national income produced by the industry in the county

$$\sum_{a=1}^{20} I_a = I_i$$

I_i = sum of national income produced by the industry of the whole country

W_a = wages paid in the industry of the given county

$$\sum_{a=1}^{20} W_a = W_i$$

W_i = sum of wages paid in the industry of the whole country

C_{as} = share of industrial fixed assets in the given county of industrial fixed assets of the whole country

1.5 = an empirical coefficient derived and used by the Hungarian Central Statistical Office and includes wages plus a social insurance contribution made by enterprises to government for each worker.

County income

At the beginning of the 1970s there were significant differences in the level of national income per capita by the counties in Hungary. In four counties (Szabolcs-Szatmar and Bacs-Kiskun county in the Hungarian Great Plain, and Somogy and Tolna county in South-Transdanubia) the national income per capita was below half of the value of the national average. In a further six counties it did not reach 75 per cent of the national average. These underdeveloped counties were located mainly in the Hungarian Great Plain and in the Western borderland of the country: in Vas and in Zala county (Figure 12.4).

By way of contrast, at that time in Budapest and in Komarom county the level of income per capita was one and a half times the national average. The difference between the two extremes was more than five times.

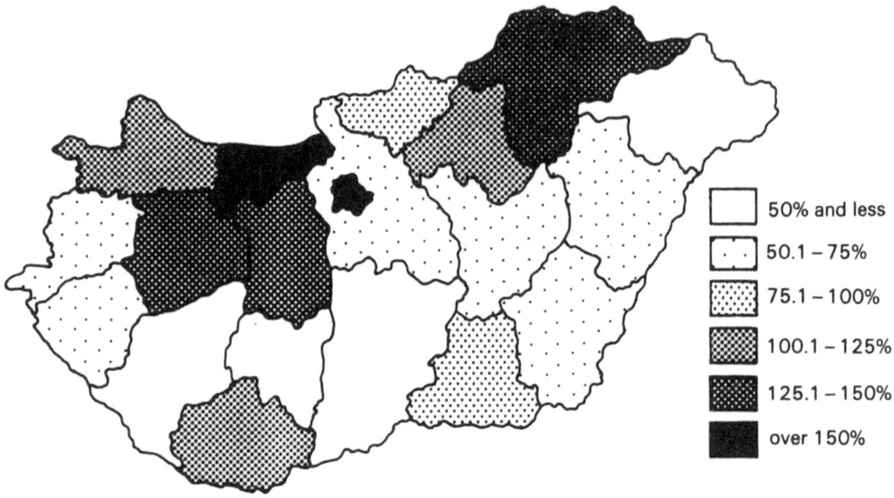

Figure 12.4: Per capita industrial national income production by counties, compared to the national average, 1972.

Between 1972-1984 the national income in industry more than doubled. The most underdeveloped counties mentioned above developed very dynamically, while the pace of development in the county of Komarom and in Budapest failed to reach 85 per cent of the national growth pace. As a consequence of these differences, a strong convergent process can be observed at the county level. The deviation between the two extremes narrowed to a difference of only three times by 1984.

In spite of the rapid development rate of the underdeveloped counties, the level of national income per capita in the Hungarian Great Plain remained relatively low while the situation improved mainly in South-Transdanubia (Figure 12.5).

Relationship between county income performance and R & D activity

As stated before, in Hungary the level of concentration of research and development activities is very high in and around Budapest. In spite of this fact, the pace of development of income per head was below the national average in Budapest and in county Pest.

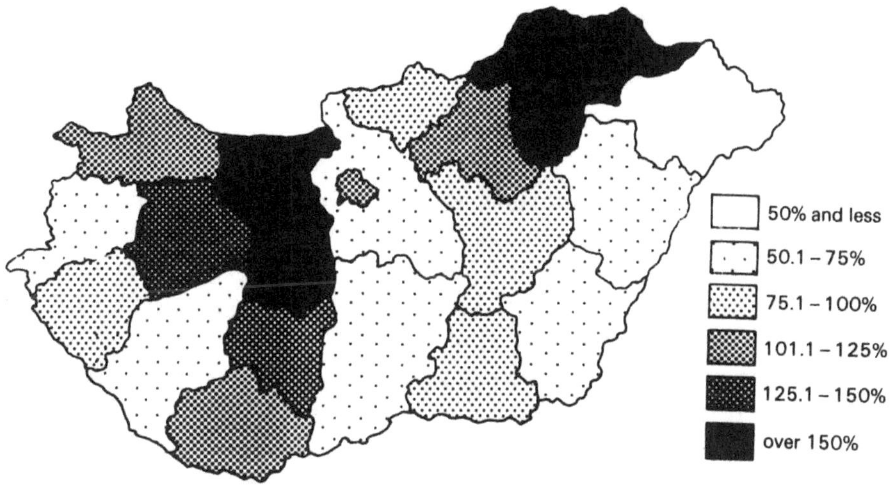

Figure 12.5: Per capita industrial national income production by counties, compared to the national average, 1984.

At the same time, in South-Transdanubia, where the level of R & D activities is very low (the proportion of scientific and technical workers employed in industry was less than two-thirds of the national average in 1986 in county Baranya and Somogy, and less than 90 per cent of the average in Tolna and Zala counties), the level of income per capita developed dynamically.

In the Great Plain the situation as far as R & D capacities is concerned is very bad in Szabolcs-Szatmar, Bacs-Kiskun, Bekes and Szolnok counties (see Figure 12.2). In these counties, despite the relatively high growth pace of income, the levels of income per capita remained comparatively low. Similar tendencies can be observed in the counties containing university-towns where the level of R & D capacities is relatively high. In North-Transdanubia and in North-Hungary the further growth of the proportional contribution to national and R & D activity probably has meant that their relative advantage in the income generation has remained, or even increased to some extent.

These experiences suggest that at the county level the links between R & D institutions and income generation within local industry are very weak. The explanation may be in the fact that the research and development units are located in a relatively few centres of the county, thus the relationship between output and

R & D activity has to be investigated at a level of smaller regional units.

Sub-county income

As stated above the administrative counties have been seen recently as an increasingly inadequate basis for economic analysis. Recent work undertaken by Klekner (1988) has also shown that statistical data at the county level tends to obscure the marked differences which characterize areas within each county. This has led to the development of a new set of 143 areas for analysis within Hungary which are called 'areas of gravity'. The boundaries of these areas were based on travel to work areas around 143 major settlements (125 towns and 18 large villages) within the country. These areas can be aggregated to the county level for meso-level analysis and upon which the analysis of described in the section above was based. Certain adjustments are necessary at the gravity area level to accommodate county boundaries. An exception to this disaggregation is made for the capital, Budapest, which is treated as a single entity.

Using the same method as before we can calculate the income generation capacity of the industry by the areas of labour force gravity in Hungary.

$$I_a = W_a \times 1.5 + (I_i - W_i \times 1.5) \times C_{as} \qquad (12.2)$$

where I_a = national income produced by the industry in the area of gravity

$$\sum_{a=1}^{143} I_a = I_i$$

I_i = sum of national income produced by the industry of the whole country

W_a = wages paid in the industry of the given area of gravity

$$\sum_{a=1}^{143} W_a = W_i$$

W_i = sum of wages paid in the industry of the whole country

C_{as} = share of industrial fixed assets in the given area of gravity of industrial fixed assets of the whole country

1.5 = an empirical coefficient derived and used by the Hungarian Central Statistical Office and includes wages plus a social insurance contribution made by enterprises to government for each worker.

This method of local income estimation suggests that at the beginning of the 1970s there were significant areal differences in the level of national income per capita in Hungary. In more than the half (55 per cent) of the 143 areas national income per capita did not reach half of the national average (in almost a quarter of them not even 25 per cent of the national average was achieved). These depressed areas were located mainly in the Hungarian Great Plain, in the eastern part of the country, and in the mid zone of Transdanubia (i.e. in that part of Hungary which lies between the Danube, the Drava and the Western frontier of the country). Areas of low national income per head were also to be found surrounding the Budapest agglomeration. However, a considerable part of the population in these apparently deprived areas around Budapest works in the capital, so income generated by them, boosts the figures for the Capital's income.

By way of contrast, in 1972 there were also 22 areas in the country which had reached more than one and a half times the national average income per head and which were located mainly in the Northern axis of the country. In the neighbourhood of several of these highly developed areas, there were also areas experiencing below the national average level of income per head, which shows that the location of industry at the local rather than the national level has not been responsive to policy intervention, despite the significant sums of money invested in promoting dispersal to a wider area surrounding nodes of development. As a consequence, these underdeveloped areas surrounding an advanced region may currently act as an obstacle to further dynamic development of the core, because past activities have attracted away the rich human capital and the fixed capital leaving the economically backward areas in a very weak position in modern development terms. One of the characteristics of local area development in Hungary appears to be the existence of these more dynamic nodes surrounded by largely rural and low income generating areas supplying the labour force to the node.

By the end of the 1970s, the areas belonging to the two extremes showed considerable convergence in terms of income potential. However, one third of all areas still did not achieve half of the national average income per capita, while 17 areas realized more than one and a half times the average value. A positive feature of the late 1970s, however, was that more than half of the areas which were the least developed in 1972 had reached a higher category of income per head.

Deviation between the two extreme categories of high and low income per head narrowed further in the 1980s. Currently, the most underdeveloped areas are still located mainly in the east, in the north-east borderland, in the middle zone of the Great Plain, and in the region of Transdanubia. Most advanced areas are located in the traditional industrial areas, in North-Hungary, in North-Transdanubia, and along the Danube. When taking into account income generated in Budapest and her surrounding areas together, the income per capita in the whole area currently exceeds the national average by 30 per cent. The overall pattern of areal distribution of income has therefore not changed dramatically in the past twenty years.

The previous discussion is based around relative performance between areas, but it should be noted that, over the past 15 years, most underdeveloped areas have experienced rising per capita income which is more than three times higher at the end of the period than at the beginning, compared with the national performance where the increase was only 220 per cent. Unfortunately, there are also several underdeveloped areas where industrial development has not taken place and in some areas it has even deteriorated. These latter areas are frequently located near other highly industrialized nodes and, as a consequence of this, did not share in the more generalized industrial advance of the nation (for instance North-Hungary). Some areas, however, are able to compensate for a lack of industrialization through the development of services, particularly in the form of tourism. These tourist centres are obviously located in the more picturesque regions: for example on Lake Balaton and at the Danube Bend. There remains, however, a set of areas which have not benefited fully from the national advance or policy intervention designed to help them.

Over the period taken as a whole (1972-84) a general rise in national income was observed which is estimated to have led to a more even distribution between the gravity areas of the country, without leading to a dramatic change in the pattern of relatively rich and poor areas of gravity. Dividing the period into two parts suggests that the rapid development of backward areas occurred mostly in the 1970s. This was a consequence of the **extensive** development of industry in this period, which was stimulated by both general economic conditions and extension of the economy plus regional policy. In combination these forces led to significant industrial developments in the underdeveloped areas of the Great Plain, in several industrialized areas (e.g. in North-Transdanubia), and in areas surrounding some of the larger cities (e.g. Budapest, Pecs and Gyor). The previously dynamic industrial development slowed down, with the world recession seriously affecting the economy and particularly those parts based upon coal mining, metallurgy, and the chemical industries.

The 1970s also saw extension of industrial development through the regional decentralization of processing industries, which led to an increase in the economic advance of underdeveloped areas where the pace of development of 37 areas of gravity rose up to 1.5 times the average area development, leading to a decrease in the degree of local area differences in income per head.

However, this levelling process slowed down significantly in the 1980s. Dynamic development (50 per cent higher than average) has been experienced in only a few areas previously classified as underdeveloped and these tended to result from specific large scale investments.

For most areas, development has taken place at about the national pace in recent years. But some areas which were already underdeveloped fell further behind and there the pace of change was less than 85 per cent of the national average. Developed areas with an average pace of development more or less kept their lead. Due to these processes, the previous convergence slowed down considerably during the 1980s.

Technological Change and Local Area Development

It has been attempted in the previous sections of this chapter to illustrate the changing levels of industrial income generated at the national and subnational levels in Hungary. The analysis shows that there are significant variations between areas in income per capita. The work here also indicates the strong concentration of research activities across a broad spectrum of higher education, enterprise and other research units in a relatively few centres of the country. This section tries to bring together these observations to see if there is any relationship between them.

Comparison shows that relatively few R & D centres are located in those areas with the lowest levels of national income generation. Collectively these areas under-employ scientific and technical workers in local industry when compared with the national average. Disaggregating research effort to the sectoral level suggests that the research based in Universities added to the technological development potential of these areas and that this was generally associated with a higher level of national income and income growth derived from industries in the local areas; University towns and their surrounding areas tend to enjoy higher levels of income per capita than the median area. However this last conclusion needs to be modified to accommodate the focus of research in each institution. If the focus of research in an area is upon agriculture, then this is less likely to be associated with high growth or high levels of local income.

Evidence from other research (e.g. Oakey, 1984) suggests that links between institutions of higher education, local innovation and output may be rather tenuous in that relatively few firms claim that links with local Universities and other institutions are a major factor in their business. The effects of this weak relationship may be observed in the fact that, while the areas in Hungary surrounding University towns have relatively high levels of income generation, the **growth** of income in these same areas over recent years has not differed significantly from the national average. Table 12.6 indicates that over the period 1972 to 1984 the University towns as a whole performed close to or at the national average in terms of income growth per capita. While the University towns, and by implication their research bases, do not appear to have led to a significantly higher level of income growth than elsewhere, it may at least have maintained that growth close to the national performance in a way that was denied to other locations.

Industrially oriented research, whether in enterprises or in centres devoted to its pursuit, could perhaps be expected to have a more positive influence on local economic performance. Fragmentary evidence suggests that where research in local enterprises was high then this did lead to a higher rate of economic development. This conclusion fits well with other research (e.g. OECD, 1986).

One of the advantages claimed to be associated with technological advance is gains in labour and industrial productivity generally (output per unit of labour input), which in turn may influence local income and growth potential. In the early 1970s labour productivity was low in relation to the national average in about one

Table 12.6
Relative growth of national labour productivity and national capita industrial income in University town areas of gravity

Areas of University Towns	Relative growth of the productivity of labour in industry (% of national growth level)		Relative growth of the per capita industrial national income (% of national growth level)	
	1972-1980	1980-1984	1972-1980	1980-1984
Miskolc	98	96	99	96
Debrecen	103	100	113	100
Szeged	106	100	92	98
Pecs	86	95	89	95
Veszprem	98	100	85	98
Sopron	98	95	99	97
Keszthely	99	105	114	108
Godollo	100	101	90	97
Budapest	105	101	85	89
National average	100	100	100	100

Source: Klekner (1988)

third of the areas of the country, made up primarily of the underdeveloped or low income areas of the Great Plain, as well as western and southern parts of Transdanubia. The highest levels of productivity were located within the more traditionally industrialized zones of North-Transdanubia and North-Hungary, where incomes have in general also been relatively high. One of the observed differences between these two types of area is that the underdeveloped areas appeared to lack investment in modern equipment and new technologies (see Bartke, 1987).

Some closing of the productivity gap between the more and less prosperous areas of the country was achieved in the 1970s due to the strong policy support for industrial growth and its extension to the underdeveloped areas. However these advances were shortlived and in the 1980s industrial investment declined in line with the national and world recession and as a result labour productivity in the problem regions stagnated or fell back. At the same time, a movement towards **intensive** (as opposed to extensive) industrial development in those areas with a long tradition of industrial activity led to rises in productivity in those areas. The productivity gap between areas therefore increased once again.

While local income and progress can be positively associated with rising labour productivity (see for example, Camagni and Cappelin, 1985) it can also be associated with capital investment. Capital investment supports the activities of workers, enhancing their human and intellectual skills, but also tends initially and over time to embody new technology with positive effects on growth of income and output (e.g. Solow, 1957; Denison and Poullier 1968). Both gross, and particularly net, increases in the stock of fixed assets and the level of capital equipment can play a decisive role in the generation of national and local income. Both gross and net investment can also introduce more modern technology to the areas of investment.

Analysis of the establishment data indicates that capital equipment per employee is more strongly associated with the observed levels and change in local income than is labour productivity or the existence of research workers, research units or R & D expenditure. Thus, at the beginning of the 1970s, the level of capital equipment per employee was well below the national average in all of the underdeveloped areas of the country, while the highest levels were observed in the traditionally industrialized counties, mainly in the North. What little investment there was in the underdeveloped areas tended towards low technology associated with low levels of labour productivity.

In several underdeveloped areas, in the north-eastern part of the country and in some areas of South-Transdanubia, despite their undeniable problems, the level of capital equipment has fallen back. Similar tendencies could be observed in some developed areas too: in North-Hungary and in North-Transdanubia. At the same time, the relative backwardness of South Great Plain lessened to some degree. The shortage of capital investment perhaps helps to explain why in this period some parts of the underdeveloped counties failed to obtain more advanced means of production or advanced machines and equipment with lower labour requirements, so their relative backwardness increased on most fronts. In some of the developed areas, industry developed by extension rather than intensification, thus their level of capital equipment per employee showed little advance.

Conclusions

The Hungarian economy has shown some economic advance over the last twenty years with the highest growth of national income experienced in the 1970s, but as a result of world and other recessionary pressures has perhaps done less well in the second half of the period. Most counties and areas of gravity have benefited from this growth in income, but area disparities in income have narrowed very little over time. Budapest continues to dominate the national economy with the other prosperous counties and areas of gravity tending to concentrate within the northern half of the country. By way of contrast, depressed areas are located mostly in the east, on the Great Plain, and the south and western borders of the country. An interesting feature of the Hungarian economy is the number of relatively small but rich nodes of income generation surrounded by relatively low income generation areas irrespective of location in the country. This level of analysis has only become available as a result of innovative work at the Institute for Economic Planning (see Klekner, 1988).

The technological effort of the country is in all respects highly concentrated in the capital, Budapest, and perhaps, from an economic development point of view, the most significant and highest concentration is in industrial research effort. The other major centres of research effort are clustered for the most part in the University towns. For the majority of these towns the level of income per capita is, and has been over time, close to the national average and, owing to the skewness in the location of economic activity in Hungary (i.e. concentration in and around Budapest), is probably much higher than the modal class of areas. Without direct evidence we can only speculate that the existence of considerable research effort in these towns has had any particular beneficial effects on the local economy or its inhabitants, except in that they are associated with relative economic prosperity.

The relation between income, income growth, research and labour productivity is also tenuous, even though the spatial differences in labour productivity have varied over time, narrowing in the 1970s but widening again in the 1980s. Perhaps the most significant relationship between rising national income at the local level and change in the economy has come from **intensification**, as opposed to **extension**, of the economy. This has led to industrial investment: investment which includes the introduction of embodied and disembodied technological progress into industrial establishments, enterprises and local areas or counties. Analysis of the establishment data indicated that capital equipment per employee was more strongly associated with the level and rate of change of income per head than was labour productivity or research effort in an area.

The conclusion is reached that much more evidence is required, at least in the Hungarian case, to indicate a relationship between technological capacity and economic activity and growth at the local level. However, a growing concern over intensification and openness of the economy may show that technology is likely to play an increasingly important role in international competitiveness. As a result, national and local area technology policies, including the effective location and integration of these activities into the economy, may need to be advanced.

References

Bartke, I. (1987) Changes in Territorial Efficiency of Industry, Tervgazdasagi Kozlemenyek. No.1.

Bartke, I. and Nemes Nagy, J.(1988) Regional pattern of R & D activities, OT, TGI, Budapest.

Camagni, R. and Capellin, R. (1985) Sectoral Productivity and Regional Policy, Documents 9-825-5535-6, Commission of the European Communities, Brussels.

C.S.O. (1988) MO14, HMSO, London.

Denison, E. and Poullier, J. (1968) Why Growth Rates Differ, Allen and Unwin, London.

Klekner, P. (1988) Regional differences in income generation capacities of industry, OT. TGI. Budapest.

Oakey, R.P. (1984) High Technology Small Firms, Frances Pinter, London.

O.E.C.D. (1988) Main Science and Technology Indicators, OECD, Paris.

Solow, R.M. (1957) Technical Change and the Aggregate Production Function, Review of Economics and Statistics, 39, 312-320.

Appendix 12.1

List of regions, counties and areas in Hungary

Regions	Counties	Areas
Central part:	Pest :	14
	Budapest :	1
North-Hungary:	Borsod-Abauj-Zemplen:	10
	Heves :	5
	Nograd :	6
North-Transdanubia :	Fejer :	6
	Gyor-Sopron :	5
	Komarom :	7
	Vas :	7
	Veszprem	9
South-Transdanubia :	Baranya :	5
	Somogy :	8
	Tolna :	5
	Zala :	6
North-Great Plain :	Hajdu-Bihar:	7
	Szabolcs-Szatmar :	7
	Szolnok :	9
South-Great Plain :	Bacs-Kiskun :	10
	Bekes :	9
	Csongrad :	7
Country: 6 regions;	19 counties + Budapest;	143 areas

13. LOCATIONAL AND TECHNOLOGICAL FACTORS IN RURAL INDUSTRIALIZATION; A COMPARATIVE INVESTIGATION

Daniel Felsenstein and Raphael Bar-El

Introduction

One of the major problems facing the stimulation of industrial development in rural areas is that of the integration of rural industrial activity into the general pattern of national industrial growth. In the wide-ranging debate on the relationship between industrialization and development, the conventional staged approach to industrial development posits a primordial stage in which surpluses are created in the primary sector. This is followed by an industrialization stage based on import substitution and after that a stage based on export orientation (Balassa, 1980). An alternative approach sees industrial development based on the development of the agricultural sector and the creation of agricultural surpluses around which agriculture-based industry, such as the industrial processing of agricultural produce, can develop (Adelman, 1984). In both these approaches, the rural industrialization that results is invariably integrated into the local rather than the national economy.

In these general schema, rural areas are assumed to benefit from the resultant industrial growth due to the 'trickle-down' or 'filtering' process (Erickson, 1976; Howells, 1983) or any other equilibrium-seeking mechanism via which industry reaches the periphery. However this process of industrial decentralization does not necessarily imply rural industrialization for two reasons. First, the net result of decentralization is usually the industrialization of small, second-tier cities in the periphery rather than industrialization of rural settlements (Ho, 1979; Gwynne, 1986).

Second, the 'trickle-down' effect and the concomitant dispersal of manufacturing activity usually results in the establishment of branch plants producing standardized products in the periphery. At relatively high levels of economic development, this form of rural industrialization does not satisfy the employment and income aspirations of an increasingly educated and skilled rural labour force. This mis-match between the demand for industrial technology by the local labour force and that supplied in practice, is symptomatic of a situation whereby rural industrialization is used purely as an instrument for achieving the general development goal of employment creation (Bar-El, Erickson and Nesher, 1987). However, in a situation where the demand is for a particular level of industrial activity, the development goal then becomes one of trying to attract production technology to rural areas that would not otherwise find its way there. This implies that regional development based on the industrialization of rural areas needs to aspire to those types of industrial activity level that are integrated into the national pattern of industrial development. This in turn, however, means a reduction in the number of areas of industrial activity in which the rural sector has some form of comparative advantage.

This situation arises because industrial growth trends in more advanced economies are invariably tied up with the development of technologically advanced products and the implementation of sophisticated production processes. The choice of industrial technology for rural areas is based on the fact that, for any given industrial activity, the production function can assume different rates of substitution between factors of production. The production technology chosen is based on the comparative advantage that the area has to offer in terms of production factors, which in turn leads to a comparative advantage in terms of products. Production technology can thus be adapted to local needs if the production function allows for substitution between factors. The result is a general pattern of rural industrialization that at the early stages of development has strong links to the local (agricultural) economy. However, where production functions are rigid, the choice becomes one of selecting between different industrial activities and not of choosing a particular production technology within a given activity (Bar-El, Erickson and Nesher, 1987).

The choice of industrial activities that are to be integrated into the national economy is however constrained in two respects. First, constraints exist whose roots lie in the nature of the rural areas themselves and second, constraints emanate from the nature of the desired industries.

Problems associated with the nature of the rural environment relate to first, the locational remoteness of the rural areas. This implies additional costs in transport and communications and further expenses associated with the (likely) sub-standard provision of this infrastructure. Second, labour constraints in a rural area can be far-reaching, affecting both the quantity and quality of labour supply and subsequently the all-important know-how that is embodied in human capital (Malecki, 1988). Third, and as a result, the supply of entrepreneurship is likely to be limited, which in turn can result in constraints in capital supply. All these factors will go a long way in dictating the character of rural industrialization in terms of firm size, the ability to reach scale economies in a rural setting, the level of capital and/or labour intensity of industry, etc.

Constraints associated with the nature of technologically-based industries relate foremost to the spatial behaviour of these industries. Much evidence points to their tendency towards spatial concentration in a few choice locations (Markusen, Hall and Glasmeier, 1986; Scott, 1988). The spatial margins of profitability are very narrow in technological industries resulting in the growth of a few select industrial agglomerations that are then reinforced by the inter-firm linkages and information networks that they develop. At the margin however, and especially in new industries where patterns of inter-firm linkages have not yet consolidated (e.g. biotechnology), there is still a pattern of some spatial dispersal (Oakey and Cooper, 1988).

The underlying assumption here is that successful rural industrialization has to be integrated into the national pattern of industrial growth and in relatively advanced economies this means aspiring to technologically-based industry. However, various constraints lie in the way of achieving this objective and successful rural industrialization is based on the choice of those production technologies within a given level of technological activity that are least affected by rural constraints.

This chapter, based on empirical evidence from 200 firms in Israel, proceeds in the following stages. To set the scene, a brief review of technological trends in rural industrialization in Israel is presented. This is followed by an outline of the methodology employed which involves comparing urban and rural firms across a wide variety of industrial characteristics. The existence of a significant urban-rural difference for a particular characteristic implies a constraint to the existence of that characteristic in the rural area and a barrier to the integration of rural industry into the national pattern of industrial growth. Finally, having established the existence of significant urban-rural differences, an attempt is made to determine whether these differences still exist when controlling for the effects of technological factors. We test for the existence of a 'substitution effect' whereby, within a given level of technological activity, rural industry can choose that type of activity and those industrial characteristics that will be least affected by the constraints of operating in a rural area.

Technological Trends in Rural Industry in Israel

At the outset it is important to note that the rural industrialization experience in Israel is not uniform. While the term 'rural' here is taken to refer to any industrial activity carried out in peripheral or remote, non-urban settlements, the rural industrialization experience is generally conditioned by the different forms of rural settlement that exist. A detailed account of this experience exists elsewhere (Bar-El, 1987) and this short discussion will deal with technological trends in the three main forms of rural settlement; the kibbutz, the moshav and the non-agricultural village.

The Arab village, while part of the rural scene, is excluded from this discussion due to a form of industrialization and a level of general development that are very different to those of the types of rural settlement mentioned above. It is characterized by an organizational structure that is essentially individualist and often based on kinship patterns, in contrast to the high level of co-operation that characterizes much of the Jewish rural sector. Furthermore, the professional and educational skill levels of the Arab labour force inhibit the potential development of an indigenous rural industrial sector. Improvements in the educational system have made for a narrowing of the skill gap between Jewish and Arab labour, although the level of available human capital is still estimated as one of the major constraints in the industrialization of the Arab village (Czamanski and Meyer-Brodnitz, 1987).

The relative share of rural industry in total national industry varies greatly across branches, reflecting differing degrees of integration across industrial activities. As a whole, rural industry accounts for some eight per cent of national industrial employment, 13 per cent of output and 14 per cent of exports (Trattner, 1988). On an individual branch basis the picture is far less uniform (Table 13.1). In certain traditionally 'low-technology' branches such as food, textile and wood products, rural industry is over-represented, while in many of those sectors popularly perceived as technologically intensive (chemicals, electrical and electronic products, etc.) rural industry is under-represented. However, it should be noted that these aggregate figures are heavily weighted by the influence of kibbutz industry which, out of the three settlement types outlined above, accounts for some 85 per cent of all employment and nearly 90 per cent of all industrial output. In all forms of rural settlement, aside from the general constraints imposed by a peripheral location and by the specific requirement of technologically based industries, a further set of barriers to industrialization exist in the form of structural constraints imposed by the ideological and organizational structure of the settlement. The effect of these varies to a differing degree across rural settlements, but in all of them it imposes certain restrictions on firm size and the ability to reach scale economies, and on labour composition and the ability to create an in-house technological capability. These constraints are therefore 'given' and impact on the type, and technological level, of industrial activity to which rural firms can aspire. As such, they dictate certain industrial characteristics of rural firms such as size, capital intensity, skill level of labour and so on. These industrial characteristics can be conceptualized as 'rigid' in relation to location, in that they are predetermined, to an extent, by the type of settlement in which the firm is found. Alternately, where these structural constraints do not apply, or have little effect, the resultant firm characteristics can be conceptualised as 'flexible' in relation to location.

Table 13.1

The Composition of Israels Rural Industry in Relation to National Industry, by Selected Branches, 1984[1]

BRANCH	EMPLOYMENT		OUTPUT		EXPORTS		LOCATION[2] QUOTIENT	RURAL EMP. AS % OF NAT. EMP.
	Nat.	Rural	Nat.	Rural	Nat.	Rural		
Total[3]	303	22	4563	586	4005	572	1.0	8.1
Percentage of Total:								
Food	15.0	34.0	20.6	47.1	9.3	30.1	2.27	16.5
Textiles	16.6	4.6	9.6	14.8	9.3	40.2	0.28	2.1
Wood	4.8	5.8	2.8	2.4	0.6	1.8	1.21	9.0
Plastic and Rubber	4.1	15.4	4.9	11.2	3.5	13.7	3.76	27.0
Chemicals	6.1	1.9	10.8	1.2	18.7	0.9	0.31	2.3
Metals	20.5	18.0	17.5	8.1	19.5	9.1	0.88	6.4
Electrical & Electronic Products	11.9	4.4	12.5	2.0	16.9	1.5	0.37	2.7
Transport Equipment	6.5	0.0	5.3	0.0	7.4	0.0	0.00	0.0
Miscellaneous	2.4	10.9	2.6	10.9	7.8	2.7	4.54	32.1

Notes: 1 Data exclude Moshavim and Non-Agricultural Villages.

2 Computed as: $\frac{\text{\% Employment Share of Branch - Rural}}{\text{\% Employment Share of Branch - National}}$

3 Absolute Terms: Employment (Thousands); Output (Million Shekels); Export (Million $).

Source: Bar-El et al. (1989).

The kibbutz (pl: kibbutzim) is perhaps the most organizationally and ideologically constrained form of rural settlement. As a collective with joint ownership of all means of production and assets and with an aversion to the hiring of outside labour, the kibbutz finds that its limited labour pool does not generally offer the level of human capital needed for technologically intensive production. Thus the entry of kibbutz industries into those sectors producing 'high technology' products has not been particularly fortuitous. Rather the kibbutz has chosen to invest in those forms of activity that are generally capital intensive and depend on skilled rather than scientific labour (such as the plastic and rubber products sector).

Industry on the kibbutz was historically introduced as an adjunct to agricultural activity and as an employment solution for various sub-groups in the community. These included second generation members that demanded more varied and challenging employment and ageing members that demanded less strenuous activity (Don and Leviatan, 1987). As a result, the kibbutz strives towards a 'socio-technological' solution to the question of quality of employment. One aspect of this has been the introduction of sophisticated production processes such as pick-and-place and handling systems to replace monotonous activities (Rosner, 1988). While the uptake of these process innovations is generally higher in the kibbutz sector than in industry at large, their overall impact is still limited.

Anywhere between 15 and 35 per cent of the existing 300 kibbutz plants are engaged in a given year in R & D projects (Kibbutz Industries Association, 1984). Many of these are involved in the modification of existing products whose technology has been acquired externally (such as in the metals and plastics sectors). Rather less kibbutzim have the in-house technological capability of developing and producing innovative products. As expected, there have been some excursions into agriculture-related activities in the fields of agro- and biotechnology. In these instances it is usually the simpler and lower risk end of these activities that is adopted; in biotechnology this finds expression in a number of firms engaged in the growing of plant tissue cultures for example, rather than genetic engineering.

The kibbutzim are increasingly using inter-kibbutz co-operation as a way for circumventing certain of the constraints imposed by their unique organizational structure. These include organizational innovations in the field of joint marketing and sales, production co-operation in order to overcome the problem of scale economies and inter-kibbutz R & D as a risk-dispersing strategy. In this respect the kibbutzim, as part of a unified movement, have an advantage over other forms of rural settlement. They can more easily mobilize their overall organization in order to embark on innovative schemes in organization and production than can other forms of settlement (Kipnis and Meir, 1983). In addition they also have the financial, credit and support systems and the prestige and political clout that are needed to accompany any process of innovative change.

In contrast to the kibbutz, the moshav (pl: moshavim) as a co-operative community of family-run small-holdings faces fewer organizational and ideological constraints. While services are provided collectively, ownership of the means of production is individualistic. On the one hand, because of the size of the individual unit of production, the moshav suffers more than the kibbutz from labour and capital constraints. On the other hand, none of the ideological objections to hiring outside labour are in force.

The moshav industrial experience has been limited. Much of it amounts to not much more than labour-intensive workshops in traditional sectors (food, furniture, leather, etc.). Hard data on the moshav sector is not available, but its relative share in the national economy has been estimated as 0.3 per cent of employment and 0.2 per cent of output (Trattner, 1988). Those moshav-based firms that are engaged in full-scale industrial activity operate generally capital intensive plants employing semi- or unskilled labour (Schwartz, 1987). Like the kibbutz, industry was adopted in the moshav as an employment solution due to the contraction of work possibilities in agriculture and in order to provide for the needs of an increasingly educated second generation. While much moshav industry is centred on the processing of agricultural produce, unlike the kibbutz there has been little attempt to introduce process innovations, such as handling systems, in these activities.

Production of innovative products is also missing in the moshav economy. The uncertainty and market fluctuations that go with this type of production call for sources of capital that are usually unavailable to the moshav enterprise (venture capital, etc.). Furthermore, the size constraint of the moshav does not allow for the operation of the concurrent processes of fast growth and rapid organizational change that is needed in order to reach optimal size, and that is often the hallmark of firms producing innovative products.

The non-agricultural village is an umbrella term to describe various forms of rural settlement whose common denominator is an economy based at the outset on non-farm activities and a population comprised of former city-dwellers seeking a rural quality of life. Some of these villages (eg. 'industrial villages') are organized around the moshav pattern, viz. the family as the unit of production and collective ownership of the means of production. As such they suffer from some of the constraints operating on the moshav outlined above. Other however operate on a purely atomistic basis (e.g. 'community settlements') and constraints here hardly exist in the economic sphere.

Both types of non-agricultural village have been initiated for the geo-political purpose of ensuring the settlement of sparsely populated areas, but the level of commitment to the initiation of economic activity varies greatly. For the industrial village, industrial activity is the a priori basis for its existence. The community settlement in contrast shares no such commitment and some of them function as purely residential settlements for a population that finds employment in the city. The human capital composition of both these forms of settlement is relatively high with a high proportion of young professional, academically-trained families.

Industrial activity based in this form of settlement, despite its short history and the dearth of data available, points to the influence of the available labour force as a potent factor in the conditioning of the form of rural industrialization. A case study of an agglomeration of these villages in the north of Israel has pointed to the variations in levels of sophistication of production that exist across the various settlements and also to the relative concentration of advanced services and information-based activities (Bentolila, 1987). These include software houses, architectural consultants, computerized graphics and technical writing services, etc. Product innovation and R & D are also undertaken based on either in-house capabilities or collaboration with local academic institutions and external consultants. While the small size of these settlements operates as a constraint on the pool of labour available, their open organizational structure allows for labour mobility between settlements themselves and even between the settlements and the nearby cities. There is also some evidence of the creation of regional industrial parks in order to create external agglomeration economies necessary to assist the firms' survival. A further step along the road to liberalization is the encouragement, in the community settlements at least, of external private industrial investment.

Data and Methodology

The empirical analysis below is based on a sample of rural and urban firms in Israel and is based on data collected via a postal questionnaire. The sampling design is structured so as to equally represent both rural and urban independent firms and firms with different levels of technology (see below). As such, the sample does not aspire to be representative of Israeli industry as a whole. For each firm, data are presented relating to the main industrial characteristics of the production process. These characteristics are subdivided for the purpose of analysis into three groupings; variables relating to the inputs, outputs and spatial linkages of each firm (Appendix 13.1). It should be noted that in this classification R & D investment, raw materials expenses and the work-roles played in the production process by groups (R & D, production, administration), are taken as characterizing the composition of outputs of the firm. Taken together they give us some indication as to the level of sophistication of the product and whether it has a greater or lesser component of new technology.

The sample of responding firms is cross-classified by type of location (form of settlement) from which the firm operates and the technological profile (characterising the different components of technology existing in the firm) to which it belongs (Table 13.2).

The classification by type of location makes the basic distinction between rural and urban firms. These sectors are further subdivided into four groups:

(1) 'Old' Rural Sector; all firms in this group are kibbutz or moshav enterprises that exist on settlements that were set up prior to 1967 (the Six Day War). In nearly all instances industrial activities were started as an adjunct to the agricultural activities that formed the mainstay of the economies of these settlements. Today however, in some settlements, industry has surpassed agriculture as the income and employment backbone of the local economy (80 firms).

(2) 'New' Rural Sector; firms in this category are located in rural settlements established since 1967. These settlements are in the main non-agricultural villages along with some kibbutzim and are characterized by a more concerted effort than in the past for the a priori promotion of industry over agriculture. They are also less co-operative in nature than has been the tradition in rural settlement, displaying a variety of organizational forms ranging from the collective to the private enterprise village (34 firms).

(3) Development Towns; the firms found in this category are all located in small, relatively peripheral towns that have been established since the 1950s as part of the government effort to achieve a balanced spatial distribution of population. The location of these towns can be in rural areas and they often have rural settlements in their orbit. However, class, ethnic and economic differences separate them from the rural sector (46 firms).

(4) Urban Centres; firms found in the large, centrally located urban centres of the country, particularly the Tel Aviv and Haifa conurbations. The vast bulk of industrial activity, at all technological levels, takes place in these urban centres (37 firms).

The reason for the above classification is that the methodology employed involves the comparison of rural and urban firms across an array of industrial characteristics. The existence of a significant rural-urban difference for a

Table 13.2

Cross Classification of Firms in the Sample by Type of Location and Technological Profile

TYPE OF LOCATION	TECHNOLOGICAL PROFILE: LOW LEVEL OF TECHNOLOGICAL COMPONENTS	SOPHISTICATED PRODUCTION PROCESS	SKILLED LABOUR	ACADEMIC EMP. AND R & D INVESTMENT	TOTAL
´Old´ Rural Sector	6	33	19	22	80
´New´ Rural Sector	5	8	8	13	34
Development Town	10	19	10	7	46
Urban Centre	2	9	10	16	37
TOTAL	23	69	47	58	197

particular characteristic indicates the 'rigidity' of that factor in relation to location and implies a constraint to the existence of that characteristic in the rural area. These constraints are structural and are usually a direct function of the organizational make-up of the rural sector. Thus their effect is not easily mitigated, even by the 'right' policy recommendations. The existence of 'rigid' industrial characteristics is taken as pointing to a particular barrier to the integration of rural industry into the national pattern of industrial growth. Conversely, the absence of a significant difference between types of location for a particular characteristic indicates its 'flexibility', suggesting that it can exist in all forms of location. The subdivision of rural firms into 'old' and 'new', representing the change in rural industry over time, also has wider implications viz. the type of organizational structure, level of co-operation and so on.

Having established the existence of significant rural-urban differences across a variety of characteristics, we then attempt to answer the question; do these differences still exist when controlling for the effects of technological factors? To what extent do the rural-urban differences stem from a different composition of technological profiles between urban and rural firms and to what extent can the rural sector, within a given level of technological activity, find 'niche' activities that can accommodate the constraints that go with a rural location?

To this end a further classification of the firms is employed, this time into 'technological profiles'. These profiles endeavour to measure the technological intensity of the firm based on four defining variables; proportion of academic labour, proportion of skilled labour, sophistication of the production process and level of R & D investment. It is felt that taken together these four variables express the various components of technology that combine to account for the technological level of the firm. If each one of these four variables takes on a dichotomous value, sixteen possible combinations of profiles are created. The methodology used to construct the technological profiles involves assigning each firm to one of the various theoretical profiles on the basis of these defining variables. Using a clustering algorithm, the various profiles can be combined to give four major technological groupings; firms with a low level of technological components; firms whose technological level is expressed via the level of academic labour and R & D investment employed; firms whose technological level is embodied in skilled labour and firms whose technological level is expressed via sophisticated production processes. Empirical testing of these technological profiles at the level of the industrial branch has been conducted elsewhere (Felsenstein and Bar-El, 1989) and the profiles have been shown to have consistent industrial characteristics over and above their common technological characteristics. Here however we utilise the notion of technological profile at the firm level.

The four technological profiles employed here are:

(1) Low Level of Technological Components; firms assigned to this profile have scored low on all four of the defining variables (23 firms).

(2) Sophisticated Production Process; the common feature of firms in this category is the use of sophisticated equipment in the manufacturing process (69 firms).

(3) Skilled labour; firms in this category have above average proportions of skilled labour as their main identifying feature (47 firms).

(4) Academic Employment and R & D Investment; above average levels of academic labour and R & D investment indicate the main technological features of this group of firms (58 firms).

The methodology employed here uses analyses of variance (one and two-way) in order to determine the level of sensitivity of the various industrial characteristics in relation to type of location and technological profile. A significant difference between average scores, as indicated by F-ratios, is taken to indicate the sensitivity or 'rigidity' of the factor in that it displays a particular connection to a location or technological profile.

Differences in the Characteristics of Rural and Urban Industry

An analysis of the characteristics of the inputs to the production process shows a clear division between the 'old' and 'new' rural sectors across a variety of variables (Table 13.3). The former come close to the overall pattern displayed by Israeli industry in all that is to do with average firm size, capital-intensity and composition of the labour force. However the picture is rather different when it comes to the firms of the 'new' rural sector. They are on average smaller, younger and less capital-intensive than the other rural firms. The main, and potentially most telling, difference however lies in the structure of the labour force. 'New' rural sector firms on average employ more academic labour and less skilled labour than all other firms, pointing to a labour force that is based more on formal 'schooling' than informal, on-the-job 'skilling'. Yet while, in this respect at least, the 'new' rural sector seems to be integrated into technological trends in industry, the wage level for academic labour in this sector is way below that of urban industry. This could point to over-supply of academic labour in the 'new' rural sector serving to suppress wages and to the general mis-match in supply and demand for suitable quality work that seems to be inherent in rural industry. This mis-match could also point to the sub-optimalization of the available labour force whereby rural employment is based on

Table 13.3
Mean Values and F-Ratios for Industrial Characteristics by Type of Location

VARIABLE	MEAN VALUES OVERALL MEAN	'OLD' RURAL SECTOR	'NEW' RURAL SECTOR	DEVELOPMENT TOWN	URBAN CENTRE	F-RATIOS AND PROBABILITY
Inputs						
Emp. per Firm	79.9	61.8	14.8	90.7	158.8	4.46 *
Capital per Emp.	36.2	46.1	18.2	36.2	34.5	1.98
Year of Establishment	71.14	69.5	81.1	72.1	65.5	11.25 *
% Female Emp.	25.5	26.8	36.6	20.0	19.1	6.81 *
% Academic Emp.	14.3	14.3	28.2	6.1	13.3	10.81 *
% Technicians	9.6	11.3	12.2	6.6	8.4	2.53 *
% Skilled Lab.	29.0	31.7	17.5	28.5	33.7	3.18 *
% Unskilled Lab.	45.8	43.6	41.6	54.5	42.7	2.04
% Acads High Wage	75.1	59.5	60.1	79.1	90.3	3.69 *
Sophisticated Prod Tech (%)	17.2	19.5	18.9	8.4	22.8	1.49
University Collaboration	.69	.86	.69	.42	.65	3.81 *
Outputs						
R & D Invest as % of sales	6.1	5.4	10.3	2.9	6.4	2.21
Product Innov.	1.85	1.91	1.93	1.56	2.02	3.00 *
Changes in Demand	2.94	3.04	2.93	2.69	3.05	.99
Market Behaviour forecast	1.17	1.47	1.07	.84	1.03	2.32
Raw Materials Expenses (%)	38.8	41.2	31.1	54.6	34.6	4.37 *
% Production Employment	71.4	69.7	66.4	76.8	73.5	2.75 *
% R & D Employment	7.2	7.3	11.7	3.2	7.4	3.08 *
% Admin. Employment	21.3	22.9	21.8	19.8	19.0	1.02
Spatial Linkages						
% Acads - Non Local	27.1	10.7	24.5	55.9	26.5	8.79 *
% Technicians - Non Local	14.8	3.1	19.7	34.9	28.9	5.11 *
% Skilled Lab - Non Local	15.7	5.6	11.3	26.5	22.8	3.02 *
% Unskilled Lab - Non Local	12.5	5.6	18.5	11.6	23.1	4.02 *
Location - Main Raw Material Source	.53	.65	.33	.33	.69	7.30 *
Location - Main Market	.95	1.0	.91	.87	.97	4.11 *
% Export	25.4	31.6	18.4	19.2	24.1	1.73

Note: * $p < 0.05$.

over-qualified labour engaged in work tasks that do not allow for the full utilization of their skills. As expected, rural firms are also distinguished from urban-based operations by the relative size of the female labour force.

These differences in the composition of human capital play themselves out in differences in output. Products emanating from firms in the 'new' rural sector would seem to come close or even exceed the urban sector in indices relating to level of innovation and R & D investment. These products are characterized by a relatively low use of primary raw materials and probably a heavier dependency on products from the secondary sector and a greater relative dependency on labour as an input than on raw materials. These firms also operate in markets with average levels of fluctuations and risk. There is also some indication that products from this sector are less capital or consumption goods oriented and more oriented towards services and information, as evidenced by the low overall percentage of production labour. Thus the composition of outputs from the 'new' rural sector in some instances mirrors trends taking place in the urban economy and in others even seems to surpass them. Output from firms of the 'old' rural sector, in contrast, falls below outputs from the urban sector in some instances (R & D innovation, dependence on secondary materials) and mirror this sector in others.

In relation to the spatial linkages of the firms in the different types of location we again find differences between the two rural sectors. The 'old' rural sector firms, as expected, seem to use their organizational and co-operative structure to realise economies of scale and scope that location in a rural setting would generally deny them. Thus the extent to which they receive inputs from outside the local area, service non-local markets and export abroad resembles the pattern of the urban firms. In contrast, the 'new' rural sector's pattern of extra-local linkages comes close to the limited pattern displayed by the development towns.

On the issue of spatial linkages, one further finding relates to the spatial extent of labour markets for firms in different types of location. As shown above, rural industry is often characterized by a mis-match between supply and demand, especially for labour at the highest levels. As a result, the extent of the spatial labour market at this level is generally limited. At lower levels, (technicians, skilled and unskilled labour) the difference in the operation of ideological constraints between the 'old' and 'new' rural sectors comes to the fore. The former exhibits very narrow spatial labour markets as a result of self-imposed restrictions on hiring outside labour. The latter is less restricted ideologically and as such exhibits wider spatial labour markets, although these still are narrower than those of the development towns and urban centres.

The Technological Factor and the 'Flexibility' of Industrial Characteristics

The difference in industrial characteristics outlined above raise the question of the extent to which these differences are conditioned solely by the particular technological profile of the firms in rural and urban areas or whether they reflect the flexibility of industrial characteristics and the adaptation of these different characteristics to the conditions existing in the rural environment for any given technological profile. If we control for the effects of these technological factors will the differences between rural and urban firms still exist? If significant differences are still found we are led to the conclusion that this characteristic is 'rigid' and is a sine qua non for a given technological profile. Rural industry in such a case should choose industrial activities from a different technological profile that are more 'flexible' and likely to be adapted to rural conditions.

To test this proposition we use analysis of variance with two independent variables; type of location and technological profile. This gives us the net effect of each independent variable on the industrial characteristics and also the interaction between the independent variables indicating a lack of 'clean' effect on the dependent variable. For the purpose of this analysis, the list of industrial characteristics (Appendix 13.1) has been limited to those found most important in the previous section (14 variables).

The results from Table 13.4 illustrate that amongst the input variables some can be considered 'rigid' while others are more 'flexible'. Predominant amongst the former category is capital intensity. This characteristic would seem to be dictated by choice of technological profile. The implications of this for rural industry are clear. In order to circumvent the barriers that exist to setting up capital-intensive operations, rural firms need to choose technological activities of a lower level of capital intensity rather than trying to accommodate less capitally-intensive production technologies into existing activities.

Other input variables are also influenced by technological profile, but they are more likely to be 'flexible' and adapted to the conditions of the rural environment. For example, in order to get around the size constraint that exists in rural firms, the results show that it is possible within technological profiles to raise the level of employment by increasing the size of the female labour force (a characteristic that continues to separate rural from urban firms even after neutralizing the effect of technology). In all probability, rural and urban firms continue to be differentiated on the basis of levels of academic employment in any given technological profile even though this is not corroborated statistically (because of the high interaction effect between location and technological profile).

Table 13.4
Results of Two-way Analysis of Variance: Industrial Characteristics by Type of Location and Technological Profile

VARIABLE	F-RATIOS AND PROBABILITY			OVERALL F-RATIO OF THE MODEL
	TYPE OF LOCATION	TECHNOLOGICAL PROFILE	INTERACTION	
Inputs				
Emp. per Firm	3.49 *	.40	.32	.97
Cap. per Emp.	2.03	5.81 *	.87	2.09 *
% Female Emp.	6.19 *	11.10 *	1.48	4.35 *
% Academic Emp.	8.71 *	21.43 *	2.15 *	7.31 *
% Skilled Lab.	2.36	1.84	.60	1.20
University Collaboration	3.75 *	4.77 *	1.36	2.52 *
Outputs				
Product Innov.	1.28	4.75 *	1.27	1.97 *
Raw Material Expenses (%)	3.06 *	9.57 *	1.42	3.38 *
% Production Employment	2.45	4.91 *	1.26	2.23 *
% R & D Employment	2.24	13.61 *	2.43 *	4.63 *
Spatial Linkages				
% Acads - Non Local	8.26 *	.33	.77	2.18 *
% Skilled Lab - Non Local	2.53	.90	2.13 *	1.96 *
Raw Materials - Non Local	7.02 *	1.21	.48	1.94 *
Markets - Non Local	5.13 *	2.16	1.65	2.45 *

Note: * $p < 0.05$.

Output characteristics are uniformly determined by choice of technological profile. The innovation level of the firm, the nature of the product (service or manufacturing oriented) and the level of dependence on raw materials or secondary products are in the main dictated by choice of technological profile, with hardly any margin for maneuverability within a given technological profile to take into account the constraints of location.

In contrast to the output characteristics, the variables relating to spatial linkages in any given technological profile, are unanimously influenced by the type of location from which the firm operates. Thus the difference between urban and rural firms in the employment of non-local academic labour is reiterated. The structure of the other spatial linkages (the amount of raw materials bought non-locally and the size of the non-local markets serviced by the firm) illustrates the adaptation of industrial characteristics to the specific locational conditions in any given technological profile. In the 'new' rural sector, the constraints of geographic remoteness cause a reduction in external linkages and an increasing dependence on linkages in the region. In the 'old' rural sector in contrast, the developed co-operative organizational structure with its centralized administration allows for the overcoming of the constraints imposed by remoteness and the existence of spatial linkage patterns similar to those of urban firms.

Conclusions

The above discussion has suggested the existence of a 'substitution effect' whereby, within a given level of technological activity, rural industry can choose a type of industrial activity whose main characteristics are flexible in relation to location. However, the integration of rural industry into national technological growth patterns also serves to highlight various processes that are likely to influence the further development of rural industrialization.

Within the rural sector itself, the distinction between 'old' and 'new' rural firms has illustrated that what the 'old' rural firms gain in terms of organization they loose in terms of dynamism. While the reliance on a developed, nationally co-ordinated and co-operative organizational structure has enabled 'old' rural firms to realize scale economies, operate larger, more capitally-intensive plants and in this respect move towards an industrial structure found in the urban sector, it nevertheless seems to have anchored them into a mould based on classic manufacturing production. This has been accompanied by a certain reticence to move into more innovative operations coupled with objective difficulties, especially in terms of suitably trained labour.

The converse is true for firms in the new rural sector. What they lose in organization that make up for in dynamism. The smaller, more innovative firms of this sector have entered fields that are suited to their small size and small capital resources, such as information and advanced services. However, potentially these size and capital constraints pose a very real threat to the ability of these firms to scale up production in the face of increases in demand. These firms operate in markets characterized by very short product and process cycles that demand the ability to concurrently increase output and effect organizational innovations. Size and capital constraints coupled with marketing barriers and problems of distance from service and information networks all pose a very real question mark over the ability of these firms to be able to sustain fast growth in the future.

Ironically, the over-supply of academic labour that was noted above to exist in the case of 'new' rural sector firms might also work to their detriment. The rural sector as a whole is unable to capitalize on this relative over-supply due to the fact that firms for whom the availability of highly educated labour is their primary location factor tend to favour urban areas, because of their larger absolute supply of this resource. Thus, the size constraint imposes a large opportunity cost, acting as a barrier that inhibits rural settlements from exploiting their surpluses of highly educated labour. This in effect could limit their opportunities for partaking in national industrial growth trends, constraining them to more traditional industrial activities.

References

Adelman, I. (1984) Beyond export-led growth, World Development, 12 (9), 937-944.

Balassa, B. (1980) The Process of Industrial Development and Alternative Development Strategies, World Bank Staff Working Paper, No.438.

Bar-El, R. (ed.) (1987) Rural Industrialization in Israel, Westview, Boulder.

Bar-El, R., Erickson, E. and Nesher, A. (1987) Rural industrialization in Israel; concluding considerations, in Bar-El, R. (ed.) Rural Industrialization in Israel, Westview, Boulder, 169-190.

Bar-El, R., Felsenstein, D., Bentolila, D., Spiegel, D. and Kedar, F. (1989) Technological Trends in Rural Industry in Israel, Bialik Institute, Jerusalem (Hebrew).

Bentolila, D. (1987) The non-agricultural village, in Bar-El, R. (ed.) Rural Industrialization in Israel, Westview, Boulder, 05-142.

Czamanski, D. and Meyer-Brodnitz, M.B. (1987) Industrialization in Arab villages in Israel, in Bar-El, R. (ed.) Rural Industrialization in Israel, Westview, Boulder, 143-168.

Don, Y. and Leviatan, U. (1987) Kibbutz industrialization, in Bar-El, R. (ed.) Rural Industrialization in Israel, Westview, Boulder, 1-56.

Erickson, E. (1976) The 'filtering-down' process; industrial location in a non-metropolitan area, Professional Geographer, 28, 4-60.

Felsenstein, D. and Bar-El, R. (1989) Measuring the technological intensity of the industrial branch; a methodological and empirical approach, Research Policy, 18, 239-252.

Gwynne, R.N. (1986) Industrialization and Urbanization in Latin America, John Hopkins University Press, Baltimore, Maryland.

Ho, S.P.S. (1979) Decentralized industrialization and rural development; evidence from Taiwan, Economic Development and Cultural Change, 28, 77-96.

Howells, J.R.L. (1983) Filter-down theory; location and technology in the UK pharmaceutical industry, Environment and Planning A, 15, 147-164.

Kibbutz Industries Association (1984) Annual Report 1983, Tel Aviv.

Kipnis, B. and Meir, A. (1983) The Kibbutz industrial system; a unique rural-industry community in Israel, in Hamilton, F.E.I. and Linge, G.J.R. (eds.) Spatial Analysis, Industry and the Industrial Environment, Vol.2, Wiley, Chichester.

Malecki, E.J. (1988) The challenge of rural development, Environment and Planning A, 20, 567-568.

Markusen, A., Hall, P. and Glasmeier, A. (1986) High-Tech America: The What, How and Why of the Sunrise Industries, Allen and Unwin, Boston.

Oakey, R.P. and Cooper, S.Y. (1988) High technology small firm location, entrepreneurship and the potential for peripherally located science parks, paper presented at 28th Annual Congress of the European R.S.A., Stockholm.

Rosner, M. (1988) High Tech in Kibbutz Industries, Report No.73, The Kibbutz University Centre, University of Haifa, Haifa.

Schwartz, M., Bar-El, R., Nesher, A. and Finkel, R. (1987) Moshav-Based Industry, in Bar-El, R. (ed.) Rural Industrialization in Israel, Westview, Boulder, 57-104.

Scott, A.J. (1988) Flexible production systems and regional development; the rise of new industrial spaces in North America and Western Europe, International Journal of Urban and Regional Research, 12,171-185.

Trattner, E. (1988) Rural Industrialization in Israel : Its Importance, Achievements and Problems, Research Paper, Settlement Study Centre, Rehovot, (Hebrew).

Appendix 13.1
Definition of Variables[1]

VARIABLE	DESCRIPTION
Inputs	
Employment per Firm	
Capital per Employee ($ Th)	
Year of Establishment	
% Female Labour as % of Total Employment	
% Academic Labour as % of Total Employment	
Technicians as % of Total Employment	
% Skilled Labour as % of Total Employment	
% Unskilled Labour as % of Total Employment	
% Academics High Wage	% Academic Employment > 20% above average national wage.
% of Production Using Sophisticated Production Technology (CNC, Laser Technology, Pick-and Place and handling systems)	
University Collaboration	0 = No Collaboration; 1 = Infrequent; 2 = Frequent.
Outputs	
R & D Investment as % of Sales	
Product Innovation	0 = no product innovation by firm; 1 = modification of existing products; 2 = product developed and produced by firm.
Changes in Demand	Over last 3 years: 1 = unstable; 2 = declining; 3 = stable; 4 = rising.
Market Behaviour Forecast	Over next 3 years: 1 = sales unstable; 2 = sales stabilize; 3 = sales rise.
Expenses on Raw Material as (%) of Total Firm Expenses	
% Employed in Production out of Total Employment	
% Employed in R & D out of Total Employment	
% Employed in Administration out of Total Employment	
Spatial Linkages	
% Academic Employment from Outside the Area (radius 30 km)	
% Technicians from Outside the Area (radius 30 km)	
% Skilled Labour from Outside the Area (radius 30 km)	
% Unskilled Labour from Outside the Area (radius 30 km)	
Location of Main Raw Material Source	0 = in local area (radius 30 km); 1 = otherwise.
Location of Main Market	0 = in local area (radius 30 km); 1 = otherwise.
Exports as % of Total Sales	

Note: 1 - all data relates to 1985/6

14. REGIONAL DIFFERENCES AND DETERMINANTS OF ENTREPRENEURIAL INNOVATION - EMPIRICAL RESULTS OF AN AUSTRIAN CASE STUDY

Franz Tödtling

Introduction

The changing macro-economic conditions accompanied by a changing international division of labour have led to reinforced entrepreneurial competition in industrialized countries since the early seventies and as a consequence speeded up industrial restructuring. In the course of this restructuring process many firms have tried to increase their competitiveness by taking one or more of the following strategies or measures (Massey and Meegan, 1982; Todtling, 1986; Stohr, 1987a):

- **cost cutting measures** such as the reduction/closure of unprofitable lines of production, the concentration of production into the more efficient plants within the enterprise, the relocation to cheaper locations, the introduction of a more efficient form of organizing production ('intensification') or the introduction of cost saving new technology;

- **search for new markets** via increased market penetration or extension, the modification of existing products or the introduction of new products;

- **flexible adjustment to specific demands** (e.g. subcontracting or 'custom tailoring').

Technological change and innovation (in the broad notion of Schumpeter) plays an important role in most of these strategies and thus seems to be of general importance for the improvement of the competitive position of plants.

In the Austrian context there are two spatial levels to be considered with regard to problems of entrepreneurial innovation. For the country as a whole there appears to be a certain 'backwardness' compared to other industrialized countries (low levels of R & D, unfavourable balance of trade in patents to licenses, above average share of mature and basic products in the economy (Volk, 1987)) and at the regional level there has been some indication that striking differences exist between localities with regard to entrepreneurial innovation (Todtling, 1987). Up to now, however, there has been no systematic analysis of this latter aspect.

The remainder of this chapter will be devoted to this regional aspect by focusing on the **regional** differences in entrepreneurial innovation, particularly the introduction of new products and processes at the plant level. Firstly, theoretical aspects and concepts which indicate explanatory factors will be discussed. Secondly, some empirical evidence from five Austrian regions will be presented based on interview data in selected industries. Finally, this empirical evidence will be related to:

- relevant structural characteristics of plants;
- innovation activities (introduction of new products and processes);
- results of a PROBIT-analysis with regard to determining factors.

Regional Differences in Entrepreneurial Innovation - Theoretical Aspects

To date there has been no consistent theory of regional innovation, but there are several concepts which indicate explanatory factors. In general these explanatory approaches relate to the context outlined in Figure 14.1 (Ewers et al., 1980; Thwaites, 1982; Cappellin, 1983; Brugger and Stuckey, 1987; Todtling, 1988).

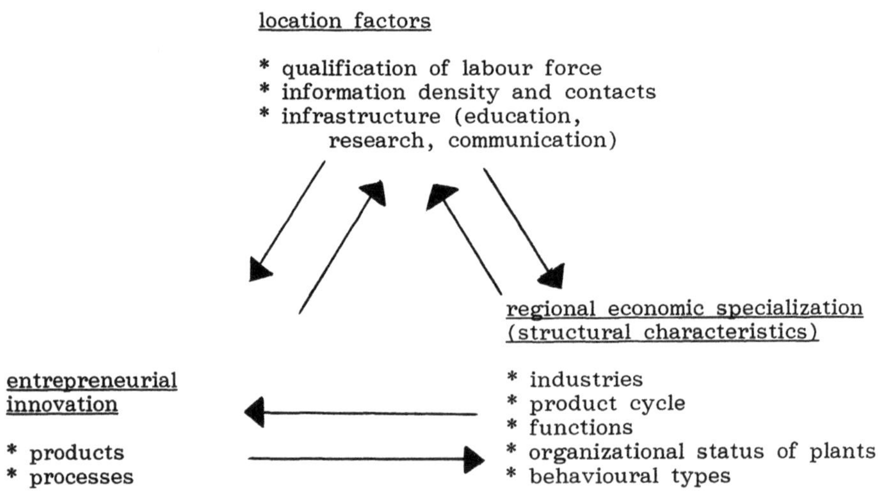

Figure 14.1: The Regional Context for Entrepreneurial Innovation

Early approaches to innovation diffusion tended to place considerable stress on the importance of location factors where the diffusion process was seen as being mainly determined by the transmission of information between suppliers or early adopters and potential adopters, depending on factors such as the hierarchical or geographical distance between them or the size of agglomeration (Hagerstrand, 1967; Pred, 1977). These approaches have neglected economic factors - internal and external to the plant - and thus have little explanatory power. In the following a brief survey of more recent approaches and concepts is given, indicating relevant factors of influence which refer to both locational factors as well as structural characteristics of plants (see Figure 14.2).

Locational factors of entrepreneurial innovation

The following locational factors have been considered in the literature as relevant to entrepreneurial innovation (Ewers et al., 1980; Malecki, 1983; Aydalot, 1986; Stohr, 1987b; Brugger and Stuckey, 1987; Davelaar and Nijkamp, 1987):

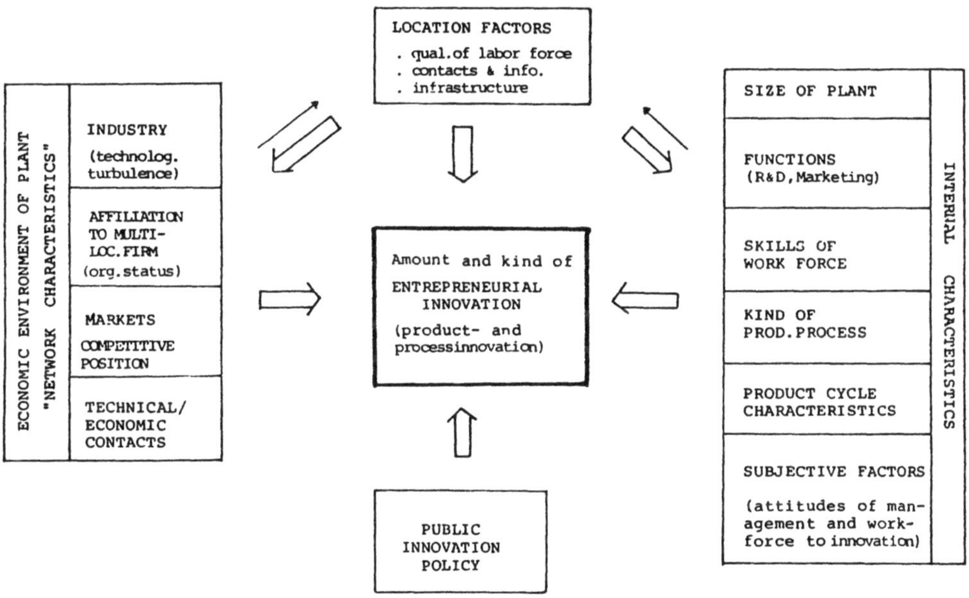

Figure 14.2: Factors Influencing Entrepreneurial Innovation

- Factors related to the **qualification of the labour force** such as the availability of qualified labour (particularly technical and economically well trained personnel); the quality of the educational system (technical and management schools and universities) and the general quality of living for highly skilled labour (cultural and leisure facilities, environmental quality);

- **Information density and contact potential**: these are mainly determined by the access to research facilities and universities, the access to private and public services for information transmission (producer services, 'diffusion agencies' for specific technologies) as well as the density of economic activities and the linkages existing between them;

- **'Innovation-relevant' infrastructure** such as facilities for telecommunication as well as for rapid transport (allowing personal contacts: air connection, rapid train).

The **availability of venture capital** has also been considered as a locational factor for innovation, but more recent studies have pointed out that this factor is more important at the national rather than at the local or regional level (Ewers et al., 1980; Mouven and Nijkamp, 1985; Stohr, 1987b).

Some authors have argued that it is not the availability of single location factors, but the **interaction of all of them** which is important (i.e. synergy effects or 'milieu': see Aydalot, 1986; Perrin, 1986; Stohr, 1987b; Aydalot and Keeble, 1988). These milieux-effects have to do in part with the 'technological capabilities' of a region, which result from informal contacts and processes, learning by watching, doing and using as well as 'tacit knowledge' of the people in the region (Malecki, 1988). The innovation process is also thought to be influenced by socio-political and cultural factors (Aydalot, 1986; Maillat and Vasserot, 1986). However, the effects which result from synergetic interface and through the contributions of a local 'milieu', including cultural factors - although probably very important - are difficult to operationalize and to measure in the empirical studies.

It should also be pointed out that the relative weight of the above stated locational factors differs for various kinds of innovation as is suggested in the product cycle theory. In general they are considered more important for product innovations (the more radical the innovation the higher the locational requirements) and less important for minor modifications of products and for process innovations (which are very often incorporated into purchased machinery). The diffusion of process innovations is also considered to depend to a considerable degree on specific information networks, 'diffusion agencies' or service networks (Brown, 1981; Mudespacher, 1987).

Regional specialization of plants and innovation

Besides the direct impact of location factors on entrepreneurial innovation, the existing regional specialization (structural characteristics of plants) can to a large degree determine local innovative activity. The following concepts have relevance in this context (Ewers et al., 1980; Thwaites, 1982; Cappellin, 1983; Brugger and Stuckey, 1987; Todtling, 1988; see also Figure 14.1):

(a) **Sectoral characteristics**: Sectors or 'industries' are the relevant economic environment of the plant/firm and they differ with regard to the growth of demand, the degree of competition and of market concentration (Scherer, 1980; Porter, 1980). They also differ in terms of technological opportunities and dynamics and thus in the average amount of R & D spent and the general importance of innovation to their progress. For example, high technological dynamics have been observed in past years in sectors penetrated by micro-electronics (Freeman, 1982). For Austria it has been shown that a considerable share of regional differences in R & D activity is due to the sectoral composition of regional economies (Todtling, 1987).

(b) **Product cycle characteristics**: Its is argued that products go through a life cycle (stages) in the course of which markets, as well as factor requirements, and the optimal location change (Vernon, 1966; Erickson and Leinbach, 1979; Suarez-Villa, 1983). In the course of the cycle the typical innovation pattern (kind and intensity of innovative activity, as well as its location) also changes (Utterback, 1979). Furthermore there is a strong relationship between product and process innovations. The spatial pattern of innovation is expected to be concentrated in the early stage of the product cycle (product innovations) and more dispersed in the growth and mature stages, in which product modification and process innovation dominate. The final stage of the cycle is characterized by standardization and a lack of further innovative activity.

(c) The concept of the **functional specialization of plants considers** activities outside the direct production activities, particularly the 'boundary spanning' functions of R & D and market-related functions which are of central importance to the innovation process and also for regional development. Regional differences in innovation are seen mainly as a consequence of a spatial concentration of these functions (Ewers et al., 1980; Bade, 1984; Brugger and Stuckey, 1987).

(d) **Organizational characteristics of plants**: The affiliation of plants to larger multi-regional enterprises may furthermore be influential in the innovation process. These plants on the one hand have access to resources of the larger enterprise (financial resources, R & D, market networks, planning departments and information networks) and on the other hand they are subject to an internal division of labour in the firm, which may be detrimental, particularly to peripheral branch plants (Watts, 1981; Thwaites, 1982; Gibbs and Edwards, 1983; Massey, 1984; Todtling, 1984; Holst, 1987). In general expectations are that the branch plants will have low product innovation, but with regard to new production processes they are considered to be earlier adopters than single plant enterprises due to their integration into a larger enterprise. Single plant enterprises on the other hand are much more dependent on the quality of their immediate locational environment with regard to innovation activities. It is argued that such firms face disadvantages in this context, particularly in old industrial areas and peripheral rural areas where the innovation support environment is thought to be weaker (Thwaites and Oakey, 1985; Goddard et al., 1986; Grabher, 1988).

(e) **Firm size and the size of the plant** have been considered as influential factors on entrepreneurial innovation since the writings of Schumpeter (Schumpeter, 1934; Freeman, 1982; Kleine, 1983). While firm size is reflected in the organizational aspects discussed above, the size of the plant may have an additional influence, although the arguments are ambivalent. On the one hand large plants have more financial and other resources for introducing costly new products and processes (Gibbs and Edwards, 1983; Rees et al., 1984), on the other hand it is argued that they have more bureaucratic organizations and may be lacking intensive lateral and informal information flows which are considered important in the innovation process (Burns and Stalker, 1968; Sorge, 1986).

(f) **Skills of the work force** have already been mentioned in the locational context; their importance, however, becomes manifest internally to the plant. They reflect the level of 'competence' in the plant which can be regarded as a precondition both for the introduction of new products as well as of new processes (Andersson and Johannsson, 1984; Johannsson and Larsson, 1986). These skills are frequently historically determined and are based upon existing products and equipment. Furthermore, the introduction of specific new technologies depends on the **existing production processes** in the plant: for example, CNC, because of its flexibility, is particularly suited for medium-sized production runs, but rather less so for unit- or for mass-production (Gibbs and Edwards, 1983).

(g) **Firm strategy and innovation**: Some authors like Freeman (1982), as well as behavioural geographers (Taylor and Thrift, 1983; Taylor, 1983), have stressed the fact that firms differ (explicitly or implicitly) in the strategic position which they take within the economy. It is argued that only for some types of firms is innovation and technological change of crucial importance (for example, those which pursue 'offensive' or 'defensive' innovation strategies and to a lesser extent for those pursuing imitative strategies). For other segments, having for example, 'dependent', 'traditional' or 'opportunistic' strategies, it is of little relevance. In the following empirical analysis this behavioural dimension was operationalized through the identification and use of several variables: dependence on large customers (subcontracting), extension of the market (export-rate), competitive position of the firm (technological advantages versus price advantages), the amount of technical-economic contacts and the subjective attitudes of management to innovation.

Besides the locational and structural factors, the entrepreneurial innovation process is influenced by **public innovation policy**. As in most market economies, Austria tries to promote and stimulate innovation activities in the economy by giving state support or advice to firms, or by promoting the development and diffusion of specific technologies. As a result, in Austria there are several programmes and institutions which provide financial support for the development and introduction of new products: within the 'TOP' programme, soft loans for the introduction of new products are given; a fund for research and development ('FFF') provides subsidies to the R & D activities of firms; there are also institutions concerned with information transmission in the innovation process (at the federal level, for example, 'Innovationsagentur'); and there exist programmes for the development and application of specific technologies (microelectronics, biotechnology). Firms which can obtain support from these programmes may find it easier to overcome financial and other bottlenecks in the innovation process. However, up to now, there has been only a limited regional dimension to these public innovation programmes.

Figure 14.2 summarizes the factors suggested as influencing the level and pace of regional innovation which have been discussed above. The innovation activities of plants can be seen - with reference to organizational science and behavioural theory - as being influenced by the relations of the plant to its relevant 'environment' (economic, institutional / political, locational) as well as by its internal characteristics. Because of the regional and theoretical perspective chosen, the main focus here is on these structural factors of plants. Subjective factors such as management style and attitudes of managers and employees, or factors such as the more differentiated internal organization of plants and enterprises are not addressed directly.

The Austrian Case Study

Background of the empirical study

The following results are based on interviews in 149 plants manufacturing in the metal and steel, machinery, electrical and electronics industries located in five regions in Austria. These sectors were selected on the premise that technological change - in the form of micro-electronics penetration - plays a major role in their development. The regions were selected according to their a priori assumed differences with regard to locational and structural preconditions for innovation (see description below and Figure 14.3):

- A case study area within Austria's largest agglomeration Vienna (Wien-Sud);

- A more dispersed but dynamic industrialized area (Vocklabruck);

1) Wien-Südd (Liesing, Industriezentrum NO-Süd
2) Vöcklabruck-Gmunden
3) Obersteiermark
4) Süd- und Oststeiermark
5) Wald- und Weinviertel

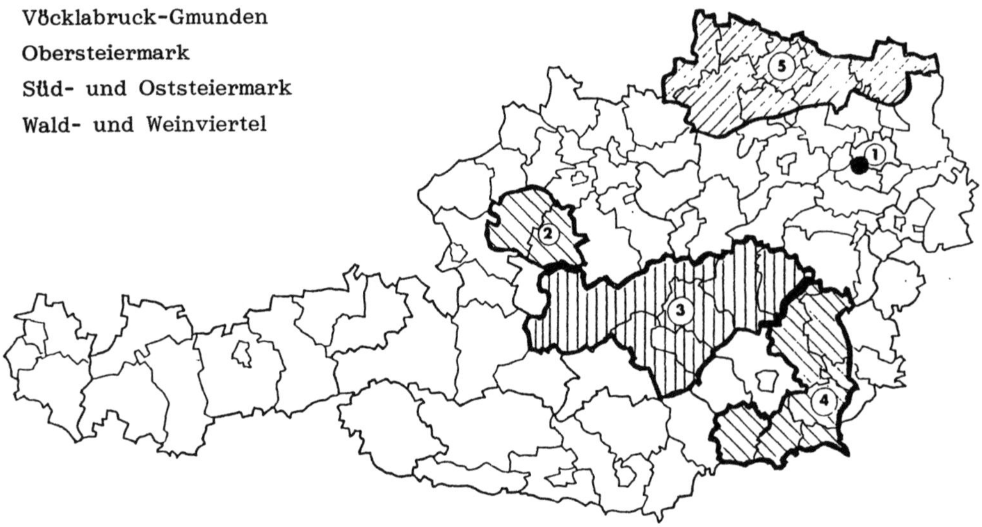

Figure 14.3: Austria: Investigated Areas

- An 'old' industrialized area (Obersteiermark); and

- Two peripheral rural areas (Sudsteiermark and Waldviertel).

In the investigated sectors and regions all plants which had more than 20 employees were asked for an interview, 149 plants (more than half) with about 30,500 employees finally agreed, representing from 45 per cent to 70 per cent of the employment in the respective sectors and regions. The data were collected by means of interviews with the managers or owners of the plants in the spring and summer 1987. The interviews were conducted by personnel from the University of Economics and Business Administration, Vienna, the author (Sudsteiermark), H. Hofer (Waldviertel), A. Strasser (Vocklabruck), M. Weinrother (Wien-Sud) as well as H. Glatz from the Institute of Advanced Studies (Obersteiermark). They were based on a questionnaire covering the following aspects:

- basic characteristics and trends within the plant (age, organizational status/ownership, employment, sales);
- qualifications of the workforce and functions performed;
- characteristics of product (age, product cycle), markets (spatial extension, concentration of customers) and production processes (batch size);
- introduction of new products and processes (types, motives, objectives, sources and problems);
- type and spatial extension of technical-economic contacts;
- satisfaction with location factors (in general and with respect to innovation).

In the region of Obersteiermark due to time and resource constraints a shorter questionnaire had to be used and as a result not all of these factors were included with the same level of detail.

Characteristics of the investigated regions and plants

The investigated regions differ, as a consequence of the selection procedure, with regard to general locational and economic preconditions for innovation, such as the qualification of the labour force, the contact and information potential, infrastructure, sectoral structure and overall dynamism. They also show considerable differences in the structural features of the investigated plants within the selected sectors in each region with respect to size, organizational status and ownership, functions performed, qualification of the workforce, age of products, extension and concentration of markets and types of production processes used as well as the intensity of technical-economic contacts. Constraints on the length of this chapter mean, however, that only a summary of the characteristics of the region, as well as of the structural features of the investigated plants can be presented here.

i) Wien-Sud

This case study area consists of the industrial areas Liesing and Industriezentrum-Sud in the south of the Vienna agglomeration. As the capital of the country, the agglomeration is well endowed with the above stated locational factors for innovation such as the availability of highly qualified labour, research centres and universities, technical and management schools and producer services.

Regarding the investigated plants (27 plants, 5950 employees), the machinery industry and the electrotechnical industry have a high share of employment. There are, compared to the other regions, relatively more subsidiaries (mostly of foreign firms), main plants of multi-regional firms and indigenous plants and, in general, the plants are older, larger and organizationally more differentiated. They have more R & D as well as marketing functions (in terms of full-time equivalents) and a clearly better qualified labour force (higher shares of qualified commercial and technical personnel). Moreover, the technical contacts are more frequent as well as more specialized.

ii) Vocklabruck

This area - consisting of the districts of Vocklabruck and Gmunden - is a relatively dynamic industrialized area with a dispersed settlement structure populated by small and medium sized plants. The area contains no large regional centre, but has good accessibility to the agglomerations Linz and Salzburg and also to the dynamic region of Munich and to the European market.

The investigated plants (37 plants, 6240 employees) belong more often than the sample average to the machinery industry and are also more often indigenous, main plants of multiregional firms or subsidiaries, with only a few branch plants. Similar to Wien-Sud the plants have more functions such as planning, marketing or R & D, a higher qualification of the work force (technicians, commercial staff) and also more technical contacts than the plants in the rural areas and the old industrial Obersteiermark.

iii) Obersteiermark

This is an 'old' industrial area, based on iron and steel with a high incidence of state owned enterprises. It is fairly mono-structured, has a considerable share of its employment in branch plants of large firms and a lack of producer services. The region faces economic decline and increasing unemployment.

Because of the tradition of iron and steel in the region, the investigated plants (29 plants, 6940 employees) are - compared to the other regions - also more often in metal working and steel products. Because of the exclusion of the two large state owned enterprises (branch plants of VOEST-ALPINE and VEW) a high share of the plants in the sample is formally autonomous, but many of them are subcontractors to Austrian and German firms and some are in public ownership. Most of these firms also have a low level of R & D and marketing activities. Concerning the qualification of the employees there is a lack of higher order commercial and technical staff but - in contrast to the rural areas - a high share of skilled blue collar workers. Some of the larger plants are in crisis, experiencing considerable restructuring, and have recently lost a considerable amount of their employment.

iv) Sudsteiermark

This is a rural area, peripheral to the national and European markets, but with good accessibility to the regional centre Graz. The latter is a medium sized agglomeration of about 300,000 people, well endowed with schools, a university, producer services and information networks. The investigated area has a typical 'rural' industrial structure (apparel, electrical products, some traditional industries for the regional market). There has been some industrial dynamism in the past 20 years, partly due to the establishment of branch plants and (very often foreign) subsidiaries, which in many cases have been supported by regional policy.

The investigated plants (29 plants, 7720 employees) show in many respects similar features to those characterizing Waldviertel (see below) where there are two different segments. On the one hand, small indigenous plants, very often in metalworking and machinery; on the other hand larger externally controlled plants, very often in the electrotechnical industry. Both segments lack important functions (planning, R & D, marketing) and they have a low qualification of the labour force. Compared to the Waldviertel there is, however, a stronger penetration by extra-regional and foreign firms and a higher share of plants in 'modern' electronic industries. Due to this fact, there is more mass-production, a higher level of productivity, shorter product cycles and a higher share of technicians than in the Waldviertel. The plants also show more technical contacts, very often with the regional centre of Graz.

v) Waldviertel

Like Sudsteiermark, Waldviertel is also a rural area, but if anything is even more peripheral to the national and Community markets. Large parts of this area have, in contrast to the Sudsteiermark, a very low accessibility to larger centres (Wien, Linz) and there is generally a low density of activities. The region has some older industries (textiles, glass, stone, wood) but has also attracted new plants in the 1960s and early 1970s (mainly labour intensive industries such as apparel and electrical products).

The investigated plants (27 plants, 3570 employees) consist - as in the Sudsteiermark - of two rather different segments. On the one hand, small plants in metal and steel products or (agricultural) machinery are represented more strongly. They are very often older indigenous plants and subcontractors to larger enterprises outside the region, and in general have a low productivity. On the other hand, there are some larger plants in the electrotechnical industry which absorb almost two-thirds of the employment of the investigated plants. The latter are mostly branch plants in the more mature stage of the product cycle and are subject to a well articulated division of labour within the respective firms. Both segments lack important functions such as R & D, marketing and planning and consequently they also have a very low qualification of the labour force (low shares of technical staff and management personnel and high shares of unskilled female workers). Furthermore their level of technical contacts is low.

Regional differences in innovation activity

In the empirical survey managers of the plants were asked to indicate the innovations which had taken place in the years 1981-1986. 'Innovations' were broadly defined to include the opening up of new markets with existing products (market expansion or penetration), the introduction of new products (both 'new to the market' as well as 'new to the plant'), the modification of products and the introduction of new processes (in production and in the office). Selected results from the surveys are presented in the following sections.

i) Product innovations

About four fifths of the plants stated that they carried out some kind of product innovation. This rather high share is due to the length of the period in which innovation could take place (five years) and that all kinds of new products were included. The differences between regions with regard to this general indicator were not very strong. More significant differentiation could be seen for those product innovations which were considered by the respective managers as 'new to the market' (see Table 14.1):

In Vienna two-thirds of the plants reported this kind of product innovation. The plants were found to be generally of large size, employing 94 per cent of the respective work force. The plants of the industrialized region of Vocklabruck as well as the rural Sudsteiermark were fairly active in this regard. In the case of Vienna and Vocklabruck the higher product innovation activity is in line with the better preconditions which were presented above (both locationally as well as with regard to the structural characteristics of the plants). In the Sudsteiermark the high product innovation activity was not expected a priori, with both subsidiaries and branches from multilocational enterprises (mainly in the electrotechnical/ electronics industry with typically short product cycles) as well as small indigenous enterprises (mainly in the machinery sector) involved. These latter firms very often were former producers of agricultural machinery which benefited from the technical skills of the owner manager and the work force. These skills are often informal

Table 14.1
Regional differences in product innovation

	Product innovation 'new to the market'	
	per cent of plants	per cent of employment
Wien-Sud	67	94
Vocklabruck	49	66
Sudsteiermark	55	35
Waldviertel	30	60
Obersteiermark	17	7
All areas	44	49
Contingency coefficient	0.33	0.52
Probability of χ^2	0.001	0.001

qualifications resulting from training on the job. They also seem to benefit from the strength of the regional centre (Graz) in the engineering sector and the technical research infra-structure (technical university, research centres, enterprises).

The level of product innovation was, however, found to be very low in the old industrial area of Obersteiermark (only 17 per cent of the plants with just seven per cent of the employment) as well as in the peripheral rural Waldviertel (30 per cent of the plants). In the latter region the innovative plants are large (mainly electrotechnical industry) and employ 60 per cent of the labour force.

Table 14.2

Regional variations in the modification of existing products

	Modification of products	
	per cent of plants	per cent of employment
Wien-Sud	70	89
Vocklabruck	78	89
Sudsteiermark	66	72
Waldviertel	63	69
Obersteiermark	28	33
All areas	62	71
Contingency coefficient	0.34	0.45
Probability of χ^2	0.001	0.001

In the case of the modification of existing products the regional differences are also significant and the ranking of the regions stays about the same. The plants in Wien-Sud and Vocklabruck are again more active (more than 70 per cent of the plants with about 90 per cent of the employment). Those in the old industrial area (Obersteiermark) are clearly below average (Table 14.2). This is obviously the more usual route to innovation for most plants outside the Obersteiermark region, including those located in the peripheral rural area of Waldviertel.

These results for the product innovation activity are fairly consistent with the data about the age structure of the products (number of years in production programme). The youngest product structure was not surprisingly found in the 'innovative regions' Wien-Sud, Sudsteiermark and Vocklabruck, the oldest in Obersteiermark and Waldviertel.

ii) Introduction of new processes

Results from the surveys suggest that there was very little and insignificant regional differentiation with regard to the introduction of new production processes in general. In all regions, between 78 per cent and 83 per cent of the plants claimed to have introduced some kind of new production processes over the five years to 1986. Plants in the old industrial area (Obersteiermark), in contrast to the low product innovation activity, adopted process technology at the average rate for the sample (Table 14.3).

Greater regional differentiation in the adoption of processes were, however, found with regard to specific technologies such as CNC and particularly CAD. These technologies were proportionately introduced to a higher degree in Wien-Sud

Table 14.3

Regional pattern of process technology adoption

	Introduction of new production processes (per cent of plants)		
	general production processes	CNC*	CAD*
Wien-Sud	78	44	41
Vocklabruck	78	38	11
Sudsteiermark	83	38	14
Waldviertel	78	19	11
Obersteiermark	79	-	-
All areas	79	35	18
Contingency coefficient	0.05	0.19	0.30
Probability of χ^2	0.99	0.21	0.008

* These were not asked about in the region of Obersteiermark.

(44 per cent and 41 per cent of the plants) than elsewhere. The adoption of CNC was clearly below average in the region of Waldviertel, and adoption of CAD was below average in all three regions outside of the capital. The regional differences in the level of diffusion was thus much more pronounced in the case of the newer CAD technology than in the older CNC (Table 14.3).

Results on the introduction of information technology in the office also appear to show some regional differentiation (Table 14.4): with regard to the introduction of computers the peripheral Waldviertal and the Obersteiermark appear to be below average. Statistical tests, however, suggest that these differences are not significant. Similar findings characterize the results for the new communication technologies where again the Obersteiermark is below average.

A further step of the analysis was to investigate those factors associated with the innovation activities. This was approached in two ways:

- questions were asked of respondents with regard to (subjectively indicated) sources and obstacles to innovation;

- further statistical methods (cross-tabulation statistics as well as a multivariate probit analysis) were applied. Some of the results of the latter PROBIT analyses are summarized in the following section.

Table 14.4

Regional variation in the introduction of information technology

	Information technology in the office (per cent of plants)	
	computers	new communication technology
Wien-Sud	63	78
Vocklabruck	70	51
Sudsteiermark	76	79
Waldviertel	52	70
Obersteiermark	52	28
All areas	63	65
Contingency coefficient	0.20	0.25
Probability of χ^2	0.21	0.038

Results from the PROBIT-analysis

There have been only a few attempts to analyse entrepreneurial innovation activities at the regional level in a multivariate framework (see, for example, Alderman et al., 1982; Johannsson and Larson, 1986; Wrigley and Brouwer, 1986). In the present study a PROBIT-analysis was applied to identify the probability of the introduction of selected innovations at the plant level depending on a set of explanatory variables (see Todtling, 1988)[1].

The PROBIT-analysis - as applied here - has many similarities with a conventional multivariate regression analysis. The basic difference between them is that PROBIT-analysis takes into account the fact that the dependent variable is a binary one and cannot take values along a continuum. Binary dependent variables create problems for conventional regression analysis (e.g. heteroscedasticity: see

Ben-Akiva and Lerman, 1985). In the PROBIT-analysis of the present study each plant is seen to have the alternative to introduce a specific innovation or not. The probability of introduction is determined by attributes which are specific to the alternative (these are in the present model captured by an alternative-specific constant) and by characteristics of the decision making unit (plant).

The latter explanatory factors are the internal and external characteristics of the plant which have been discussed above (see Figure 14.4). In the estimation procedure the explanatory variables were all introduced at the beginning simultaneously in the larger set as indicated by Figure 14.4, then those which contributed only a little to the total explanation were systematically eliminated by a stewise process. The final model represents the best fit to the data, taking into account the degrees of freedom (largest corrected G^2). In the estimation a maximum-likelihood procedure has been applied. The search process has been started at a parameter vector which accounts for the observed frequencies of specific

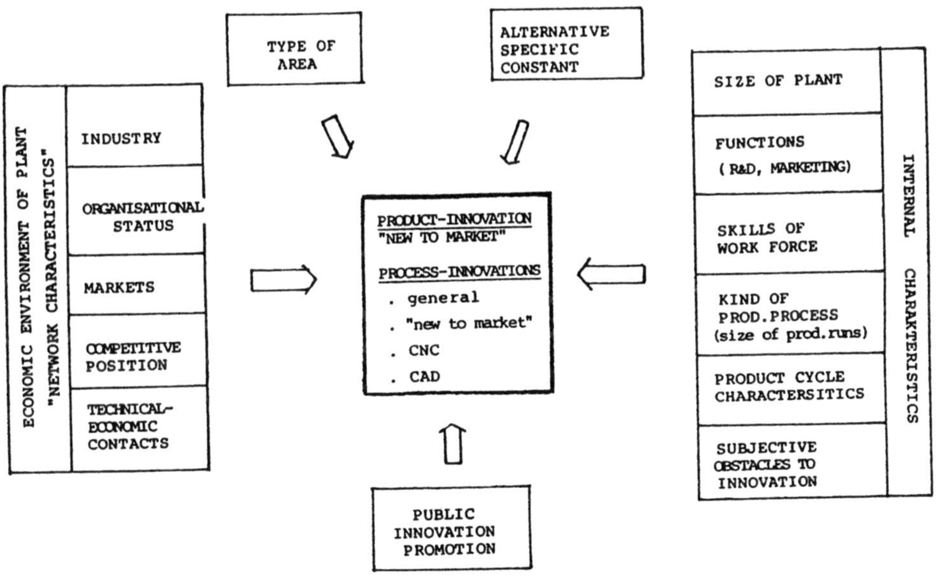

Figure 14.4: Scheme of Investigated Relations in the Probit Analysis

innovations. Thus, the G^2-values in Figures 14.5 to 14.7 only refer to the additional explanatory power of the structural variables (for details see Maier, 1987; Todtling, 1988). Although the number of cases (149) is not large and of course better results could be obtained from a larger sample, the number is sufficient for this kind of analysis. The calculations cover several kinds of innovation (product innovations 'new to the market', specific process innovations), but only a summary of this analysis can be provided here.

i) Product innovations 'new to the market' (Figure 14.5)

From the external factors the location of the plant in the old industrial Obersteiermark (measured by a dummy variable) was significantly negative. In this area the conditions for product innovation thus appear to be worse than in the other regions, even after controlling for the influence of other factors. It seems that this result is only partly to do with locational factors such as contact potential, access to universities and research institutions, since these are even worse in the rural areas. The interviews suggested that the less tangible socio-economic conditions such as the existence of rigid and bureaucratic organizations (both in enterprises and unions), conservative (in the literal sense) attitudes of management and workforce as well as the traditional expectation of outside help (government) were more important.

Status as an indigenous plant also had a negative influence on this kind of product innovation. This may be due to the fact, that these are usually smaller firms which lack financial and other resources for product innovation and cannot rely on a network of a larger enterprise.

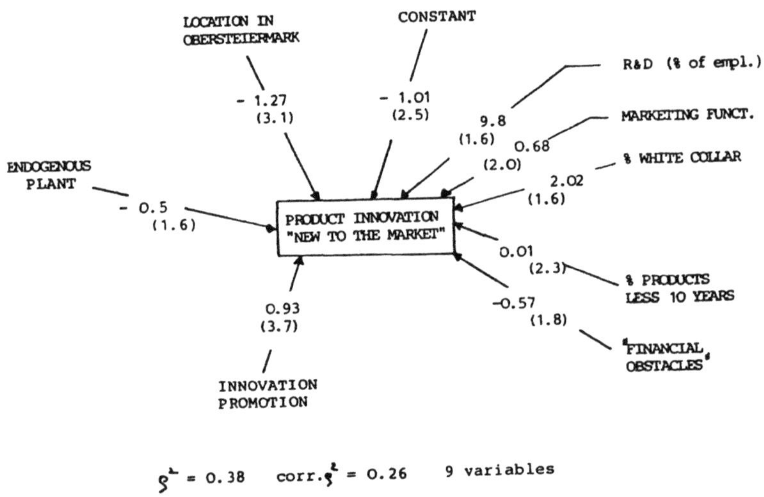

Figure 14.5: Product Innovation "New to Market" (Coefficients and T-values)

The fact that the plant has received support from public innovation policy had a positive influence on this kind of innovation (not on others). In most cases this support was financial aid for the development of products and/or the start up of production. The results indicate that the public support was helpful in overcoming financial bottlenecks in the innovation process and that the impact of these programmes is at least in the intended direction. The role of financial bottlenecks as a limit for innovation projects also appears in the negative effect of the internal factor 'financial obstacles to innovation' (Figure 14.5).

In another version of parameter estimation, in which the plants of the Obersteiermark were excluded due to missing data, the amount of technical-economic contacts had an additional positive influence indicating that a strong outward looking behaviour of the plant is favourable for product innovation.

From the internal factors the 'boundary spanning' functions marketing (existence of activity) and R & D (per cent of employment in full-time equivalents) as well as the share of white collar employees had significant positive influences. Both the positive influence of the above mentioned external contacts and of the internal 'boundary spanning functions' and skills thus support the view that product innovations occur particularly at the interface of external links (information flows) and internal capabilities (functions and skills). Furthermore a youthful **production programme** (resulting from product innovations of the previous period) had a positive influence, indicating the continuous character and the existence of learning effects and routines inherent in the innovation process (Nelson and Winter, 1982; Lambooy, 1986).

ii) CNC and CAD (Figures 14.6 and 14.7)

CAD as the younger and less diffused technology is better explained by the model than CNC. For both technologies the size of the plant as well as export-intensity are significant factors. The positive effect of size probably has to do with the relatively high cost of the introduction as well as threshold effects (Gibbs and Edwards, 1983). The effect of export-intensity probably comes from stronger competition in external markets as well as those plants having generally better information, leading to a more up to date technological standard of the plant.

In the case of CNC (Figure 14.6) in addition to the above factors, the kind of production process (length of the production runs) was a significant factor: unit production on the one hand and mass production on the other had negative influences. Both of these batch sizes are considered as suboptimal for CNC technology. Furthermore CNC was introduced at a lower level in the electrotechnical industry as well as in branch plants. In the electrotechnical industry there is less possibility for application in the production process than in the metal-working and machinery industry. In branch plants the lack of decision making power and/or the division of labour in the firm might lead to an adoption lag (Gibbs and Edwards, 1983). A good profit situation in the plant was favourable for introduction of CNC,

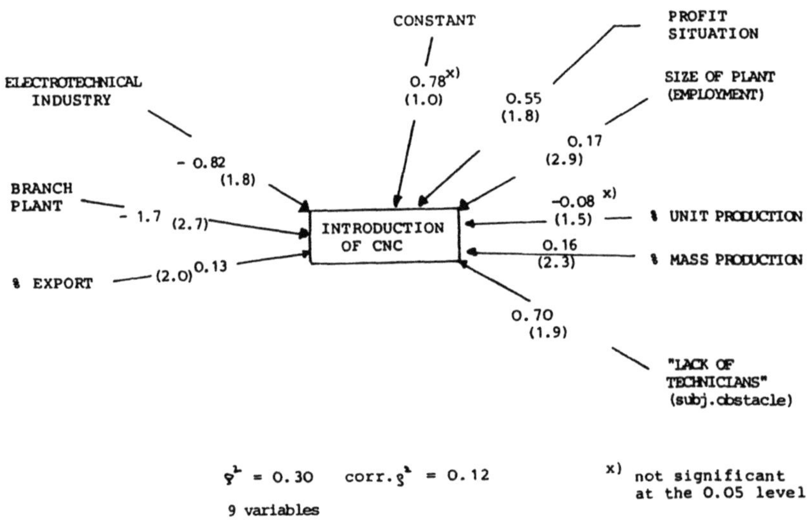

Figure 14.6: Introduction of CNC (Coefficients and T-values)

probably due to the fact that costs and risks involved are taken more readily in a buoyant economic context.

In the case of CAD (Figure 14.7), further significant positive influences came from the qualification of the work force (higher level commercial employees, other white collar employees) and an already existing technological advantage in the competitive position of the plant - again indicating previous innovation activity and thus the continuous character of the innovation process (Nelson and Winter, 1982).

Summary

In the investigated sectors and case study regions there were - due to the historically evolved regional specialization of plants - quite clear differences both with regard to preconditions for innovations (locational factors, structural characteristics of plants), as well as the innovation activity itself. For product innovations classified as 'new to the market' as well as for specific process innovations (CAD, less for CNC) the regional differences were particularly strong. They were also noticeable for new information technologies (computers and communication technologies). In general, a higher innovation activity could be observed for the area in the agglomeration of Vienna, partly also for Volklabruck and Sudsteiermark. These activities were rather low for the old industrial area of Obersteiermark and to a lesser extent the peripheral and rural Waldviertel.

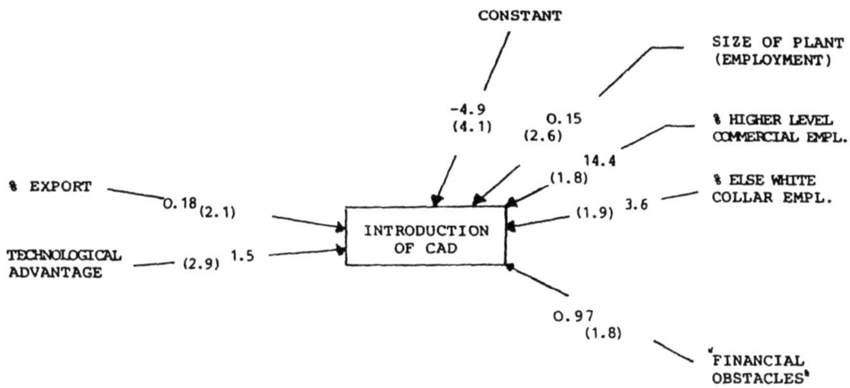

$\rho^2 = 0.44$ corr. $\rho^2 = 0.25$ 7 variables

Figure 14.7: Introduction of CAD (Coefficients and T-values)

The most important explanatory factors for the innovation activity found in the PROBIT-analysis, were selected structural and behavioural features of the plants. There were significant internal factors such as the boundary-spanning functions performed (R & D and marketing), the qualification of the labour force and the kind of production process (length of production runs). Also external factors ('firm-environment' or 'network characteristics') such as the export activity, the kind of competitive position, the information activity and the industry affiliation turned out to be important. Locational factors have some direct effect on the innovation activities of plants (particularly via the labour market), but the stronger impact is via the above stated structural features of the plants, which may be due to the historically evolved regional specialization. In other words, some of the plants which are not innovating may have adjusted their production programme, functions performed, skills of the work force and technology to the lower demands of their regional environment. They have subjectively decided that they have 'no need for innovation', either because they have other survival strategies (subcontracting, flexibility and accuracy with regard to delivery dates, low wages) or because they do not realize it as a problem. Therefore such respondents usually saw no locational problem for innovation. Of course the relation of a plant to its regional environment is not one way. The plant also shapes the regional environment by its factor demand (e.g. the skill of its labour force) and by the strength of its linkages to other regional firms (e.g. the demand for producer services) and thus influences the locational conditions for innovation in the medium and long run.

Due to this long run interrelated adjustment process and the stability of structural features and behavioural patterns it appears, in general, quite difficult to change the innovative behaviour of plants and firms in regions with low innovation activities. The stimulation of plants to increase their innovation activities just by the supply of finance or information (transfer and consulting activities) is probably not sufficient. It seems important and necessary to integrate a regional innovation policy more with the long run structural improvement of regional economies. The latter implies more selectivity and a change in the regional investment promotion: both the subsidization of new plants as well as the expansion of existing plants should be linked much more to structural criteria (e.g. level of skill, functions performed, potential linkages to regional firms). The promotion of investment should also be related more to 'non-material investment' (human capital, organizational improvement, functions outside production) than in the past. It also implies an improvement of locational conditions such as the qualifications of the labour force, the access to producer services and information centres (transfer agencies, research centres) and the availability of innovation relevant infrastructure (telecommunications, rapid transport) in those regions.

Notes

1 Gunther Maier gave advice with regard to methodological aspects and wrote the necessary computer programs, linked to the SAS statistical model (Maier, 1987).

References

Alderman, N., Goddard, J.B. and Thwaites, A.T. (1982) Regional and urban perspectives on industrial innovation: applications of logit and cluster analysis to survey data. Discussion Paper 42, Centre for Urban and Regional

Development Studies, University of Newcastle upon Tyne.

Andersson, A.E. and Johannsson, B. (1984) Knowledge intensity and product cycles in metropolitan regions, IIASA, Metropolitan Study 8, Laxenburg.

Aydalot, P. (ed.) (1986) Milieux Innovateurs en Europe, GREMI, Paris.

Aydalot, P. and Keeble, D. (eds.) (1988) High Technology Industry and Innovative Environments: The European Experience, Routledge, London.

Bade, F.J. (1984) Die funktionale Struktur der Wirtschaft und ihre raumliche Arbeitsteilung, Deutsches Insititut fur Wirtschaftsforschnung (DIW), Berlin.

Ben-akiva, M. and Lerman, R. (1985) Discrete Choice Analysis : Theory and Application to Predict Travel Demand, MIT Press, Cambridge, Mass.

Brown, L.A. (1981) Innovation Diffusion : A New Perspective, Methuen, London.

Brugger, E.A. and Stuckey, B. (1987) Regional economic structure and innovative behaviour in Switzerland, Regional Studies, 21, 241-254.

Burns, T. and Stalker, G.M. (1968) The Management of Innovation, Tavistock Publications, London.

Cappellin, R. (1983) Productivity growth and technological change in a regional perspective, Giornale degli Economisti ed Annali di Economia, 42, 459-482.

Davelaar, E.J. and Nijkamp, P. (1987) The urban incubation hypothesis: Old wine into new bottles? Mitteilungen des Arbeitskreises fur Neue Methoden in der Regionalforschung, Vol.17, Wien.

Erickson, R.A. and Leinbach, T.R. (1979) Characteristics of Branch Plants Attracted to Non-metropolitan Areas, in Lonsdale and Seyler, Non-metropolitan Industrialization, Wiley, New York.

Ewers, H.J., Wettmann, R., Kleine, J., Krist, N. and Bade, F.J. (1980) Innovationsorientierte Regionalpolitik, Schriftenreihe des Bundesministers fur Raumordnung, Bauwesen und Stadtebau, Nr.6.042, Bonn-Bad Godesberg.

Freeman, C. (1982) The Economics of Industrial Innovation, MIT Press, Cambridge, Mass.

Gibbs, D.C. and Edwards, A. (1983) Some preliminary evidence for the interregional diffusion of selected process innovations, in Gillespie, A. (ed.), Technological Change and Regional Development, London Papers in Regional Science, Pion Limited, London.

Goddard, J.B., Thwaites, A.T. and Gibbs, D. (1986) The regional dimension to technological change in Great Britain, in, Amin, A. and Goddard, J.B. (eds.) Technological Change, Industrial Restructuring and Regional Development, Allen and Unwin, London.

Grabher, G. (1988) De-Industrialisierung oder Neo-Industrialisierung? Innovationsprozesse und Innovationspolitik in traditionellen Industrieregionen. Edition Sigma, WZB Berlin.

Hagerstrand, T. (1967) Innovation diffusion as a spatial process, University of Chicago Press, Chicago.

Holst, M. (1987) External control and innovation activities in manufacturing industry, paper presented at the 27th European Congress of the RSA, Athens.

Johannsson, B. and Larsson, J. (1986) Characteristics of the firm and the speed of technological innovation, CERUM working paper, University of Umea.

Kleine, J. (1983) Investitionsverhalten bei Prozess innovationen - Ein Beitrag zur mikrookonomischen Diffusionsforschung, Campus Verlag, WZB Berlin.

Lambooy, J.G. (1986) Regional development trajectories and small enterprises: The case of the Amsterdam region, in, Aydalot, P. (ed.) Milieux Innovateurs en Europe, GREMI, Paris.

Maier, G. (1987) Die Schatzung diskreter Entscheidungsmodelle mit Hilfe der SAS Prozeduren BROBIT and MNLOGIT., IIR-Discussion 27/2. Institut fur Raumordnung, WU Wien.

Maillat, D. and Vasserot, J.Y. (1986) Les millieux innovateurs - le cas de l'Arc Jurassien suisse, in Aydalot, P. (ed.) Milieux Innovateurs en Europe, GREMI, Paris.

Malecki, E.J. (1983) Technology and regional development: a survey, International Regional Science Review, 8(2), 89-125.

Malecki, E.J. (1988) Research and development and technology transfer in economic development: the role of regional technological capability, paper presented at the European Summer Institute of the RSA, Arco.

Massey, D. (1984) Spatial Divisions of Labour: Social Structures and the Geography of Production, McMillan, London.

Massey, D. and Meegan, R. (1982) The Anatomy of Job Loss: The How, Why and Where of Employment Decline, Methuen, London.

Mouven, A. and Nijkamp, P. (1985) Knowledge centres as strategic tools in regional policy, paper presented at the colloquium 'Technologies Nouvelles: Condition de Renouveau des Regions en Crise', Bruxelles.

Mudespacher, A. (1987) Adoptionsverhalten der Schweizer Wirtschaft und regionale Aspekte der Diffusion der Neuerungen der Telematik, Jahrbuch der Regionalwissenschaft 8/1987, 106-134.

Nelson, R.R. and Winter, S.G. (1982) An evolutionary theory of economic change, Harvard University Press, Cambridge, Mass.

Perrin, J.C. (1986) Les synergies locales - elements de theorie et d'analyse, in Aydalot, P. (ed.) Milieux Innovateurs en Europe, GREMI, Paris.

Porter, M.E. (1980) Competitive Strategy: Techniques for Analysing Industries and Competitors, The Free Press, New York.

Pred, A. (1977) City Systems in Advanced Economies, Hutchinson, London.

Rees, J., Briggs, R. and Oakey, R. (1984) The adoption of new technology in the American machinery industry, Regional Studies, 18, 489-504.

Scherer, F.M. (1980) Industrial Market Structure and Economic Performance, Rand McNally, Chicago.

Schumpeter, J. (1934) The Theory of Economic Development, Harvard University Press, Cambridge Mass.

Sorge, A. (1986) Informationstechnik und Arbeit im sozialen Prozess: Arbeitsorganisation, Qualifikation und Produktivkraftentwicklung, Campus, Frankfurt.

Stohr, W. (1987a) The spatial division of labour and entrepreneurial strategies, in Muegge, H. and Stohr, W. (eds.) International Economic Restructuring and the Regional Community, Avebury, Aldershot.

Stohr, W. (1987b) Regional Innovation Complexes, Papers of the Regional Science Association, 59, 29-44.

Suarez-villa, L. (1983) Dynamics of manufacturing and the theory of location, unpublished paper, University of California, Unpublished paper, Irvine.

Taylor, M. (1983) Technological change and the segmented economy, in Gillespie, A. (ed.) Technological Change and Regional Development, London Papers of Regional Science, Pion, London.

Taylor, M. and Thrift, N. (1983) Business organization, segmentation and location, Regional Studies, 17, 445-465.

Thwaites, A. (1982) Some evidence of regional variations in the introduction and diffusion of industrial products and processes within British manufacturing industry, Regional Studies, 16, 371-382.

Thwaites, A. and Oakey, R. (eds.) (1985) The Regional Economic Impact of Technological Change, Francis Pinter, London.

Todtling, F. (1984) Organizational characteristics of plants in core and peripheral regions of Austria, Regional Studies, 18, 397-412.

Todtling, F. (1986) Betriebliche Reorganisation und regionale Industrie-Entwicklung bei veranderten Rahmenbedingungen 1973-1981 in Osterreich, IIR-Discussion 32, WU-Wien.

Todtling, F. (1987) The regional pattern of industrial R and D in Austria: Sectoral, organizational and locational determinants, Revue d'Economie Regionale et Urbaine, 2, 239-256.

Todtling, F. (1988) Regionale Unterschiede und Einflussgrossen der betrieblichen Innovation - Ergebnisse regionaler Fallstudien fur Osterreich, IIR-Forschung Nr.13, Wirtschaftsuniversitat Wien.

Utterback, J.M. (1979) The dynamics of product and process innovation in industry in, Hill, C.T. and Utterback, J.M. (eds.), Technological Innovation for a Dynamic Economy, Pergamon Press, New York.

Vernon, R. (1966) International investment and international trade in product cycle, Quarterly Journal of Economics, 80, 190-207.

Volk, E. (1987) Die Innovationsaktivitaten der osterreichischen Industrie - Technologie- und Innovationstest 1985. Osterreichisches Institut fur Wirtschaftsforschung, Wien.

Watts, H.D. (1981) The Branch Plant Economy - A Study of External Control, Longman, London.

Wrigley, N. and Brouwer, F. (1986) Qualitative statistical models for regional analysis, in Nijkamp, P. (ed.) Handbook of Regional and Urban Economics, Vol.1, Elsevier, New York.

Appendix 14.1

List of Variables in the Probit-Models

Dependent Variables (binary variables)

product innovation 'new to the market': plant has introduced from 1981 to 1986 one or more product innovations considered as 'new to the market'.

introduction of CNC: plant has introduced CNC-technology.

introduction of CAD: plant has introduced CAD-technology.

Explanatory Variables:

location in Obersteiermark: dummy variable for location of the plant in the case-study area of Obersteiermark.

indigenous plant: dummy variable for lack of organizational or ownership links of the plant to extra-regional firm(s).

branch plant: dummy variable for plant with extra-regional headquarter.

electrotechnical industry: dummy variable for plant belonging to electrotechnical or electronics industry.

% export: percentage of exported sales

technological advantage: dummy variable for indication of a competitive advantage via 'technology'.

innovation promotion: dummy variable for plant having received financial support from public innovation policy.

size of plant: number of employees of the plant.

profit situation: dummy variable for indication of 'good to high profits' for past 5 years.

R & D: percent of employment concerned with R & D activities (full-time equivalents).

marketing: dummy variable for indication of full performance of marketing activities.

% higher level commercial employment: percentage of employees with medium to high level commercial education.

% white collar: percentage of other white collar employees.

% unit production: percent turnover of unit production.

% mass production: percent turnover of mass production.

% products less 10 years: percent turnover of products which are less than 10 years in the production program of plant.

financial obstacles: dummy variable for indication of financial obstacles in the innovation process.

lack of technicians: dummy variable for indication of problems with regard to the availability of technical personnel in the innovation process.

15. INNOVATION DIFFUSION, INNOVATION ACQUISITION AND PATH DEPENDENCY IN TRADITIONAL SECTORS: AN EMPIRICAL INVESTIGATION FOR THE VENETO REGION OF ITALY

Fiorenza Belussi

Introduction

This work will examine the processes of innovation diffusion and innovation acquisition within a regional context characterized by the overwhelming presence of firms in traditional sectors, offering some elements of analysis and interpretation of the manner in which technology is generally acquired and utilized. Two main analytical traditions have provided major theoretical contributions to the understanding of the role of technology development in firms: 'innovation diffusion' literature and 'technical change' economics. An enormously rich amount of theoretical developments, fieldwork and analyses have been carried out in the last decades by many scholars. In spite of this, two main criticisms can be made of their work.

First, with regard to the innovation diffusion literature a major limitation is that most analyses are exclusively focused on individual product-process diffusion and on their path of evolution over time which are not included within frameworks incorporating the **real** process of firms' long term technology developments. Traditional analyses on innovation diffusion can be principally criticized not only for the uncertain assumptions underlying the S-shaped diffusion patterns[2], but, more seriously, because the majority tend to overlook the fundamental problem that processes and products during the diffusion process are not unchangeable, and the population of potential and actual adopters is not stable (Gold, 1979; Freeman, Clark and Soete, 1982; Soete, 1985; Freeman and Perez, 1986).[3]

Adoption, in other words, is not **simply** diffusion. It is certainly more. Each implementation by an organization of a new technology requires a strategic 'moving around', 'improving upon', 'slight changing'. Several factors like uncertainty, technological appropriability and opportunity, interactions with the external environments, and the need for organizational and managerial changes challenge the process of firm innovation adoption. Thanks to the work of scholars such as Rosenberg, Mowery, von Tunzelman and others, within the historical literature on technical progress, and because of the growing importance of innovation studies elaborated within economic analysis by Freeman, Pavitt, Nelson and Winter, Metcalfe, Dosi and others, we certainly now possess a better understanding of the given characteristics of technology:

- where a significant amount of firms' resources are spent in searching, exploring, discovering, experimenting, developing and looking out (Dosi, 1988a);

- being technology inter-related, complex, firm specific, cumulative and partially tacit (Pavitt, 1986);

- with technological trajectories of firms being strongly influenced by path-dependency characteristics (Langlois, 1987); pushed by the creation of technological fusions and 'break-throughs' (Kodoma, 1986) and shaped by learning activities: learning by doing (Arrow, 1962), by using (Rosenberg, 1982), by learning (Stiglitz, 1986), by failing (Maidique and Zirgler, 1985), by interacting (von Hippel, 1976; Lundvail, 1985; Metcalfe, 1981; Aoki, 1984 and 1986; Sako, 1987), and by marketing (Belussi, 1987);

- where firms' behaviour is constantly marked by building up routines, iterations, heuristic and sequential procedures, and selections (Nelson and Winter, 1982; Dosi et al., 1988); and

- where technology involves the firms' knowledge assets (Winter, 1987, p.170); a great deal of tangible and intangible, articulable and non-articulable, observable and unobservable know-how.

On the other hand, adoption is **not just** a process which involves a 'technological donor' and an 'absorptive recipient firm'. And diffusion is not a football match where two or more players (innovators) make a pass with the ball (innovation). As discussed by Dodgson (1989)[4], contemporary technology extends the boundaries of a single firm. Few firms, in fact, have sufficient in-house resources and expertise to generate novel technologies. Most often, technology is developed in conjunction with, or in response to, activities of other firms, in customers-suppliers-partners-competitors linkages. On the other hand few, if any, firms are passive adopters: as is well known, an enormous quantity of effort is normally required in order to apply new technology to existing processes. Progressive modifications occur during the adoption phase which can also, for instance, lead to a firm technologically retrofitting in the automation path (see Cainarca et al, 1987). Continuous innovations (Baba, 1988) and learning are the fundamental components of industrial dynamics. A basic implication of this view is that along the innovative cycle, process and product innovations have been transformed through a process of continuity and competition in 'post innovation performance' (Georghiou et al, 1986; Scherer, 1984).

In addition, traditional theory in the main tends to place the process of spreading technical change in a static framework, where recipient firms are all potential users with the same level of capability, knowledge, information, etc. Time is the only explanatory variable which in the end will equalize all 'actors' involved in the process. However, these assumptions are clearly unrealistic. In the real world the existence of a cumulative pattern of technical change, with leads and lags, with technological asymmetries, differences in learning, imperfect degrees of information about markets, oligopolistic powers, corporate strategies, etc., causes a sequential development of **firms' specific** natural (Nelson and Winter, 1977, 1982) and technological (Dosi, 1982) trajectories. In other words, firms on the frontier will not await the end of the diffusion process to start innovating; as for imitators, they have an 'open exit' choice on how quickly to follow, catch up or even overtake innovators.[5]

We can now come to some provisional conclusions. First, the innovation diffusion process, the mechanism through which changes are generated and spread in the system, is a more complex process than those reviewed by 'diffusionists'. Second, this process interacts deeply with many other characteristics of a firm: ability, capability, potentiality, path of growth of the firm. In addition, as has been showed convincingly by an increasing number of studies, gains which each firm can extract from that technology may imply very different levels of productivity. The outcomes may depend upon a firms' production capability (Cohen and Zysman, 1987; Abernathy et al, 1983), strategic competitive position in the market (Porter, 1980) global strategy (Porter, 1985; Spence and Hazard, 1988; Hout et al, 1982); on firms' co-specialized assets (Teece, 1986); product complementarities (Richardson, 1972), organizational flexibility (Piore and Sabel, 1984; Sabel and Zeitlin, 1982; Friedman, 1988), and on its ability to incorporate the new technology (Freeman, 1987). However, diffusion analysis per se does not reveal firms' potential technological capability.[6] Thirdly, firms are not isolated, especially in the context of technology. As discussed by many authors - among these Sweeney (1987), Anderson (1985), Russo (1985), Brusco (1982, 1986), Sabel et al. (1987), Garofoli (1988), Kamann and Nijkamp (1988) and Lazerson (1988) - technology transfer depends critically on the external environment of firms.

As regards the technological change literature, our main attention shifts to the economics of R & D on which subject there is now an extensive body of literature.[7] The question to be asked is how technological progress is generated and who are the subjects who benefit from it? More importantly what are the typical loci of creation and transmission of useful knowledge? In the 1960s Nelson, Peck and Kalechek found that 'industrial research and development probably accounted for significantly more than half of the total national effort to advance technological knowledge' (Nelson et al, 1967, p.45). The question today is whether or not the body of scientific and technical knowledge is still primarily generated in research and development laboratories? Answering this question raises, of course, the very important issue of how good R & D indicators are as proxies for technological capacity (or innovativeness) of a firm, region and country? Analyses which have explored sources and causes of technological change have brought to light several different knowledge producing and knowledge using regimes.

First, there is a general pool of knowledge generated within the system by Universities, Government Agencies and non-profit centres. Knowledge produced by these institutions is the most generally available but is still subject to constraints and perhaps even increasing constraint. Some of these constraints are based on transmission costs between different areas and countries. Firms located in the same country or in the physical proximity of research institutions are clearly in a favourable position: the more widely information is diffused, the higher the risks of 'entropic' dissipation and 'redundancies'. At the same time, the more nodes are involved the more risks there are of information degeneration (Premus, 1988). Direct linkages between firms and research centres stimulate a growing R & D efficiency. Therefore in science-based firms these contacts might play a fundamental role: this is why we encounter Science Parks, Silicon Valley and the Route 128 (Senker, 1985), and why the European and US laboratories of giant firms in the electronics and chemical industries are now migrating to Japan. However, the lack of internal R & D in a firm is not necessarily a sign of technological weakness. Firms can rely on external specialized centres or networking of R & D results.[8]

Second, each firm can support within their own R & D departments, the effort needed to advance in selected technological fields, which may or may not be provided with exclusivity and overall control of the fields. Control and exclusivity can be dependent on the system of protection of intellectual property, that is to say, to the different appropriability regimes, as described by Levin (1984): patents, industrial secrets, lead times, learning curves, marketing and services activities. It is important to observe here that firms are also confronted with more than one alternative: they can co-operate with other firms (upstream and downstream in the production cycle) to form 'strategic alliances' (Mowery, 1988), or subcontract external research projects, or even acquire small innovative firms in relevant areas of research (Granstrand and Sjolander, 1988). Third, firms may acquire technology knowledge (totally or, more frequent, partially) externally, through:

- 'reverse engineering' techniques (Pavitt, 1984);

- the purchase of licenses;

- technology embodied in new machinery - it can be a single item or as often occurs in Third World countries a 'turn-key' system (Bell, 1988);

- all forms of learning: these activities can be viewed as an inter-firm coordination among internal departments or as activities which bind the internal department with the external environment (see particularly Rosenberg, 1976);

- relationships with customers, suppliers, subcontractors (von Hippel, 1976, 1979, 1988; Lundvall, 1985, 1986; Freeman and Lundvall, 1988);

- recruitment/training activities of the workforce (Senker, 1986; Freeman, 1987).

Fourth, technical knowledge might also be developed in non-R & D departments. In spite of the difficulties related to the recording of these activities (Freeman, 1974; Pavitt et al, 1987; Kleinknecht, 1987; Santarelli, 1988), this appears to form an increasingly important modality of development of innovations, within manufacturing and could be defined as 'implicit R & D'. It is clear that 'implicit R & D' activities involve firms in continuous innovation, learning, reverse engineering, and incremental innovations.

The foregoing discussion suggests that formal R & D activity is not the only way to acquire productive knowledge, but one of the many complex options existing in a whole range of creative strategies (see also Kay, 1979; Cohen and Levinthal, 1988). Despite the fact that generally R & D expenditure is considered the fundamental tool for generating productive knowledge for firms, R & D can play another important role in enhancing the ability to assimilate and exploit existing information. R & D expenditure therefore should only be seen as indicative of a firm's information generating and assimilation efforts; it does not measure the total knowledge gained (externally, by cooperation, by imitation, by networking, etc.) by the firm, and as a result provides no direct link with the results obtained in terms of innovation and imitation capacity.

Indeed R & D performance depends on the degree of uncertainty and bounded rationality of firms, on their past technological experience (R & D performance is strongly path dependent: what you are going to do is related to what you are now doing) on learning ability, on R & D appropriability and opportunity. It has to be noted here that the level of technological opportunity shapes the range of firms' chances of gaining from R & D activity. In 'innovation producing' sectors (Scherer, 1982; Pavitt, 1984) R & D plays a fundamental role of advancing the frontiers of knowledge, whereas in 'innovation using' sectors, 'implicit R & D', incremental improvements, learning processes are perhaps the required skills of the firm.

To conclude: the analysis of the different knowledge producing and knowledge using regimes cannot be limited to R & D investigations. A large body of research on all issues raised above is required.

In regional economics and in geography there has been generally more attention paid to the **initial** and to the **final** steps of the 'innovative cycle': in other words to the issue of the spatial location of R & D (how concentrated? in which firms or sectors?) and to the feature of diffusion (Hagerstrand, 1967; Davies, 1979; Brown, 1981). However, competition among firms, areas, regions and nations is fundamentally marked by the 'maximum efficiency'[9] with which firms, areas, regions and nations are able to explore all three dimensions of the 'innovative cycle': invention, innovation and diffusion. In regional studies, a recent field of research has tried to bridge the gap between innovation related topics and diffusion studies.[10] The present study work takes up and tries to extend this stream of analysis. Using empirical data this chapter provides, within a regional context of analysis of Italian manufacturing industry, a detailed picture of the complex inter-dependency between the adaptation of exogenously derived technological change (i.e. innovation diffusion) and the internal innovative capability of firms (which is dependent upon firms' internal strength and 'in-house' innovative competence).

An empirical survey of Veneto industry was carried out through interviews using a semi-structured questionnaire addressed to entrepreneurs and managers. The data collected includes objective information, aspects referring to the firm's characteristics, and the opinions of the firm's leaders. The interviews focused both on the firms and on their competitive environment. A sample of 122 firms in nearly 300 lines of business, involving about 35,000 employees was identified. The interviewed firms represented nearly 10 per cent of all industry in the Veneto region in 1981 (firms with more than 50 employees). Firms selected in the sample are representative of the sectoral and size variability of the region's industry. Previous interviews regarding innovation diffusion and firms' innovative capability, etc., confirmed the representativeness of these data. For analytical purposes the sample and analyses are divided into traditional and non-traditional sectors. The basic dichotomy pivots around sectors with low technological content, based on staple products whose production techniques are well known and spread among countries, and sectors more 'advanced' with, generally speaking, a much higher technological content, whose development is more recent and can be traced starting from significant scientific and technological discoveries of the eighteenth century.

The remainder of this chapter attempts to place this survey in the context of the wider literature on technological change. The next section deals with the conventional indicator of R & D as a measure of innovative activity, while the following section offers an analysis of other sources and channels of technological advance. The chapter goes on to discuss the importance of innovation acquisition throughout the diffusion process and then throws light on the output indicators of innovative activity in firms. We discuss the importance of the adoption of organizational innovations and analyse the importance of non-technological factors for the firms' competitiveness. Finally, literal and empirical evidence is drawn together to form conclusions.

R & D As An Indicator of Innovative Activity

The development of the so-called peripheral areas or 'Third Italy' has been widely regarded as the major novelty of the Italian economy in the 1970s. However, despite the fact that growth indicators for the 1970s and for the first half of 1980s have shown a positive performance (Anastasia, 1985) by nearly all regions included in the peripheral belt which surrounds the older and more developed industrialized 'industrial triangle'[11], the conventional indicators of innovative activity such as R & D and patents have remained at a very low level (Belussi, 1985; Archibugi, 1987; Santarelli, 1988; Santarelli and Sterlacchini, 1988; Dosi and Moggi, 1986). At first glance this could be considered as 'proof' of the low innovativeness of firms and sectors located in such areas. This is not the opinion of the author of this chapter. On the contrary, the research presented below will try to show that innovativeness is high among firms in the Veneto region, but that the characteristics of sources, nature and determinants of innovation activity in such 'peripheral areas' goes largely unmeasured. The following discussion will be related to the features of different regimes of knowledge production and use, described above. The main point to emerge from the research is that innovation activity does not depend exclusively on internally generated R & D activity. This is especially true for traditional sectors. Indeed, traditional sector firms base their activity on staple products and processes, which until recently were not thought to require a marked participation in R & D activity. An additional measurement problem is revealed if we take into account the size distribution of the predominantly small firms belonging to these sectors. In such firms where the division of labour among the various productive functions is not so clear as in larger units a detailed account of all resources spent on innovative activity is impossible and as a result undervalued. The expectation is that technological advance in Veneto based manufacturing firms within traditional sectors will be low if traditional measures of innovation are used. As Table 15.1 illustrates, the empirical survey reveals a marginal, and sectorally varied presence of formal R & D activities in traditional industries,but a relatively high presence of 'implicit' R & D activities which the research revealed were linked to engineering and manufacturing departments, oriented both to design functions and to the improvement of existing equipment and machinery.

Table 15.1

Institutionalized R & D and 'implicit' R & D

	Presence of R & D %	Presence of 'Implicit' R & D %
Timber and wood products	7.7	76.9
Food	12.5	75.0
Textile	54.6	45.4
Clothing	29.4	52.9
Footwear	14.3	57.1
Non-ferrous metals	83.3	16.7
Total : Traditional sectors	**30.0**	**56.5**
Iron and Steel	80.0	20.0
Metal Goods and engineering	60.0	34.3
Chemical industry	50.0	25.0
Paper products and printing	20.0	40.0
Total : Non Traditional sector	**56.6**	**32.0**

The propensity towards formalized R & D activities is clearly less significant in traditional than in non-traditional sectors. We found a strong association between the presence of R & D functions, the level of R & D expenditure and the size of the firm. Data presented in Table 15.1 confirms the relevance of implicit R & D which we found in 80 per cent of the cases in which there was no formalized R & D activity recorded. In the majority of cases, the personnel charged with this kind of 'implicit'R & D were also directly involved in production functions, such as quality-control tasks and responsible for testing new products and new technology. In many cases, their real function was to form an interface between the technological needs of the firm and the supply of innovations from external sources. As we shall see further, in some firms it was the entrepreneur himself who undertook the majority of the innovative functions. The interviews revealed a consistent contribution to the innovative activity of the firm coming from production departments, as well as from technicians and some blue-collar workers. We also found some evidence that in small firms, where the situation was characterized by a less formal division of labour, the environment appeared much more receptive to capturing new ideas and suggestions stemming from the shop floor.

Sources and Channels of Innovative Activity

Respondents to the empirical survey were asked to indicate which were the primary sources determining the firm's innovative performance in products and processes. Table 15.2 shows the relatively marginal importance of the internal R & D factor as an initiator of innovations. Sources most frequently mentioned - both in traditional and in non-traditional sectors - were 'customers' stimuli' (chosen by 50.0 per cent of the firms in the sample), and 'suppliers' stimuli' (39.0 per cent of cases). Nevertheless market stimuli do not come in terms of intangible economic signals but embodied in 'concrete' forms. Consequently, innovative activity by the firms appears to be stimulated less by fluctuations of the trade cycle (as described by Schmookler, 1966), than by co-operative relations[12] among firms and among 'producers' and 'users' (according to von Hippel, 1976).

Other significant sources of innovations were endogenous to the firms, namely internal engineering departments (40.0 per cent) and their entrepreneurs (30.8 per cent). In contrast the purchasing of foreign patents and licences (8.3 per cent) play a much smaller role in overall innovative activity. In a certain number of firms there was evidence of 'reverse engineering' activity (15.8 per cent), but rarely were firms purely imitators, tending instead to include improvements in their designs. Finally, we observed that the activities of university and public research centres did not appear significant. Nearly all sectors commented on the absence of positive and continuing linkages between Italian Universities and industry. Within traditional sectors, the entrepreneur is very influential on innovative activity, whereas, in non-traditional sectors, R & D departments and engineering activities appear more important.

Innovation Acquisition Through Diffusion

The adoption of technological innovations

Conventional indicators of industrial innovation have tended to concentrate on the inventive/innovative phase while the effort here is to extend the analysis to include the subsequent diffusion process. Data in Table 15.3 refer to the product, material and process innovations introduced in the period 1974-84. Three main points are identified. First of all, it has to be noted that in all three types, the performance achieved by firms is substantial: in 89.3 per cent of the cases, firms have introduced process innovations, in 47.9 per cent technological and non-technological innovations in materials, and in 42.1 per cent technological innovations in products. Second, the data show a high degree of integration among the various types of innovations. Most firms have adopted more than one type of

Table 15.2
Sources of Firm's Innovations in Products and Processes (variables non mutually exclusive)

	Entrepreneur %	R&D %	Engineering %	Foreign patents %	Reverse engine-ering %	Stimuli of sup-pliers %	Stimuli of cust-omers %	University %	Other %	no.
Timber and wood products	53.8	-	46.2	15.4	7.7	61.5	53.8	-	-	(13)
Food	37.5	25.0	-	-	12.5	50.0	50.0	12.5	-	(8)
Textile	20.0	20.0	30.0	-	20.0	40.0	70.0	-	10.0	(10)
Clothing	41.2	23.5	52.9	5.9	17.6	35.5	35.3	-	11.8	(17)
Footwear	50.0	-	28.6	-	21.4	28.6	50.0	-	-	(14)
Non-ferrous metals	16.7	16.7	33.3	-	16.7	50.0	16.7	-	-	(6)
Total : Traditional sectors	**39.7**	**13.2**	**35.3**	**4.4**	**16.2**	**42.6**	**47.1**	**1.5**	**4.4**	**(68)**
Iron and steel	20.0	40.0	40.0	20.0	20.0	40.0	60.0	20.0	-	(5)
Metal goods and engineering	20.6	41.2	55.9	8.8	17.6	23.5	58.8	8.8	2.9	(34)
Chemical industry	12.5	50.0	37.5	25.0	12.5	37.5	50.0	12.5	-	(8)
Paper products and printing	20.0	-	-	20.0	-	100.0	20.0	20.0	-	(5)
Total : Non traditional sectors	**19.2**	**38.5**	**46.1**	**11.5**	**15.4**	**34.6**	**53.8**	**11.5**	**7.6**	**(52)**

Table 15.3
Firms' Innovative Activity During the Period 1974-84

	Product innovations	New materials		Process innovations		
		Non High-tech	High-tech	Non High-tech	High-tech	Total
	%	%	%	%	%	no.
Timber and wood products	30.8	46.2	-	92.3	46.2	(13)
Food	57.1	14.3	-	87.5	14.3	(8)
Textile	18.2	18.2	27.3	100.0	45.5	(11)
Clothing	35.3	11.8	11.8	82.4	82.4	(17)
Footwear	14.3	14.3	35.7	71.4	50.0	(14)
Non-ferrous metals	50.0	33.3	33.3	66.6	100.0	(6)
Total : Traditional sector	**30.9**	**21.7**	**15.9**	**86.9**	**44.9**	**(69)**
Iron and steel	60.0	40.6	20.0	80.0	60.0	(5)
Metal goods and engineering	65.7	14.3	42.9	94.3	57.2	(35)
Chemical industry	50.0	25.0	50.0	100.0	75.0	(8)
Paper products and printing	-	-	40.0	80.0	40.0	(5)
Total : Non traditional sectors	**56.6**	**17.0**	**43.4**	**92.5**	**50.9**	**(53)**

innovation: process innovations associated with materials innovation appears as the combination most often represented in the sample (45.4 per cent), together with the combination of product and process innovations (40.5 per cent). In about one-quarter of analysed firms one can observe the presence of all three types of innovations. Third, although traditional sectors score quite well in the adoption of all kinds of innovations, the data show that there are differences between traditional and non-traditional sectors: the former introduced fewer than average technological product innovations, but average amounts of process innovations and new materials. The latter introduced a larger number of types of innovation.

Inter-firm diffusion

To understand the extent of innovative activity it would be inappropriate to limit our analysis to the simple presence of absence of a specific or certain range of innovations. Indeed, what affects the firm's performance is much more the extent of changes directly or indirectly produced by the introduction of innovations. For instance, Mansfield (1988), studying the diffusion of robotics in Japan and in the United States, found that the most striking difference between the two countries was not the inter-firm but the intra-firm diffusion rate. The gap in productivity between the two countries can perhaps be better explained using the intra-firm rather than the interfirm diffusion variable. To explore this point further, data were collected in Veneto on the proportion of total output which in the period 1980-84 was obtained with new or radically changed machinery. With varying intensity (see Table 15.4), firms appear to have been strongly engaged in the modernization and restructuring of the production process. It is interesting to observe that the pattern followed by traditional and non-traditional sectors is more or less the same; conversely, small firms (those with less than 250) employees perform much better than large firms (those with more than 250 employees) in their modernization and restructuring of the production process.

The timing of diffusion

In addition to the adoption of any technique is the **rate** at which firms have incorporated technological progress into their production processes. For the same innovations discussed above the survey attempted to record the time-lag of adoption of each firm from the first Italian adopter of the specific technology. Data collected show vividly the existence of a short time-lag for those innovations which have been introduced.

Time-lags of less than a year were declared by 76 per cent of firms regarding product innovation, by 68.8 per cent regarding process innovation, and by 89.3 per cent regarding the utilization of new materials. In many cases we found that

Table 15.4

Firms' Output Realized with New or Radically Modified Equipment and Machinery in the Period 1981-84

Proportion of Output	0-10% %	11-20% %	21-30% %	31-50% %	More than 50% %	total %	no.
Traditional sectors	27.0	9.5	7.9	15.9	39.7	100.0	(63)
Non traditional sectors	19.6	13.7	15.7	17.7	33.3	100.0	(51)
Small firms (up to 250 employees)	21.0	11.1	11.1	12.3	44.4	100.0	(81)
Large firms (250 plus)	30.3	12.1	12.1	27.3	18.2	100.0	(33)

innovation acquisition was not just a passive adoption. Knowledge about innovations was not generated inside the firm but collected from external sources. However, firms were able to implement these innovations in a 'creative context' to meet the firms' specific needs. The ability to look for and discover profitable innovations was linked to the general high level of knowledge about technology which we found existing in each firm. More interesting, we found that about 20 per cent of firms (more than 35 per cent in the case of mechanical engineering firms) were at the same time partially producing their process machinery in-house. In all industrial sectors, and in both traditional and non-traditional firms, the time-lags in adoption appear to be relatively short. These results differ from many existing empirical studies (Freeman, 1982; Nabseth and Ray, 1974; Ray, 1984), which have underlined the long duration of the process of innovation and diffusion. The explanation may lie in the fact that within the survey sample many more incremental changes are recorded than perhaps is the case in earlier studies. The previous studies also cover periods before the widespread use of information technology which in the 1980s may have both increased the stock of knowledge of appropriate technologies and relevant competition and the pace with which knowledge is disseminated.[13] However, the data show that there still exist important time-lags in the diffusion of process innovations. This result is consistent with mainstream innovation literature which points out the existence of various barriers to the adoption of some new technologies, including finance, tacit knowledge, technological cumulativeness, and industrial secrecy (Pavitt, 1984; Dosi, 1984; Nelson and Winter, 1982; Levin et al, 1984).

Innovation diffusion of microelectronics technology: the path of firms towards automation

Despite the difficulty of collecting, interpreting and comparing data on the use of microelectronic technology in manufacturing industry, the level and pace of diffusion of these types of technology in firms represents an interesting proxy for innovative capacity of the firm more generally. Data displayed in Table 15.5, referring to innovations previously adopted and in use before 1987, show the extent to which these processes have been adopted by the firms in the sample. They reveal a surprisingly high level of introduction of numerous forms of micro-electronic based technology in both traditional and non-traditional sectors. Nearly half of the sample has introduced numerical control machinery in the productive process (on the contrary only in 21.0 per cent of cases were more systemic forms of production process discovered in the form of centralized numerical control machinery). In one third of the sample we found automated process control and a similar result emerges for quality control technology and for the introduction of robots. Smaller percentages still were observed for more complex technologies, such as CAD-CAM (18.9 per cent), automation of lay-out (16.4 per cent), and automation of warehousing functions (13.1 per cent). Chemical, metal goods and mechanical engineering, paper products and printing, and iron and steel appear to lead in the implementation of automation technologies. Nevertheless, even among firms in traditional sectors one can note the presence of some innovative 'peaks': firms in the clothing industry in CAD-CAM technology; firms in the timber and wood products industry in the automation of lay-out; firms in the textile industry in numerical control machinery; and firms in the food industry in quality control technology. The data also suggest that size of firm may be a more satisfactory explanation of the level of adoption of technology than is sector. The number of applications of each technology by firms also indicates a similar pattern of sectoral and size variation.

The Output Indicators of Innovative Activity

Patenting activity

Patenting activity has been widely used by academics as a significant indicator of innovative activity (e.g. Pavitt, 1985). However, in traditional sectors where the nature and pace of technological advance may be relatively slow this indicator may not be appropriate.

The picture presented in Table 15.6 is quite striking: in the sample, only 25 per cent of the firms belonging to traditional sectors have patented in the period; by contrast, in non-traditional sectors, this percentage rises to 58.5 per cent.

Table 15.5
Adoption of Microelectronics Technology by Type of Product and Process - User Firms

	Cad-cam %	Process control %	Single m.t. %	Centralised m.t. %	Robots %	Automation of warehousing %	Automation of layout %	Quality control %	no.
Timber and wood products	7.7	-	46.2	7.7	30.8	-	30.8	7.7	(13)
Food	12.5	37.5	62.5	25.0	25.0	25.0	12.5	50.0	(8)
Textile	27.3	54.5	72.7	18.2	27.3	-	18.2	36.4	(11)
Clothing	35.3	35.3	35.6	17.6	17.6	11.8	17.6	5.9	(17)
Footwear	7.1	7.1	21.4	14.3	7.1	7.1	7.1	14.3	(14)
Non-ferrous metals	-	16.7	66.7	16.7	50.3	16.7	33.3	33.3	(6)
Total : Traditional Sectors	**17.4**	**24.6**	**46.4**	**15.9**	**27.5**	**7.2**	**18.8**	**20.3**	**(69)**
Iron and steel	-	80.0	60.0	40.0	40.0	20.0	40.0	60.0	(5)
Metal goods	25.7	42.9	57.1	25.7	42.9	17.1	8.6	42.9	(35)
Chemical industry	25.0	75.0	37.5	25.0	25.0	37.5	25.0	62.5	(8)
Paper products and printing	-	20.0	60.0	40.0	60.0	-	-	20.0	(5)
Total : Non Traditional Sectors	**20.7**	**49.1**	**54.7**	**28.3**	**35.8**	**20.7**	**13.2**	**45.3**	**(53)**
Small firms (up to 250 employees)	12.5	21.6	48.9	13.6	26.1	4.5	11.4	19.3	(88)
Large firms (250 plus)	35.5	70.6	50.0	41.2	44.1	35.3	29.4	61.8	(34)
TOTAL	**18.9**	**35.2**	**50.0**	**21.3**	**31.1**	**13.1**	**16.4**	**31.1**	**(122)**

When considering the number of patents recorded, the survey clearly reveals that in specialized supplier firms (essentially mechanical engineering firms) patenting activities play a more pronounced role than in some of the other sectors. The data also show that as firm size increases so does the propensity to patent. The previous debate and analysis clearly suggests that patents as a measure of innovative activity tends to underestimate the technological performance of firms, particularly those manufacturing in traditional sectors.

Table 15.6

Patenting Activity by Sector and Firm Size

	Presence of Patenting % of firms
Sector	
Traditional	25.0
Non-Traditional	58.5
Size	
1 - 100 employees	26.3
101 - 250 employees	40.9
251 - 500 employees	50.0
500 plus employees	60.9

Product redesign

In the literature product innovation is often regarded as an homogeneous concept, whereas in reality it describes a spectrum of activities ranging from fundamental technological change to minor modification and customisation. Taking the broadest view of product innovation surveyed firms were requested to provide information on their performance of the past two years. The results show that 80 per cent of the sample had introduced at least one or two new products, 60 per cent of firms introduced between three and ten, and 35.6 per cent had introduced 11 or more new products over the period in question. In the higher category are several firms belonging to the fashion sectors (clothing and footwear). About 50 per cent of firms in these two sectors have introduced more than 100 new products. In general, small firms show a performance correlated with the average of the sample.

Focusing on the qualitative aspects of these phenomena, only a small number of the firms (mainly belonging to mechanical and engineering sectors) resorted to forms of product innovation involving radical changes in the product (17.1 per cent) and accounted for less than 10 per cent of current total output in most firms. The majority of the firms in our sample (as seen in Table 15.3) pursued a 'soft' trajectory of developing incremental innovations based on product redesign.[14] By way of contrast, 88.4 per cent of the firms had introduced incremental product innovations, and for nearly half of these firms such innovations currently make up more than 50 per cent of total output. The performance of traditional sector firms is on average very high; indeed reaching over 90 per cent in clothing and footwear. Incremental change seems more important to the small, rather than the large, firms where more than 50 per cent of total output results from such innovation.

The Adoption of Organizational Innovations

Despite some historical reconstruction to provide us with a record of managerial innovations (Chandler, 1962), there has been very little attention directed to the role played by the adoption of organizational innovations in the competitiveness of individual firms or of industrial districts. It can be hypothesised that consistent growth of firm competitiveness requires constant adjustment to take advantage of new organizational techniques and new forms of management.[15] It is now widely recognized that these particular influences on productivity growth have had an important role to play in trends and changes within trade cycles. Firms in the survey were therefore asked to indicate which organizational changes were most linked with the growth of competitiveness. We found a confirmation of the relevance of organizational behaviour depending on the changes of work organization and induced by the introduction of innovations.

In the case of textile firms and, above all, of clothing firms, great importance has been ascribed to the so-called decentralization of production. Many other features have been emphasised during the interviews, such as: the adoption by the firms of a form of work organization by phases; the tendency towards a more flexible production cycle; the focusing towards dynamic scale economies of product and not of plant; the emerging of a pattern of specialization[16]; and the expansion of external subcontracting networks. These changes were strongly 'capital-labour-energy-space-time' saving.

We also tested the level of automation of firms' tertiary activities. The number of firms in which these functions are managed with the aid of a computer is consistently small. Access to information technology appears limited for many traditional sector firms, above all for the firms in timber and wood products and footwear, and in general, for smaller firms with less than 100 employees. This is consistent with findings in the UK (Foord and Gillespie, 1988).

A consistent pattern of organizational innovation in distribution was also found in the sample : nearly 40 per cent of the firms have recently set up new commercial organizations abroad, 34.4 per cent have changed channels of distribution and sales techniques, and 23.0 per cent have changed methods and structures of after-sales assistance. In addition about one third of firms have strengthened their market and advertising services. Moreover, about one quarter of firms have developed technical assistance, training, and R & D activities; about one third of firms have strengthened maintenance, and more than one half of the firms in the sample have strengthened quality control and office automation.

The Relevance of Technological and Non-Technological Factors

As the research evolved it became increasingly clear that the simple analysis of innovative processes and the measurement of inventive/innovative activity in isolation were inadequate means of identifying the real sources and causes of firm competitiveness. An adequate explanation must cover the entire activity of the firm, where a number of variables seem to influence the success and failure of firms in the market place. Entrepreneurs and managers were therefore asked to evaluate the role of technological change within a broader set of factors which were likely to affect the competitiveness of the firm. Data presented in Table 15.7 illustrate the complexity of understanding individual firm performance. The relevance of technological factors is recognised within the sample (19.6 per cent), but non-technological factors are of comparable importance. Therefore, competitiveness based on technology does not seem the dominant factor. While technological factors are more important in some non-traditional sectors, in contrast, product innovation seems more crucial in the fashion sectors, and chemical and mechanical engineering industries display an above average role of innovation in distribution. The survey findings appear to confirm the linkages between a firm's performance and a complex strategy of multiple innovative factors.

Conclusion

The results of the empirical survey suggest that the traditional indicators of innovative activity are an inadequate means of describing the inventive/innovative process and, above all, the innovative/diffusive process. Such indicators of the input and output of innovative activity would suggest the technological backwardness of the majority of firms analysed. However, the more detailed evidence collected in the survey suggest a much greater involvement in innovative activity than could be expected on the basis of R & D expenditures, R & D employment and patents.[17] For that reason, additional indicators on which to judge technological capability have been proposed and tested in the field.

These include: the firms' capacity for adopting product-process-new material innovations[18], the time-lags of diffusion, the spread of micro-electronics in production and in the firms' coordination functions, the diffusion of organizational innovations, and the presence of product redesign.

Table 15.7
Evaluation of Innovation Factors on their Influence on Firm's Competitiveness - average score

	Technological innovations %	Product innovations non tech %	Organizational innovations in production %	Organizational innovations in admin. methods %	Organizational innovations in distribution %	Total Score obtained by the item %
Timber and wood products	20.5	26.9	21.8	14.1	16.7	100.0
Food	26.1	21.7	21.7	17.4	13.1	100.0
Textile	13.9	16.4	25.3	24.0	22.2	100.0
Clothing	15.2	34.3	20.9	14.3	15.3	100.0
Footwear	9.2	32.9	19.7	15.8	22.4	100.0
Non-ferrous metals	17.8	11.1	26.7	28.8	15.5	100.0
Iron and steel	29.6	-	27.0	14.8	18.6	100.0
Metal goods and eng.	21.5	13.6	20.4	18.5	26.0	100.0
Chemical industry	27.1	6.8	16.9	22.0	27.2	100.0
Paper products and printing	24.2	12.2	21.2	21.2	21.2	100.0
TOTAL	**19.6**	**18.2**	**21.8**	**18.6**	**21.1**	**100.0**

The survey shows that the firms, particularly the traditional firms, devote considerable resources to incremental improvements (which can be represented, in part, by 'implicit' R & D activity) of adopted innovations. The results suggest that technology is **firm specific**, **cumulative**, and **tacit** (Pavitt, 1984, 1986), but in a softer way, in comparison with high-tech firms (Dosi, 1984). Intense **learning** activity, and some **technological implementation** of innovations, characterize the firms' technological performance and the competitive relations amongst firms in the market. We found that regarding both products and processes, the interactions between firms and suppliers on one hand, and firms and customers on the other, were the most important source of innovation. Following issues of forward and backward linkages with producers the importance of internal engineering activities were mentioned, as well as the creative intervention of the entrepreneur. Therefore, internal sources of innovations were indicated as being less prominent than external ones, but may have been of greater relevance in later stages of the incorporating process.

The average level of innovation diffusion in the last decade has displayed an **acceleration** of the innovative cycle, i.e. a substantial reduction of the time lag of adoption.[19] This tendency was related much more to the diffusion of new products and new materials, whereas there remain long time lags regarding process innovations. Our results do not contradict, and perhaps strengthen, those questions of appropriability of technology which have been discussed by recent economic literature (Nelson and Winter, 1982; Rosenberg, 1976, 1982; Pavitt, 1986; Levin et al, 1984; Dosi, 1984).

The data presented here suggest that in contrast with the predominant view (see, for instance, Arcangeli, Dosi and Moggi, 1987), Italian small and traditional firms are not always under-automated and under-informaticized. Thus, the specificity of traditional firms which leads to high performance in terms of economic indicators[20] does not only relate to a well-matched set of corporate strategies, work organization modes, institutional conditions etc., but also seems to include some technological capabilities. Moreover, their adoption of microelectronic technology has been rapid by the standards of other sectors and of other countries such as Britain (see for instance Northcott, 1986).

Regarding traditional sectors, the relevance of embodied technical change, as well as the autonomous capacity for generating incremental innovations in products and processes, has been highlighted. Considering all indicators analysed, traditional firms show above all a particular ability to generate innovations in the design sphere, that is, in non-technological product innovations.

Organizational innovations have played an important role in the last decade in terms of increasing firms' competitiveness. Organizational innovations in production, distribution, and in administration were estimated by those interviewed as being as important as technological innovations in the competitive process. In exploring the sources and the causes of firm competitiveness, we found that technological factors, though relevant, were only of equal importance to non-technological factors.

Finally, regarding size, our survey has revealed a clear association between large firms and the presence of R & D activities. In other dimensions of innovative activity, technological adoption lags have not proven relevant in differentiating traditional and other sectors at all, such as in the case of equipment and machinery restructuring. Small firms have been found to be more innovative in products; however, in the adoption of complex and integrated microelectronics technologies, they have lagged well behind larger firms.

Acknowledgement

I would like to acknowledge Professor C Freeman, Professor K Pavitt, Dr N von Tunzelmann, Dr N Clark, Professor R Rothwell, Dr P Senker and Dr M Dodgson, for their helpful comments, and also Professor G Dosi for his suggestions during the preparation of the questionnaire.

Notes

1. The theoretical framework was defined during a visiting period at SPRU in 1985 and further refined later. The empirical evidence is provided from the results of a preliminary research project organized by Ires Veneto and Daest (Dipartimento di analisi economica e sociale del territorio dell'Istituto Universitario di Venezia) and was carried out by the author. A representative sample of firms was selected based upon the sector and size of the industrial structure of the Veneto region of Italy.

2. The shape, in 20 years of literature, has been found to be influenced by many disparate elements. These were: perceived attributes of innovations, type of innovations decision, communication channels, extent of effort by change agents, nature of the social system (Rogers, 1982); profitability of innovation, level of profit expectation for the adopting firm, financial record of adopting firms, firm size (in Mansfield's pioneer analyses: 1961, 1963 and 1968); profitability, attitudes of management, and availability of credit (in Nabseth and Ray, 1974; and Ray, 1984); the comparison between the costs of using the old and new technology (in Salter, 1960); the size (or the 'threshold' effect) of firms (in David (1975) examples of which in his historical analysis were farms adopting mechanical reapers); the condition of supply (in Metcalfe, 1981); the environment variability (in Gold, 1981). For discussion see Stoneman (1985), Mahajan and Peterson, (1985), and von Tunzelmann (1988).

3. For example, some improvement can dramatically widen the intersectoral applicability of a technology (the 'pervasive effects' of technical change in the economy as a whole have been addressed by Freeman and Soete (1985, 1987), Freeman and Perez (1986) and by Perez (1985)) whereas, in contrast, some technologies are not fully applicable given their poor performance in terms of reliability, product quality, costs, etc.

4. See Dodgson's introductory chapter to Technology Strategy, Firm Management and Public Policy, a volume which contains many detailed studies on corporate strategy.

5. This process seems particularly important to explain the path of development of less industrialized countries (see on this point Rosenberg and Frischtak, 1985) and for the establishment of new international technological leaderships (see for instance: Freeman, 1987; Katz, 1987; Fransman and King, 1984; Malecki, 1988; and Dosi, 1988b). For a review which tackles the issues of 'long-run' economics see also Clark and Juma (1987).

6. Within a regional analysis context, this concept has been discussed extensively by Malecki (1988).

7. Many books and articles provided a good survey on this topic, particularly, Freeman (1982); Scherer (1982); Rosegger (1980); Coombs, Saviotti and Walsh (1987); Malerba (1985); Thwaites and Alderman (1988).

8. For instance, this is the case of two particularly successful firms selected in the example of our survey - Fidia Spa and Glaxo Spa. Fidia, a small-medium firm in the pharmaceutical sector recently opened up a department of research at Georgetown University in Washington, but it does not perform R & D in its headquarter in Padua. The case of Glaxo is common to all multinational enterprises. Glaxo, which is a renowned firm in the same industrial field as Fidia has a strong technological leadership in Europe (Pavitt and Patel, 1988): the R & D department in Verona is not representative at all of the total R & D efforts undertaken by the firm.

9. The 'effort of growth' must be analysed in a non static context of resource allocation. See on this topic Klein (1977), Dosi (1984), Freeman (1987), Nelson (1988), Landau and Rosenberg (1986).

10. For instance: Malecki, 1977, 1983 and 1988; Gillespie (1983); Camagni, 1985; Thwaites and Oakey (1985); Gold (1979) and Knaap and Wever (1987).

11. Conventionally, the so-called 'industrialized triangle' covers the territory of Lombardy, Piemonte and Liguria, linking together three main industrial cities: Milan, Turin and Genoa.

12. On this topic see also Aoki (1986 and 1988), Dore (1987), and Sako (1988).

13. This survey has not measured (directly or indirectly) the impact of the social context in accelerating the pace of the diffusion rate. However, other research focusing on the analysis of the 'local factors' of development (see for instance Bagnasco and Trigiglia, 1984) has elucidated for the Veneto case, the high levels of 'interactions' and 'socialization' existing among entrepreneurs, and between the entrepreneurial class and the local authorities (local government, banks, and collective organizations). The hypothesis that much information could be diffused 'informally' through the interactions of this highly cohesive group would deserve to be examined closely. My thanks to M. Dodgson for a stimulating discussion on this point.

14. The importance of redesign innovations have been stressed for a long time by the innovation literature. See for instance Rothwell and Gardiner (1983), Gardiner (1985) and Langdon and Rothwell (1985).

15. This point raises the issue of a transition from Fordist production techniques to new (post-Fordist) ways of organizing production. See Roobeek (1987), Aglietta (1979), Perez (1985), Coriat (1980), Sayer (1986), Murray (1987), Murray (1987), Moden (1983). Schonberger (1982) and Lazerson (1988). This issue is discussed in Belussi (1987 and 1988).

16. As noted by Leborgne and Lipietz (1988) as regards the influence of new technology on firms' organization 'the optimal management of integrated modular processes is likely to entail more vertical integration, but flexible automation and computerized management of flows lead to new possibilities in vertical disintegration. The new "fix" appears to be the "specialized firm", producing a restricted scope of differentiated goods (final or intermediate), with an optimal management of quality, innovation, and time-saving' (p.20).

17. These results are consistent with many other surveys. For instance, the fact that innovative activity in small firms tends to be strongly under estimated has been recently stressed by Kleinknecht (1987) and Pavitt (1986).

18. These results are consistent with the research organized by Istat-Cnr (1986) which found a consistent level of innovation diffusion in all regions characterized by peripheral development. For a further discussion see Santarelli and Sterlacchini (1988).

19. On this topic see also Alderman, Davies and Thwaites (1988) whose recent work indicates - on a wide national basis (UK) - that for innovations such as computers and CNC the rate of first time adoption is now falling.

20. For an analysis of firms' economic performance see Belussi (1988).

References

Abernathy, W. et al (1983) Industrial Renaissance: Producing a Competitive Future for America, Basic Books, New York.

Aglietta, M. (1979) A Theory of Capitalistic regulation : the US Experience, London NLB.

Anderson, A.E. (1985)Creativity and Regional Development, Papers of the Regional Science Association, 56, 5-20.

Aoki, M. (1984) The Co-operative Game Theory of the Firm, Oxford University Press, Oxford.

Aoki, M. (1986) Horizontal vs Vertical Information Structure of the Firm. American Economic Review, 76. n(5), 971-983.

Alderman, N., Davies, S. and Thwaites, A.T. (1988) Patterns of Innovation Diffusion, CURDS, University of Newcastle-upon-Tyne.

Amendola, M. and Gaffard, J.L. (1988) The Innovative Choice, Blackwell, London.

Anastasia, B. (1985) L'evoluzione dei divari di produttivita tra le regioni italiane (1973-1982). Archivio di studi urbani e regionali, 24.

Arcangeli, F., Dosi, G. and Moggi, M. (1987) Patterns of Diffusion of Electronics Technologies, DRC Discussion Paper No.44.

Arcangeli, F. and Camagni, R. (1988) The flexible automation trajectory in time and space, paper presented at the European Summer institute of the Regional Science Association 'Theories and Policies of Technological Development at Local Level', Arco, Italy.

Archibugi, D. (1987) La struttura settoriale dell'innovazione industriale in Italia. Un'analisi dei brevetti italiani negli Stati Uniti, Rivista di Politica Economica, 2, 131-176.

Arrow, K. (1962), The Economic Implications of Learning by Doing, Review of

Economic Studies, 29, 80, 155-173.

Baba, Y. (1988), Systemic Innovation: its Nature and How to Benefit From It. Mimeo, SPRU.

Bagnasco, A. and Trigiglia, C. (eds.) (1984) Società e Politica nelle Aree di Piccola Impresa. Il caso di Bassano, Arsenale editore, Venezia.

Belussi, F. (1985) Localizzazione delle attivita di R&S., del servizi alle imprese, delle reti informatiche e delle attivita brevetuali delle imprese. Archivio di Studi Urbani e Regionali, 23, 81-152.

Belussi, F. (1987) Benetton: Information Technology in Production and Distribution. A case Study of the Innovative Potential of Traditional Sectors Occasional Paper, 25, SPRU, University of Sussex.

Belussi, F. (ed.) (1988) La diffusione delle innovazioni nei settori tradizionali, Milano, Franco Angeli.

Bell, R.M. (1988) The Role of Oil Companies in Technological Innovation in the Offshore Oil Industry, Mimeo, SPRU.

Brown, L.A. (1981) Innovation Diffusion: a New Perspective, Methuen, New York.

Brusco, S. (1982) The Emilian Model: Productive Decentralization and Social Integration, Cambridge Journal of Economics, 2, 167-184.

Brusco, S. (1986) Small Firms and Industrial Districts: the Experience of Italy, in M. Bassand et al, Self Reliant Development in Europe, Gower, Aldershot.

Burns, T. and G.M. Stalker (1961), The Management of Innovation, Tavistock Pub., London.

Camagni, R. (1984) Il Robot Italiano, ed. II Sole 24 Ore, Milano.

Camagni, R. (1985) Spatial Diffusion of Pervasive Process Innovations, Papers of the Regional Science Association, 58, 83-95.

Cainarca, G.C., Colombo, M.G. and Mariotti, S. (1987) Innovazione e diffusione: il caso dell'automazione flessibile, L'Industria, 4.

Chandler, A. (1962) Strategy and Structure, MIT Press, Cambridge.

Cicciotti. E. (1982) Differenze regionali nella capacità innovativa, Politica ed Economia, 3.

Clark, N. and Juma, C. (1987) Long Run Economics: An Evolutionary Approach to Economic Growth, Pinter Publishers, London.

Choen, S.S. and Zysman, J. (1987) Manufacturing Matters: the Myth of the Post-industrial Economy, Basic Books, New York.

Cohen, W.M. and Levinthal, D.A. (1988) Innovation and Learning: The Two Faces of R&D. Implications for the Analysis of R&D Investment, mimeo. Carnegie-Mellon University, Pittsburg, Pennsylvania.

Coombs, R., Saviotti, P. and Walsh, V. (1987) Economics and Technical Change, Macmillan, London.

Coriat, B. (1980) The Restructuring of the Assembly-line: A New Economy of Time and Control of Labour, Capital and Class, 11, 34-43.

David, P. (1975) Technical Choice Innovation and Economic Growth, Cambridge University Press, Cambridge.

Davies, S. (1979) The Diffusion of Process Innovations, Cambridge University Press, Cambridge.

Dodgson, M. (ed.) (1989) Technological Strategy and the Firm, Management and Public Policy, London, Longman.

Dore, R. (1986) Flexible Rigidities, The Atheone Press, London.

Dosi, G. (1982) Technological Paradigms and Technological Trajectories: A Suggested Interpretation of the Determinants and Directions of Technical Change, Research Policy, 11, (3), 147-162.

Dosi, G. (1984) Technical Change and Industrial Transformation, Macmillan, London.

Dosi, G. (1988a) The nature of Innovative Process, in G. Dosi et al. (ed.), Technical Change and Economic Theory, Frances Pinter, London.

Dosi, G., Orsenigo, G. and Silverberg, G. (1986) Innovation, Diversity and Diffusion: a Self-organization Model, paper presented at the Venice Conference on Innovation Diffusion.

Dosi, G, and Moggi, F. (1986), La capacita innovativa italiana: alcuni confronti internazionali, inter-settoriali e inter-regionali, in F. Onida (ed.), Vincolo Estero, Struttura industriale e Credito all'Esportazione, Il Mulino, Bologna.

Fransman, M. and King, K. (eds.) (1984) Technological Capability in the Third World, St Martin Press, New York.

Freeman, C. (1982) The Economics of industrial innovation, Frances Pinter, London.

Freeman, C. (1987) Technology Policy and Economic Performance, Frances Pinter, London.

Freeman, C. (1988) The Third Kondratieff wave: age of Steel, Electrification and imperialism, mimeo, SPRU, University of Sussex.

Freeman, C., Clark, J. and Soete, L. (1982) Unemployment and Technical Innovation, Frances Pinter, London.

Freeman, C. and Soete, L. (1985) Technological unemployment, mimeo, SPRU, University of Sussex.

Freeman, C. and Soete, L. (1987) Technological Change and Full Employment, Basil Blackwell, Oxford.

Freeman, C. and Perez, C. (1986) The Diffusion of Technical Innovations and Changes of Techno-economic Paradigm, paper presented at the Venice Conference on Innovation Diffusion.

Freeman, C. and Lundvall, B.A. (eds.) (1988) Small Countries Facing the Technological Revolution, Frances Pinter, London.

Friedman, D. (1988) The Misunderstood Miracle, Industrial Development Political Change in Japan, Cornell University Press, Ithaca.

Gardiner, P. (1985) Invention, Innovation, Re-innovation and the Role of the User: a case Study of British Hovercraft Development. Technovation, 3, 167-186.

Garofoli, G. (1988) Local Patterns of Development: the Systems of Small Firms, mimeo, University of Pavia.

Georghiou, L. et al. (1986) Post-innovation Performance, Macmillan, London.

Grandstrand, O. and Sjolander, S. (1988) The Acquisition of Technology and Small Firms by Large Firms, paper presented at the Conference 'Markets for Innovation. Ownership and Control', Stockholm.

Gold, B. (1979) Productivity, Technology, and Capital, Lexington Books, Lexington MA..

Gold, B. (1981) Technological Diffusion in Industry: Research Needs and Shortcomings, Journal of Industrial Economics, 29 (3), 247-269.

Hagerstrand, T. (1967) Innovation Diffusion as a Spatial Process, The University of Chicago Press, London.

von Hippel, E. (1976) The Dominant Role of User in the Scientific Instrument Innovation Process, Research Policy, 5, 212-39.

von Hippel, E. (1979) A Customer-active Paradigm for Industrial Product idea Generation, in J. Baker (ed.) Technology, Policy, Diffusion, Macmillan, London.

von Hippel, E. (1988) The Sources of Innovation, Oxford University Press, Oxford.

Hout, T. et al. (1982) How Global Companies Win Out, Harvard Business Review, 60, (5), 98-108.

Istat-Cnr (1986) Indagine suila diffusione delle innovazioni technologiche nell'industria manifatturiera italiana, Supplemento al Bollettino Mensile di Statistica.

Kamann, D.F and Nijkamp, P. (1988) Technologensis: Incubation and Diffusion, paper presented at the European Summer Institute of the Regional Science, Arco, Italy.

Kay, N.M. (1979) The Innovating Firm, Macmillan, London.

Kaplinsky, R. (1984) Automation, Longman, London.

Katz, J.M. (ed.) (1987) Technology Generation in Latin American Manufacturing Industries, Macmillan, London.

Klein, B.H. (1970) Dynamics Economics, Cambridge University Press, Cambridge Ma.

Kleinknecht, A. (1987) Measuring R & D in Small Firms: How Much are we Missing? Journal of Industrial Economics, 37(2), 253-256.

Knaap, B.D. and Wever, E. (eds.) (1987) New Technology and Regional Development, Croom Helm, London.

Kodoma, F. (1986) Japanese Innovation in Macatronics Technology, Science and Public Policy, 13 (1), 44-51.

Landau, R. and Rosenberg, N. (1986) The Positive Sum Strategy, National Academic Press, Washington D.C.

Langdon, R. and Rothwell, R. (eds.) (1985) Design and Innovation, Frances Pinter, London.

Langlois, R.N. (1987) Disequilibrium, Path Dependency, and Organizational Form, paper presented at the Conference at Case Western Reserve University.

Lazerson, M.H. (1988) Organizational Growth of Small Firm: an Outcome of Markets and Hierarchies? American Sociological Review, 53, 330-42.

Leborgne, D. and Lipietz, A. (1987) New Technology, New Modes of Regulation: some Spatial Implications, Mimeo, Paris, Cepremap.

Levin, R.C. et al. (1984) Survey Research on R & D Appropriability and Technological Opportunity, Mimeo, Yale University.

Lundvall, B.A. (1985) Product Innovation and User-Producers Interaction, Aalborg University Press, Aalborg.

Lundvall, B.A. (1986) Innovation as an Interactive Process from User-Producer Interaction to the National System of Innovation, IFIAS paper, SPRU, University of Sussex.

Mahajan, V. and Peterson, R.A. (1985) Models for Innovation Diffusion, SAGE Pub, Beverly Hills.

Maidique, M.A. and Zirger, B.J. (1985) The New Product Learning Cycle, Research Policy, 14, 299-313.

Malecki, E.J. (1977) Firms and Innovation Diffusion: Examples from Banking, Environment and Planning A, 9, 211-223.

Malecki, E.J. (1983) Technology and Regional Development: A Survey, International Regional Science Review, vol (2), 89-125.

Malecki, E.J. (1988) Research and Development and Technological Transfer in Economic development: The Role of Regional Capability, paper presented at the European Summer Institute of the Regional Science Association, Arco. Italy.

Malerba, F. (1985) Mutamento tecnologico, imprese e mercato, in F. Onida (ed.) quoted.

Mansfield, E. (1961) Technical Change and the Rate of Imitation, Econometrica, 29, 741-766.

Mansfield, E. (1963) The Speed of Response of Firms to New Techniques, Quarterly Journal of Economics, 77, 250-311.

Mansfield, E. (1968) The Economics of Technological Change, Norton, New York.

Mansfield, E. (1988) The Diffusion of Industrial Robots in Japan and in the United States, Mimeo. University of Pennsylvania.

Mazzonis, D. and Pianta, M. (1986) An Innovation Strategy for Traditional Industries: Experience of the Italian Textile Districts of Prato and Como, mimeo, ENEA, Roma.

Mensch, G. (1979) Stalemate in Technology: Innovations Overcome Depression, Mass.. Ballinger, Cambridge.

Metcalfe, J.S. (1981) Impulse and Diffusion in the Study of Technical Change. Futures, 13 (1), 347-59.

Metcalfe, J.S. (1986) Diffusion of Innovation, mimeo, IFIAS/SPRU.

Momigliano, F. and Siniscalco, G. (1982) Note su terziarizzazione e deindustrializzazione, Moneta e Credito, 138.

Monden, Y. (1983) Toyota Production System, Industrial Engineering and Management Press, Norcross.

Mowery, D.C. (1988) Collaborative Ventures Between US and Foreign Manufacturing Firms, An Overview, in D.C. Mowery (ed), quoted.

Murray, F. (1987) Flexible specialization in the Third Italy, Capital and Class. 33, 84-95.

Murray, R. (1987) Cyprus Industrial Strategy, IDS, University of Sussex.

Nabseth, L. and Ray, G.F. (1974) The Diffusion of New Industrial Processes, Cambridge University Press, Cambridge.

Nelson, R. et al (1967) Technology, Economic Growth and Public Policy, The Brookings Institute, Washington.

Nelson, R. (1988) Institution Supporting Technical Change in the United States, in Dosi et al. (eds.), Technical Change and Economic Theory, Frances Pinter, London.

Nelson, R. and Winter, S. (1977) In Search of a Useful Theory of Innovation, Research Policy, 6 (1), 36-76.

Nelson, R. and Winter, S. (1982) An Evolutionary Theory of Technological Change, Belknap Press, Cambridge.

Northcott, J. (1986) Microelectronics in Industry: Promise and Performance, Policy Study Institute, London.

Pavitt, K. (1984) Sectoral Patterns of Technical Change: Towards a Taxonomy and a Theory, Research Policy, 13, 343-73.

Pavitt, K. (1985) Patent Statistics as an Indicator of Innovative Activities: Possibilities and Problems, Scientometrics, 7 (1), 77-99.

Pavitt, K. (1986) Chips and Trajectories: how does the Semiconductor Influence the Sources and Directions of Technical Change, in R.M. McLeod (ed.) Technology and the Human Prospect, Frances Pinter, London.

Pavitt, K., Robson, M. and Townsend, J. (1987) Technological Accumulation, Diversification and Reorganization in UK Companies, 1945-83, mimeo, SPRU/ESRC, University of Sussex.

Pavitt, K. et al. (1988) Accumulation, Diversification and Organization of Technological Activities in UK Companies 1945-83, Mimeo, SPRU, University of Sussex.

Pavitt, K. and Patel, P. (1988) Technological Activities in FR Germany and the UK: Differences and Determinants, DRC Discussion Paper, 58, SPRU, University of Sussex.

Perez, C. (1985) Microelectronics, Long Waves, and World Structural Change, World development, 3, 441-463.

Piore, M. and Sabel, C. (1984) The Second Industrial Divide, Basic Books, New York.

Pisano, G.P., Russo, N.V. and Teece, D. (1986) Joint Ventures and Collaborative Arrangements in the Telecommunications Equipments Industry, in D.C. Mowery (ed.). International Collaborative Ventures in US Manufacturing, Ballinger Pub., Cambridge Mass.

Porter, M. (1980) Competitive Strategy, The Free Press, New York.

Porter, M. (1985) Competitive Advantage, The Free Press, New York.

Poznanski, K.Z. (1986) The Extinguishing Process: a Case Study of Steel Technologies in the World Industry, Technovation, 4.

Premus, R. (1988) US Technology Transfer and Regional Policy, paper presented at the European Summer Institute of the Regional Science Association, Arco, Italy.

Ray, G.F. (1984) The Diffusion of Mature Technologies, Cambridge University Press, Cambridge.

Richardson, G.B. (1972) The Organisation of Industry, Economic Journal, 32, 883-896.

Rogers, E. (1982) Diffusion of Innovations, The Free Press, New York.

Roobeek, A.J.M. (1987) The Crisis of Fordism and the Rise of a New Technological Paradigm, Futures, April, 129-154.

Rosegger, G. (1980) The Economics of Production and Innovation, Pergamon Press, Oxford.

Rosenberg, N. (1976) Perspectives on Technology, Cambridge University Press, Cambridge.

Rosenberg, N. (1982) Inside the Black Box: Technology and Economics, Cambridge University Press, Cambridge.

Rosenberg, N. and Frischtak, C. (eds.) (1985) International Technology Transfer: Concepts, Measures and Comparisons, Praeger, New York

Rothwell, R. and Gardiner, P. (1983) The Role of Design in Product and Process Change, Design Studies, 4 (3), 161-170.

Russo, M. (1985) Technical Change and the Industrial District: The Role of Inter-firm Relations in the Growth and Transformation of Ceramic Title Production in Italy, Research Policy, 14, 329-343.

Sabel, C.F. and Zeithlin, J. (1982) Alternative storiche alla produzione di massa. Stato e Mercato, 5 (3).

Santarelli, E. (1988) Distretti multilocalizzati e innovazione nell'industria italiana, Economia Marche, 7.

Sayer, A. (1986) New Development in Manufacturing: the Just in Time System, Capital and Class, 30, 43-72.

Sako, M. (1988) Neither Markets Nor Hierarchies: a Comparative Study of Informal Networks in the Printed Circuit Board Industry, paper presented at the Conference 'Comparing Capitalist Economies: Variation in the Governance of Sectors', Wingspread, Wisconsin, USA.

Santarelli, E. and Steriacchini, A. (1988) Patterns regionali di innovazione tecnologica in Italia: brevetti su l'estero e imprese innovative, L'Industria. (forthcoming).

Scherer, F.M. (1982) Inter-industry Technology Flows in the United States. Research Policy, 11, 227-45.

Scherer, F.M. (1984) Innovation and Growth, The MIT Press, Cambridge (Mass).

Schonberger, R.J. (1982) Japanese Manufacturing Techniques, Free Press, New York.

Schmookler, J. (1966) Invention and Economic Growth, Harvard University Press, Cambridge, Mass.

Senker, J. (1985) Small High Technology Firms: Some Regional Implications, Technovation, 3, 243-262.

Senker, P. (1986) Education and Labour Markets: The Challenge of Technological Change, Mimeo, SPRU, University of Sussex.

Soete, L. (1985) International Diffusion of Technology, Industrial Development and Technological Leapfrogging, World Development, 13 (3), 109-423.

Spence, A.M. and Hazard, H.A. (1988) International Competitiveness, Ma., Ballinger, Cambridge.

Stoneman, P. (1976) Technological Diffusion and the Computer Revolution, Cambridge University Press, Cambridge.

Stoneman, P. (1985) Technological Diffusion: the Viewpoint of Economic Theory, Mimeo, University of Warwick.

Sweeny, G.P. (1987) Innovation, Entrepreneurs and Regional Development, St. Martin Press, New York.

Teece, D.J. (1986) Profiting from Technological Innovation: Implications for integrations, Collaborations, Licensing and Public Policy, Research Policy, 15, 285-305.

von Tunzelmann, N. (1986) Steam Power and Nuclear Power: the Diffusion of British Energy Technologies, Mimeo, SPRU.

von Tunzelmann, N. (1988) Diffusion of Technical Change, SPRU, University of Sussex.

Thwaites, A.T. and Alderman, N. (1988) The Location of R & D Retrospect and Prospect, paper presented at the European Summer Institute of the Regional Science Association, Arco, Italy.

Vernon, R. (1966) International Investment and International Trade in the Product Cycle. Quarterly Journal of Economics, 80, 190-207.

PART IV - POLICY ISSUES

16. REGIONAL TECHNOLOGY TRAJECTORIES AND EUROPEAN RESEARCH AND TECHNOLOGY DEVELOPMENT POLICIES

Michel Quevit

Introduction

During the 1980s, the European Community has launched a series of extensive programmes in Research and Technology Development (ESPRIT, RACE, BRITE, STAR, etc.). It is already evident, from the magnitude of the budgetary allocations made to these programmes that a movement favouring the creation of a **new European-wide space in science and technology** is emerging. The creation of a vast, unified interior market, through the application of the Single European Act in 1992, will inevitably reinforce intra-European technological exchange. Therefore, the spatial distribution of science and technology activities is a current and primary concern of regional science.

It is, however too early to reach any firm and final conclusions about the eventual evolutionary path of these ongoing processes. Meanwhile, two hypotheses are often advanced regarding the creation and nature of this new technology space. According to the first hypothesis, the increasing mobility of technology in exchange will reinforce the concentration tendency of potential RTD (Research and Technology Development) around some of the existing major European poles. The second hypothesis presumes the presence of a variety of technology paths which reflect the complexity and diversity of the processes of the generation, application, and diffusion of technology.

These two hypotheses, which at first glance seem contradictory, could in reality prove to be complementary. In fact, the work of Philippe Aydalot and GREMI, concerning the diversity of technology trajectories and their relation to innovation milieux, indicates that the relationship between production and knowledge takes very different forms depending on the environments which regulate them (Aydalot, 1986a). Thus, far from finding ourselves confronting a universal trend toward technological space homogeneity, we will be faced with a variety of spatial development dynamics caused by the technology factor.

The aim of this chapter is to contribute to a clarification of this central question of the spatial distribution of technology through a demonstration of the

Belgian situation.

We will proceed to do this in three complementary stages. First, through an analysis of the distribution of R & D (research and development) investments made by firms, we will single out the diversified routes to technology which at the spatial level induce these investments. Without constituting a universal example, the example of Belgium offers an interesting field of analysis to the extent that this country, characterized by its centrality within the Community, is dominated by logics of development which are sustained by very different contexts and environments: the logic of metropolitan development (e.g. the Brussels -Antwerp axis); the logic of the reconversion of the old industrial structure (e.g. the Liege - Charleroi - Mons axis), and the logic of rural industrialization (e.g. South Wallonia and West Flanders) (Quevit, 1986b). It is therefore of major interest, through these relatively conventional logics of space, to discern the specific trajectories induced by the technology factor. A summary of the regional technological trajectories of these areas is given in Table 16.1.

Considering that Belgium is one of the members of the Community with the highest level of per capita participation in Community RTD programmes, the study of the regional impact of certain broad programmes of the European Community (ESPRIT, RACE, BRITE, etc.) on the separated technology trajectories, will allow us to eventually identify the possible territorial impact of Community RTD policies.

Finally, through the double comparison of regional paths shaped by the technology factor, and the regional impact of European RTD programmes on Belgian territory, we will attempt to work out suggestions for RTD policies which are better adapted to the specificity of the regional technology trajectories.

Spatial Distribution of R & D Investments and Regional Technology Trajectories

According to the study (Commission of the European Communities, 1987)[1] we conducted on technology potential in Belgium, three factors seem to have played a determining role in the spatial configuration of the relationship between production and skill:

- **sectoral specialization** of firms investing in R & D;
- **the size** of the firms with a technology potential;
- **regional contribution to formation**[2] supply and demand in the field of science and technology (S & T).

As we will show later, the combination of these three factors allows us to

Table 16.1
Summary Table of Regional Technology Trajectories

Technology Trajectories	Sectoral Technology Structure		Size of Technological Enterprises		Human Potential in S & T		Technological Development Potential
	concentrated	diversified	large	SME	supply	demand	
Metropolitan regions							
- based on the large enterprise	high	low	high	low	low	high	high
- based on science	low	high	medium	high	high	high	high
Regions with a tradition of industry							
- link with mono-industry	high	low	high	low	low	medium	low
- rupture with mono-industry	medium	high	medium	medium	high	high	medium
Rural Industrialized Regions							
- indigenous capacity	low	high	low	high	almost zero	high	high
- exogenous capacity	high	low	high	low	almost zero	high	weak

distinguish between a **variety of regional technology trajectories** which goes further than the traditional distinction between metropolitan areas, regions with a tradition in industry and rural regions. This three-fold typology is of course still pertinent at the meso-economic and meso-sociological levels, because it refers to an economic, social and cultural entity marked by specific technical-industrial evolutions.

Nevertheless, the most remarkable result of our study is that, inside these entities, the push of technology leads to different technology trajectories depending on the relative importance of these three factors and/or their connections to the regional productive structure.

Variety of Technology Trajectories in Areas of Metropolitan Density

The most striking conclusion reached by studying the regional distribution of R & D investments of private firms is, of course, the very strong concentration of R & D investments in the Antwerp - Gent - Brabant triangle which comprises the two large Belgian metropolises, i.e. Brussels and Antwerp. This region by itself, totals almost three quarters of R & D investments in the entire country (Table 16.2)

The common feature of these two regions is the very strong sectoral specialization of their productive structure in technology intensive sectors such as chemicals, electronics and electrotechnics. These three sectors represent nearly 70 per cent of the R & D spending in the country. Another indicator of the central and privileged position of these regions is the superiority of their share in technology investment as compared to their real contribution to the economic growth of the country (approximately 72 per cent of R & D expenditure against 58 per cent of the GDP).

The feature common to these two metropolitan areas, that is to say, the high degree of sectorial specialization in technology potential, does not however express with sufficient precision the nature of the technology routes that each one follows. In fact, a more refined analysis incorporating the other two factors, firm size and the nature of regional human capital in science and technology, reveals very different technology trajectories for these two regions.

A technology trajectory based on the large multinational enterprise: The Antwerp Metropolis

The Antwerp metropolitan area appears to be the prototype of technology path based on the large enterprise mastering a new field, as Aydalot defined it (Aydalot,

Table 16.2

Distribution by Region of R & D Expenditure of Enterprises and Services in Proportion to R & D Personnel Employed in Company Headquarters

Regions	R & D Expenditure per cent	Regional Share in GDP per cent
Brabant	29.4	25.8
Antwerp	37.4	20.0
East-Flanders	5.1	12.3
West-Flanders	8.0	10.5
Limburg	4.4	6.5
Liege	4.9	9.7
Hainaut	8.4	10.1
Namur	2.0	3.3
Luxemburg	0.4	1.7
Belgium	100	100
Areas		
Flanders	59.3	56.7
Wallonia	20.4	27.6
Brussels	20.3	15.7
Belgium	100	100

1986b). (63 per cent of R & D investments are concentrated in very large enterprises in the fields of petrochemicals, pharmacy and electronics. They are large multinational corporations (e.g. in telecommunications: ITT which comprises nearly 13,000 people). These enterprises acquire their technological skill through the production of knowledge within the firm in specialized units which are very often detached from the regional structure. According to authors like Dupuy and Savary (1987), these enterprises are involved in spatial labour distribution. They regulate their regional activities according to the strategies and modes of the international organization of production of the industrial groups on which they depend. As the share of this region in the formation potential of science and technology shows, these enterprises do not need to be directly sustained by the production of local skills. If we look at the formation supply (i.e. the number of universities or institutions of higher education supplying training in science and technology), the province of Antwerp contributes a mere 2.7 per cent of the national supply. From the demand perspective, (number of science and technology degrees issued in the region), the share of the province of Antwerp does not exceed 14.6 per cent while it represents 17 per cent of the total Belgian population.

It is thus noteworthy that this technology path relies heavily on public intervention. In effect, the province of Antwerp is one of the regions in Belgium where, especially through industry, research contracts and public orders, firms receive the highest level of government aid in Research and Development.

A technology trajectory founded on science: Brabant

The analysis of the technology potential of the province of Brabant presents a very different configuration. Despite a concentration of R & D investments by firms similar to the one in the province of Antwerp, the technology path of this region is more directly oriented toward science. This trajectory, we should remember, is characterized by the fact that existing production is more directly linked to basic research and draws its skill from a supply external to the firm. It is deeply connected with universities and public or quasi-public research.

In this context, Brabant has four full universities (two in Brussels, one in Walloon Brabant, and one in Flemish Brabant). The region represents 58 per cent of the national formation supply, and 25 per cent of the formative demand in science and technology. Government aid is also heavily concentrated there.

Other indicators, such as the creation of patents, confirm the privileged relationship between industry and research. This region accounts for 42 per cent of the purchase and 47 per cent of the sales of patents. Furthermore, its productive fabric offers a better structural configuration between large enterprises and innovative small and medium-sized firms (SME). The large R & D enterprises, naturally, are concentrated in the sectors of electronics, metals and services. However, a very large sectorial diversification is observed in the SMEs and the intermediary enterprises, ranging between such sectors as precision instruments, research and engineering, construction and pharmacy.

It is also noteworthy that in this region the phenomenon of capital plays a determining role through the localization of the registered offices of companies, and the concentration of public research institutions. In this context, we should also draw attention to the particular situation of Brabant which, owing to the push toward a process of technopolization created by the relatively recent foundation of the University City Louvain-la-Neuve, has led to an even more science-based development.

To conclude, we argue that there does not exist a unique model of technology trajectories for regions of particular levels of metropolitan density. These areas are characterized by direct paths to technology. This direct route can be due to the strategies of large enterprises linked to multinational industrial groups, or the spatial concentration of centres for the production of skill (universities - centres of research and experimentation) or the localization processes linked to their particularly metropolitan character (e.g. the localization of the registered offices of companies).

For all that, these zones will show a concentration of technology potential which surpasses their real contribution to general economic growth, thus leading to the accentuation of regional imbalances within nations and between nations in Europe.

Diversified Technology Trajectories in Regions with a Tradition of Industry

The works of Denis Maillat and of Jean-Claude Perrin have demonstrated that the regions with traditional industrial structures are currently directing themselves towards new development dynamics and that it is false to claim that these regions are incapable of innovation (Maillat, 1986; Perrin, 1984). The work that we are at the moment conducting in the region of Liege, within the framework of GREMI studies, widely confirms this argument. As Philippe Aydalot noted, the reconversion of this industrial fabric takes place within a **logic of filiation-continuity** which does not upset the market, the product, and technology all at the same time (Aydalot, 1986b). Innovation presents a feature of filiation within the existing fabric (e.g. the shift from mechanical activity toward electronical activity, progressive integration of technology in existing products, etc.). Thus, the older industrial structure can survive major 'technological aggression' and progressively accomplish the modernization of its productive fabric.

If this logic of filiation-continuity represents appropriately the technology paths of the industrialized regions of Hainaut, Liege and Limburg, an examination of their respective technology potential reveals that there is also a variety of configurations of possible paths which we shall attempt to describe.

A trait common to all these regions is that their R & D investments take place in sectors which are technology intensive, but directly linked to a **basic industry** that polarizes the technology efforts of the region. In the region of Liege this is the metal products industry, in Hainaut, electromechanics, and in Limburg, transport material. In each one of these regions, the basic industry by itself accounts for more than 50 per cent of R & D effort by firms. This link between technological innovation and basic industry constitutes a weakness of this productive structure that confirms another trait common to these regions. That is to say, the shares they receive of regional R & D investments are inferior to their global contribution to the economic growth of the country. Thus, these regions do not benefit from a share in technology which corresponds to their contribution to the national added value.

Meanwhile, behind this common image there also exist particular technology routes which call for a differentiation in these observations.

A trajectory of filiation in continuity with a mono-industry: Hainaut and Limburg

The technology routes of Hainaut and of Limburg are strongly shaped by the strategies of several large enterprises which attempt to monopolize particular technological opportunities. In Hainaut, this is the case with enterprises in the electrotechnical sector, which in relative terms represent the most important R & D investments in the region.

These investments are the work of several large companies. In Limburg, these are the enterprises in the transport material (predominantly motor vehicles) and the electrotechnical sectors. They account for 80 per cent of R & D investments originating from large firms in the region. Such monopolization of the technological effort also relies heavily on government interventions (support for industrial research, for prototypes, etc.).

It should also be noted that these two regions have accomplished their technological transition through existing sectors, that is to say, sectors established several decades ago. In the case of Limburg, the automobile industry was established by multinational corporations during the 1960s, and in the case of Hainaut the electrotechnical sector has existed for nearly half a century. The technological paths of these two regions therefore appear to fit easily within the logic of 'filiation-continuity'.

A trajectory of filiation in rupture with the mono-industry: Liege

The technology path of the Liege region on the other hand, presents a more specific configuration. We find there a technology filiation related to the existing industry, because metallurgy represents nearly 50 per cent of total R & D investments in the region, and the majority of these investments are made by very large enterprises.

Nevertheless, it is interesting to note that, side by side with this sector, almost 50 per cent of R & D investments are made in very diversified sectors dominated by small and medium-sized firms and intermediary enterprises (precision instruments, biotechnology, electronics, food and agriculture, etc.). If it is considered that, for the most part, the previous activities of the founders of enterprises were in already established professions, it would be interesting to know if these enterprises have undergone a break with the industry providing the motive force in the region. If so, this fact would be characterized by the dependence of SMEs on the regional productive fabric. It could also be that, different from Hainaut and Limburg, the existence of a university has had a positive effect on the emergence of new enterprise founders in this region. Contrary to the other two regions, the region of Liege in fact offers a positive supply/demand ratio of degrees in science and technology.

To conclude, the regions with traditional industrial structures do not show uniformity in technology trajectories. The logic of filiation-continuity, which appears to be at the origin of technological change in these regions, could also be influenced by a break with the mono-industries which characterized their economic development for more than a century. This rupture can be sustained by the existing entrepreneurial potential, but it also requires in these regions the presence of a high potential derived from technological formation capable of creating an entrepreneurial capacity oriented towards filling gaps in technological innovation. From the perspective of industrial change, to the extent that it consumes a major share of government R & D interventions, a very heavy technological dependence on a mono-industry handicaps the efforts aimed at modernization of the regional productive structure. The situation of the Liege region is therefore one of the most interesting ones. This is because a dynamic leading to the creation of innovative SMEs, especially in the fields of microelectronics and biotechnology, appears together with a rupture with the basic mono-industry of this region (Quevit and Bodson, 1989).

Diversified Technology Trajectories in Predominantly Rural Regions

Beside the two large metropolitan areas and their traditional industrial structures, Belgium also possesses large rural areas predominantly of forestry and agriculture. Three provinces could be defined as rural: the Walloon provinces of Namur and Luxemburg and West Flanders. The province of East Flanders, although characterized by a high level of agricultural employment, is influenced too much by the metropolitan hinterland of Antwerp and Brussels to be included in this category.

It could be problematical to talk of routes to technology while referring to rural regions. In fact, the majority of these regions are situated outside the spheres of technology exchange and centres of skill. That may be one of the reasons why this category does not exist among the diverse trajectories described by Aydalot (1986b). The peripheralization of these regions in the domain of technology is in fact undeniable. These three provinces together total a mere 10 per cent of total R & D investments in the country while they contribute 15.6 per cent of the GDP.

All the same, we cannot conclude that these regions do not possess their own technology paths and do not aim at technological innovation based development. But once more, there is no unique model.

A technology trajectory based on indigenous resources: West Flanders

The province of West Flanders offers one of the most interesting examples of a predominantly rural region which, starting from the dynamics of indigenous development, succeeded in converting itself from non-innovative to innovative sectors[3]. Primarily dominated by the textile, food and agriculture sectors, this region experienced an increase in added value during the 1970s in these sectors as well as in chemicals, electronics, financial services, transport and communications. The major feature of this expansion is the existence of a dynamic network of small and medium-sized family enterprises interlinked with a local savings network. A principal cause of this dynamism is the existence of a long historical tradition of entrepreneurship which is accustomed to act independently of links with some key 'motor' industry and according to a logic of 'survival' which relies on local financial capabilities and a strongly integrated educational environment. This is especially the case in the regions of Courtrai and Westhoek.

In the domain of R & D investments, this region by itself totals eight per cent of the entire country. This is a higher figure than those of the province of Liege and almost as much as that of Hainaut. Otherwise, it presents a very diversified structure of R & D investments. These cover the domains of ferrous and non-ferrous metals, precision instruments, electronics, textiles, chemicals and food and agriculture industries. With the exception of the metallurgy sector, the majority of R & D investments are small. As for the human potential, while formation supply in science and technology is zero, this region accounts for nearly 10 per cent of the formation demand in science and technology.

A technology trajectory based on exogenous contributions: Luxemburg

Two other rural provinces, Luxemburg and Namur, reflect an entirely different pattern. These two provinces in Wallonia have a share of 2.5 per cent of R & D investments in the country while they contribute 5 per cent of the GDP. We do not encounter developed interfirm networks in these provinces, but a very dispersed body of local employers who do not seek technological innovation. Therefore, in the province of Namur the technological effort, however weak, finds its source essentially in very small firms in several fields of innovation: electronics, precision instruments, chemicals and food and agriculture. It could be that the presence of university institutions contributes to the attraction or development of new enterprises in this region. However, as the level of R & D investment is low, this hypothesis should be formulated with caution.

The technological lags are even more obvious in the province of Luxemburg. As a region based on agriculture and forestry (80 per cent of Belgian forests are found here), it has developed mostly around a very dispersed body of local employers within the traditional industries and consumer services. Although their relative growth has been stable through time, these firms remain on the border of technological innovation. The major technology effort in this region took place in some very large enterprises, in the sector of wood and paper, and some small mechanics firms. Such investments cannot bring about a solid technology path. Furthermore, if we look closely at its human capital in science and technology, this region shows a surprising paradox: while it does not possess any institute of formation in this domain, it possesses almost seven per cent of the science and technology degrees in the country with a national share of only 0.4 per cent in R & D investments.

Contrary to present opinion, the fact that rural regions lag behind in technology is not necessarily due to the lack of a highly qualified human capital, but to the absence of economic initiatives and RTD policy instruments to allow firms access to technological innovation (Quevit, 1986a). It is to be expected in this region that there is a brain-drain of highly qualified human resources who seek employment outside their region of origin.

European Community Research and Technology Development Programmes and Regional Technology Trajectories

In this section, we will deal with the problem of the regional implications of RTD programmes initiated by the European Community. We will try to identify their connection to specific technology paths developed in the previous section.

To begin with, we should be detached from the logic which governs the choice of projects. Therefore we will limit ourselves to several large programmes, such as ESPRIT, RACE and BRITE, for which we have concrete data at our disposal. We can remark by way of introduction that Belgium has an important share in these programmes. Whereas the Belgian population is only eight per cent of the European total, it receives more than 20 per cent of the entire science and technology programme budget.

The basis of extensive RTD programmes in the european community: the standard of excellence

The main target of the European Community's RTD programme is to reduce the technological backwardness of Europe in relation to competing industrialized countries, i.e. The United States and Japan. These programmes aim above all at high level **precompetitive research** and therefore encourage the participation of firms which have sufficient budgetary means for financing such research, or centres of excellence in research. In a general way, these programmes favour co-operation between enterprises, the public sector, and research bodies. The criteria of project selection are based on standards of international scientific and technological competitiveness. Thus, only firms and centres of research excellence which have reached this 'critical threshold' are supported. These pre-requisites are present in all RTD programmes, in addition to specific requirements of individual initiatives.

i) The ESPRIT programme

The ESPRIT programme aims to stimulate R & D efforts in the field of information technology. The main objective of this programme is to reinforce European Industry so that it can catch up with the technological advances of the United States and Japan within the next ten years. The programme explicitly provides for the possible participation of small and medium-sized firms. In Belgium more than half of participant firms employ less than 200 persons, while the rest concerns firms with more than 2,000 people. It should also be noted that centres of research contribute up to 50 per cent of the cost of the projects.

This distribution between small and medium-sized enterprises corresponds well with the situation found in microelectronics where, side by side with large multinational industrial groups, there exist small innovative firms which are very active and competitive in certain areas of the sector.

ii) The RACE programme (Research and Development in Advanced Communications in Europe)

The RACE programme concerns new information technologies applied in the domain of telecommunications: the creation of terminals and equipment markets, and the eventual establishment of integrated broadband communications (IBC) networks, etc.

This programme is supported by co-operation between the owners of telecommunications networks, centres of research, and enterprises (including small and medium-sized firms). It is interesting to note that within its target, the programme explicitly foresees the possibility of improved access for less favoured regions of the Community to the advantages of the development of advanced network services. In this context, this programme will, in its operational phase, allow the exploitation of modern techniques of new telecommunications for the benefit of less favoured regions and the development of their infrastructure. The RACE programme is still in its phase of definition which will be established through projects co-financed by the Community, industrialists, centres of research and network owners.

In Belgium, the participants in this programme are initially both very large scale enterprises and institutions of research. Despite the declared intention of encouraging access to communications technology in less favoured regions, the programme has in its initial phase selected mostly projects offered by enterprises the performance levels of which correspond to standards of excellence.

iii) The BRITE programme

The BRITE programme in its turn, aims to stimulate the creation of an advanced technology base favouring the traditional industries of the Community in order to upgrade them to a level of international competitiveness. With this perspective, the programme specifically revives interest in regions with traditional industrial structures.

The programme is open to firms of all sizes, to university research institutes, and centres of scientific experimentation. To be supported, projects have to meet the following conditions:

- deal with pre-competitive R & D work;
- comprise research of an innovative character;
- attain significant budgetary level (approximately one million ECU per project).

BRITE has aroused a lot of interest in traditional sectors in Belgium, such as precision instruments, construction engineering, textiles, ceramics, transport materials, etc..

It is however amazing to find out that leading enterprises in the domains of electronics, biotechnology, or research departments are marginalized in such programmes while some of these same enterprises have been selected for other

Community RTD programmes such as ESPRIT or Biotechnology.

Finally it should also be mentioned that there exist numerous projects originating from both private research centres (centres of collective research) and universities (nearly half of the projects).

The BRITE programme has nevertheless aroused an interest, especially in Wallonia, in small and medium-sized firms (more than 60 per cent of the participants). This is an encouraging situation because one of the major problems in regions with traditional industry has been the difficulties faced by SMEs. For decades they have been disadvantaged in terms of access to, and attainment of, 'critical thresholds' in technology.

Regional Impact of the European Community RTD Programmes: Some Observations

Several observations can be made from an examination of the spatial location of Belgian participants in Community RTD programmes:

a) To the extent that selection standards do not include the regional dimension, RTD programmes do not have a direct impact on the technological trajectories of regions. In fact, they have not been designed to reduce or resolve regional technological disparities. Moreover, we will not encounter many such projects in regions which do not possess either centres of excellence in research, or firms that meet the standards of pre-competitive research. This explains the fact that predominantly rural regions, such as Luxemburg, have no share in the three programmes mentioned above, and Namur only participates in some projects originating from university research institutes located in that region.

b) Nevertheless, to the extent that they consolidate research teams and centres of scientific experimentation in the regions where they are established, RTD programmes do have a direct impact on regional technology developments. Thus, the primary effect of these programmes is the reaffirmation of the technological domination of regions that already possess a high potential of basic and applied research. In Belgium, this is the case in the metropolitan regions which follow a technology path based on science or on the technology strategies of large multinational enterprises. In the three aforementioned programmes, Brabant is clearly in the lead in its share of RTD investments, followed by Antwerp.

c) Naturally, if any of the following two conditions were to be satisfied, RTD programmes could also have a positive impact on the less favoured regions in the Community:

- If there exist, in these regions, centres of research excellence: this is the case with Liege for the programme ESPRIT or Biotechnology, and with Namur for the programme ESPRIT.

- If these programmes concern sectors where regional firms are represented in R & D investments: this is the case in Hainaut and Liege for the ESPRIT programme of microelectronics, in Hainaut and West Flanders for the RACE programme, and in the traditional industries of Hainaut, Liege and Namur for the BRITE programme.

All the same, when compared to regions with high technology potential, the proportion of RTD programmes participated in by the disadvantaged regions remains very restricted.

Factors Restricting Regional Participation in Community RTD Programmes

Programmes of the European Community have certainly contributed to increases in the performance of research teams which can satisfy the pre-requisites that international scientific competitiveness demands. In the meantime, this inevitably excludes centres of research or innovative firms which cannot reach the same thresholds.

The feasible participation of the less developed regions in RTD programmes sustaining the cumulative effect of factors which interact with each other:

- Most of the programmes are oriented toward pre-competitive research which implies the formation of skilled research teams and therefore a major investment by enterprises and/or centres of research;

- The high performance level of international scientific competitiveness is rarely attained by the bulk of regional enterprises;

- In some programmes, the rule that requires a partner (firm or centre of scientific experimentation) from another Community member country. The SMEs especially do not even have experience in interfirm contacts, and certainly not at the European level.

- The widespread lack of information about these programmes in SMEs, which because of their small size remain outside the diffusion of scientific information. The nature and procedures of these programmes are usually well known to very large multi-national enterprises which encounter, at least informally, the European authorities. Small-scale firms have a striking lack of experience as far as dealing with European administrations is concerned.

- The duration of selection procedures is often too short for a SME to meet all the standards required.

- The non-participation of regional governments which are in direct contact with firms and therefore could play, in comparison with European authorities, more efficient roles in information transmission.

To conclude, the RTD programmes of the Community have a global effect of reinforcing or consolidating the existing strong points present within regions (whether these are firms or centres for scientific experimentation). However they are very rarely capable of favouring the **creation of new strong points** which could emerge from a technology potential that has not yet attained the level of excellence required, even if this regional technology potential is of high quality. This is the case for many regions and especially so with regions with traditional industrial structures.

Linkages of European Community RTD Programmes with Regional Technology Trajectories

For a Diversified and Selective Approach to Regional RTD Policies

The diversity in regional technology routes and the logic of excellence underlying Community RTD programmes points to a need for a **diversified and selective approach** to the RTD policies of the European Communities.

In fact, the analysis of the regional dimension of the Community RTD programmes, leads to three major conclusions:

a) The RTD approach founded on the standard of excellence responds to needs which cannot be met by regions that do not have high technology or science potential (international competitiveness).

b) The RTD approach should be adapted to the nature of the regional productive and technology structure. Regional technology paths indicate by their diversity that a variety of approaches to RTD is possible and desirable if the European

Communities want to maximise the business and technology potential existing within each European region and in the Community as a whole.

c) Within this view, the orientation of RTD programmes should differ depending on the technology function, that is to say, depending on whether it is a case of targeting for openings in high technology or, with the aim of creating products and conquering markets, the incorporation of existing technologies into the existing productive apparatus.

In fact, if technological innovation has become an essential force in eonomic development, then a distinction should be made in the RTD approach, and therefore also in the formulation of its programmes, between the function of **technology production** and that of **technology utilization**. Depending on the case, we will find ourselves facing very different processes of interaction between research and technology. The production of high technology is achieved through a close link between the state of science and product sophistication. It requires specific human, financial and organizational investment. It concerns firms (large or small) which in their structures are **oriented toward scientific competition** and insertion in the flow of interaction between basic and applied research.

On the other hand, a production based on the utilization function of technology primarily aims to pick up one or more existing technologies and to introduce them into the production process, thereby creating a new product. This function concerns firms which have as their main objective the creation of a new product. It employs human, financial and management instruments that are more directly **oriented toward the market**.

It can also be hoped that the European Communities initiate RTD programmes which, through specific instruments and means, are better adjusted to regional technology paths. In this respect it seems timely to favour the initiation of programmes which correspond better to regional specificity. We can classify these according to three criteria:

- the specific technology trajectory;

- the priority of the technology function;

- the orientation of the technology effort (Table 16.3).

To conclude, it certainly is not desirable that the EEC gives up its RTD programmes targeting the international level of excellence. Technological competitiveness or catching up with the United States and Japan remains a strategic

Table 16.3
Principal Orientation of RTD Programme as a Function of the Selected Technology Trajectory

RTD PROGRAMME	TECHNOLOGY TRAJECTORY	TECHNOLOGY FUNCTION		ORIENTATION OF TECHNOLOGICAL EFFORT	
		Production	Utilization and/or adaptation	International scientific Competitiveness	Market
HIGH TECHNOLOGY	excellence pole	strong	weak	strong	weak
RECONVERSION	rupture/ filiation technological recovery	medium	strong	medium	strong
ENDOGENIZATION OF DEVELOPMENT	modernization indigenous productive structure	weak	strong	weak	strong

'must'. Within this perspective, the standards of selection in current RTD programmes should not be lowered for the sake of other principles, which underpin the reduction in regional technological disparities. On the other hand, it seems to us desirable that simultaneously with present RTD programmes focusing on international technological competition, the European Communities aim at the elaboration of RTD programmes targeting regional needs, thus rendering possible the technological recovery and the technological adaptation of the productive apparatus of regions.

Suggestion of Elements Favouring an RTD Programme of Technological Recovery

Such a programme has a double objective. The first one is above all a technological recovery of potential RTD. Thus, centres of experimentation of universities or firms can attain the 'critical threshold' demanded by the selection standards of broad Community RTD programmes, such as ESPRIT, RACE, BRITE, Biotechnology, etc.

In this case, firms and centres of scientific research can be upgraded through a specific programme to this level of demarcation. Thus, they can project themselves positively in international competition. In certain regions of reconversion there do in fact exist numerous centres of research which in regard to specific openings are not far from attaining this threshold. It is particularly important that,before encouraging them to apply for these programmes, they are given the means to develop further their basic research structures.

In this type of programme the selection of projects should not be confined to a particular sector. It will thus be a multi-sector RTD programme focusing on targeted technology themes open to all firms and centres of research which have a high chance of recovery (for instance, within five years). In this context, co-operation at the European and interregional level with successful centres of research could be decisive.

The second objective of the programme concerns the endorsement of the RTD process by the **industrial valorization of research results** in the region of origin. This valorization could be accomplished by regional firms or with the contribution of external firms which agree to establish themselves in the region. This target is fundamental for establishing the links between technology research and development dynamics.

A RTD Programme Based on the Technological Adaptation of the Regional Productive Structure

The main objective of this programme is to provide existing enterprises, and especially SMEs, with the capacity to modernize their production (process and product) through the incorporation of new technologies. Since it concerns the majority of productive activities in the region, this objective is fundamental and should have the highest priority of any RTD policy favouring the regions in the Community. The main targets of this programme are:

- the mastering of the process of the diffusion and adaptation of new technologies.
- the access to market opportunities (regional, national and international).

The central axis of the access to these programmes should be the valorization of the **indigenous resources** of the region. This requires linkages between adequate financial, human and institutional instruments. The programme should take care to support the instruments which address the internal problems of the firm and its relations with its immediate and broader environment.

In fact, the insertion of an R & D division in the firm remains, for most firm managers, a risky operation because it necessitates a **strategic vision** of the development of the firm. Technological innovation is embedded in a **sequential process** which goes from the acquisition of technology through to the commercialization of the product. The mastering of the **time factor** is an essential component. Priority should therefore be given to mechanisms and instruments which favour an **integrated approach** to the growth of the firm in its environment:

- financial engineering,
- technological adaptation to the different phases of the production process,
- competence, know-how, management and organization,
- access to the market and the commercialization of products,
- interfirm relations,
- forming networks with the educational system and so forth.

Conclusions

To end this analysis, it seems that the spatial distribution of science and technology is far from leading to a uniform configuration in space. Diversified technology trajectories appear as a function of the structure of the regional socio-economic fabric and the multiple effects of technology on the productive apparatus.

A dualization of the production of technology in favour of regions with high basic and applied research potential and thereby the creation of a new bipolarization of the European space to the advantage of existing poles of excellence is to be deplored. This is what has been revealed in this article by the analysis of the regional impact of extensive European RTD programmes. The logic of these programmes being the technological recovery of Europe vis-a-vis the United States and Japan, their application inevitably strengthens the technological domination of regions with a high technology potential. Faced with international competition, such a process is certainly a necessity. Nevertheless, it should be kept in mind that the technological effort of the European Community should also lead to a **revitalization of the regional productive structures** and therefore should be supported by RTD programmes targeting the innovation capacity of firms which do not primarily aim at the production of high technology, but at the development of new products, growth in value added and the acquisition of new markets within numerous other sectors of the economy.

Finally, the real challenge to the reconstruction of Europe in this context resides in the capacity of Europe to respond to the requirements of the complexity of international technology and simultaneously to favour the formation of innovation networks, which widen the diversified technology trajectories of the less favoured regions through programmes of research and technology development designed with these targets in mind and according to the terms set by decentralized organizations.

Notes

1. This article is based on a report which was drafted within the research programme accomplished under the direction of CURDS (Centre for Urban and Regional Development Studies of the University of Newcastle upon Tyne) and NBST (National Board for Science and Technology - Dublin), for the STRIDE project of the Commission of the European Communities (General Directorate for Regional Policy).

2. The term 'formation' refers to the educational system and more specifically in this context to graduates in science and technology.

3. It should nevertheless be pointed out that the maritime steel plant SIDMAR was

installed in this region, and the R & D investments of this sector represent 29 per cent of R & D investments in the region. The installation of this plant is completely exogenous and does not play a determining role in the indigenous process described.

References

Aydalot, P. (ed.) (1986a) Milieux innovateurs en Europe, GREMI, Paris.

Aydalot, P. (1986b) Trajectoires technologiques et modeles regionaux d'innovation, XXVe Conference de l'ASRLF, Paris.

Commission of the European Communities (1987) Research and Technological Development in the less favoured regions of the Community (STRIDE), Final report, Office for Official Publications of the European Communities, Luxembourg.

Depuy, C. and Savary, J. (1987) L'impact des groupes industriels sur les systèmes productifs locaux, in Adefi-Greco (ed.), Industries et Regions, Economica, Paris, 113-140.

Maillat, D. (1986) Les PME innovatrices et la revitalisation des Regions, Revue d'Economie regionale et urbaine, 5, 688-693.

Perrin, J.C. (1984) La reconversion du bassin d'Ales - Contribution à une theorie de la dynamique locale, Revue d'Economie Regionale et Urbaine, 2, 237-256.

Quevit, M. (1986a) Le Pari de l'industrialisation rurale, ERESA, Geneve.

Quevit, M. (1986b) La dimension regionale des politiques de Recherche - Technologie et Developpement (RTD) en Belgique - Vol.1 et Vol.2, Rapport national Commission of the European Communities (unpublished).

Quevit, M. and Bodson, S. (1989) L'enterprise innovatrice dans son milieu local, in D. Maillat and J.C. Perrin (eds.) Enterprises innovatrices et reseaux locaux, ERESA, Economica, Paris.

17. DETERMINANTS OF THE RESTRUCTURING OF POLAND'S ECONOMY IN THE 1980s

Maria Ciechocinska

Introduction

One common feature of the strategies for economic growth experienced so far in Poland, regardless of the labels put up on them in the past, has been the emphasis on **extensive** growth, which continued until the mid-1970s amid an abundance of raw materials and labour. For many decades, Poland was a country relatively rich in factors of production. Therefore, in forming the spatial structure of the economy, it was possible to give precedence to recommendations arising out of ideological doctrines over those leading to economic efficiency. Stereotyped thinking in quantitative terms predominated, because it was possible to achieve almost any desired production growth simply by increasing the scale of operation.

As the 1970s progressed, however, the possibilities of continuing the **extensive** strategies of economic growth began to run out, as easily accessible deposits of raw materials that could be mined cheaply became increasingly scarce and labour shortages appeared (Thwaites and Alderman, 1988). At the same time, signs of degradation of the environment became apparent, which in some areas, such as Upper Silesia or the Legnica-Lubin copper basin assumed the proportions of an ecological crisis (Kassenberg and Rolewicz, 1985). Meanwhile the pace of technological change elsewhere raised the requirements that Polish industry had to meet if it was to compete on the world market.

Broader awareness of these phenomena came only after considerable delay and was not reflected in the national strategy for economic development. This is illustrated by the determination with which the scale of operations and the size of organizations, institutions and enterprises was developed in all areas of socio-economic life until the late 1970s.

The collapse of the economy at the beginning of the 1980s was in part caused by this adherence to inappropriate and perhaps outdated approaches to economic development. On the other hand, awareness of the need to carry out a thorough restructuring of the economy, which is a **sine qua non** for revitalization, has only gradually emerged. Such a strategy would also call for abandonment of various handy stereotypes from the past.

The Polish economy did not go through the shock connected with the so-called 'energy crisis' caused by the OPEC oil price rises in 1972. For many economies this was an incentive to develop new and more economical technologies and means of production whereas Poland just stood by watching these transformations taking place elsewhere. Neither did Poland grasp the chances offered by the introduction of new technology at that time, which began to define the subsequent trends and directions of development in the world at large.

It is true that considerable effort was made in the 1970s to modernize Polish industry with the help of foreign credits and licences, but the emphasis was again on such traditional areas as mining, metallurgy, engineering or chemicals. The authorities failed to anticipate the depth of changes in the international division of labour ushered in by the radical restructuring of the economy toward electronics, miniaturization, computerization, automation and robotization.

By way of contrast, the 1970s were a very good period for Polish science, which through its innovative work received unprecedented opportunities for international contacts. The authorities, however, failed to capitalise on the research or to create the mechanisms that would make industry genuinely interested in the introduction of new technology, which usually involves large scale costs and an element of risk. Moreover, the lack of competition and shortages on the domestic market meant that it was easy to sell even quite outdated products, while the abundance of cheap labour was largely responsible for the lack of innovative drive, despite the large number of highly skilled engineers available. A lot of damage to the economy was done by the liquidation, carried out consistently in the 1970s, of small firms which formed the so-called 'small industry'. This was a clear step backward in comparison with the trends prevalent elsewhere in the world, as many countries began to appreciate the merits of small firms, their flexibility, low costs and high efficiency.

In Poland, these moves were supposed to facilitate the introduction of new technology by state industry administration. One example of a successful operation of this kind was the commencement of production of a Fiat-licenced small car in Bielsko-Biała where, simultaneously with the building of a new plant, a number of small enterprises were modernized and transformed into branches of the car firm. However, the majority of such moves did not produce the desired results. In many instances the local industrial firms were the only producers of thousands of simple articles. One the small factories was taken over by the giants, their production profile was changed to suit the needs of the new master and away from the needs of the broader economy. Therefore, these were cases of restructuring that could be regarded as having a negative effect on the national economy where some of the negative effects of these moves appeared only some time later, compounding the difficulties associated with the reduction in productive factor availability.

The crisis, combined with the sanctions imposed on the Polish economy in the wake of the imposition of martial law in 1981, hit hardest the most modern plants, using foreign licences and co-operating on a large scale with Western firms. Paradoxically, the position of outdated plants was consolidated as a result, just as the awareness of the need for radical restructuring of industry with a view to overcoming the crisis began to increase. Nevertheless, the strategy for economic development did undergo some revision in the early 1980s and attempts were launched to introduce economic reforms that placed emphasis on **efficiency** and restructuring of the economy; **intensification** rather than further **extension**. Similar moves to reform the economy have been embarked upon in the Soviet Union, Hungary and Bulgaria. However, the extent of the differences in the efficient use of resources between East and West is perhaps best illustrated through the following example. It is paradoxical that the CMEA countries have a lower per capita net material product than developed capitalist countries while their per capita consumption of energy is identical; that of steel is 40 per cent higher and that of cement is also higher than in western industrialized countries. This example proves that the problem of restructuring also concerns internal relations within the CMEA and relations between the CMEA and the EEC, which are not elaborated upon in this article.

The Need for Modernizing Poland's Economy Through Reforms

The negative processes occurring in the Polish economy became evident to most observers in the latter half of the 1970s. However, despite public awareness of the need for change, the launching of definite reforms proved to be a bumpy road full of conflicts and tensions despite formal declarations of broad support. The problem was that public consensus was obtained on the need for economic reform without determining its content, details of its programmes and ways in which it might be introduced.

The scope of economic problems which surfaced in the early 1980s was so vast and the domestic political situation so complicated that the conditions under which reform was undertaken proved to be extremely difficult. The authorities' preference for known and proven solutions meant that initially there was no room for new technology and modernization. Nevertheless, recognition of the need for a profound restructuring of Poland's economy was explicitly taken into account in the 'Guidelines for the regional development plan up to the year 1995', prepared by the Government

Planning Commission (Komisja Planowania przy Radzie Ministrów, 1986; Dawson, 1988). However, Poland's difficult balance of payments situation, combined with internal political divisions, meant that there was no possibility of effective action.

It was initially assumed that all the planned changes would be introduced in succession under the supervision of the central planning bodies and ministries. Such a provision was included in the aforementioned planning document. However, the rather unsatisfactory experience with the implementation of reforms led to the disbanding of a number of ministries in the latter half of 1987 and curtailment of the control and supervisory powers of the surviving ones, whose staff were cut at the same time. The Government Planning Commission was practically closed down in the middle of 1988.

More recently the government has sought to set economic processes in motion while cutting bureaucratic interference down to a minimum and delegating various powers to voivodship[1] and local authorities. (A provision for the equal treatment of the state, co-operative and private sectors has also been included in economic legislation). It should be noted, though, that such delegation of powers has also occurred in the past and has never been rescinded by subsequent legislation, but their effect was neutralized by taxation and the decisions of the administration at various levels. As a result, the 'new' powers have been received without much enthusiasm by society and especially by people interested in investing in new economic activities.

Bigger hopes are pinned on the reform of the banking system, whereby the state monopoly and rigid administrative rules were replaced, as of January 1, 1989, by more flexible economic tools, i.e. variable interest on deposits, credits, etc. It is too early to predict to what extent these moves will really contribute to the stimulation of economic activity and reward entrepreneurship to the extent that individuals or firms believe that the rewards are great enough to take a risk and to make an effort. So far, trading has been more profitable than manufacturing, which means that in many areas of the economy there is a regression in manufacturing or at least stagnation and in many instances the process of disinvestment has not been arrested.

To date, it is hard to define accurately the extent and direction of the restructuring of the Polish economy, which after all is a prolonged process. The change of structure is also the outcome of dozens of choices made by politicians, managers and ordinary people on all planes of social activity. The changes taking place in microstructures, however, are not big enough to affect perceptibly the macrostructures where the scale of the effort and its ultimate effects will only be revealed over time.

The fundamental goal of these changes should be the improvement of efficiency, both on an enterprise level and at the national level. In a centrally planned economy undergoing transformation into a decentralized state economy, systemic changes must pave the way to restructuring. In Poland, the restructuring moves are strictly tied to wider economic reforms (Kruczała, 1986). This is a source of both strength and a certain weakness in the restructuring processes. The systemic changes that are being introduced were designed to stimulate desirable structural transformations. However, these changes are being introduced only gradually and modified if necessary.

In the long-term programmes encouraging structural transformations, priority in investment will be given to projects ensuring the reduction of unit consumption of energy and materials (Komisja Planowania przy Radzie Ministrów, 1986). There are also calls for a better utilization of domestically available raw materials, for example, through the development of the food processing industry. The revitalization of Poland's national economy is to be achieved through:

- An accelerated growth of selected sectors of industry or groups of products resulting from the creation of more favourable conditions for their development.

- The initiation of selected moves aimed at equipping the national economy with efficient technological processes, advanced materials and machinery. The implementation of these undertakings will create the necessary conditions for the accelerated growth of selected areas of manufacturing and in the long run will be the basis for the emergence of specialization within Polish industry.

- The execution of selected programmes designed to provide industry with advanced materials and processes. On this basis, enterprises will design their product range and production programmes. This programme encouraging structural change is aimed principally at assisting various kinds of small-scale undertaking, but selectivity will be based on efficiency criteria at enterprise level.

The restructuring of the economy will therefore take place as a result of the combined actions of the central authorities and autonomous decisions within enterprises. The former will exert their influence through the use of a whole range of economic instruments that will encourage the generation, transfer and use of science and technology in industry, including the starting of production of advanced materials and components. These will be stimulated through the system of Government orders, financial assistance, credits and other economic and financial incentives.

The Barriers to Restructuring

Understandably, a broad restructuring of the economy requires very large scale investment. This results from the necessity to replace existing and outdated machinery and technology in many industries with new plants and processes. There is also a need to provide job retraining for a high proportion of the existing workforce. A whole set of activities is involved and their scope depends on the available resources, international connections and the place a given country occupies in the international division of labour.

A large foreign debt acts as a major impediment to restructuring activities, which must consequently be spread over long periods of time. However, a number of limited stop-gap measures can be employed with a view to bringing into play all the currently available resources and maximizing their effectiveness. This goal is served, for example, by the stabilization of the domestic market, the curbing of inflation, and improving the management of production supplies. While these are only introductory steps, which will precede restructuring proper and rely on incremental rather than major charges, they are important and may determine the future shape of the restructured economy.

These restructuring efforts must also take into account a number of factors internal to the national economies rooted both in economic and social structures and in the public consciousness. This signifies the necessity to overcome physical structural barriers which are characteristic of Poland's development and the equally important psychological barriers, which are the sum total of frustrated hopes, unfulfilled expectations and psychic depression and whose emergence is facilitated by the hardships of daily life.

In individual regions, the possibilities for restructuring depend, among other things, on the current pattern of the economy, which has been shaped by the course of history, the demographic situation, (Ciechocinska, 1983) the level of training of staff (Ciechocinska, 1986) and the state of the natural environment (Kozłowski, 1985). In the case of some branches of the Polish textile industry, mining and metallurgy, the present pattern and spatial distribution of industry has a tradition going back 150 years and it is easy to name regions and plants whose history goes back even further (Ciechocinska, 1980). In any discussion of restructuring, therefore, it is necessary to identify factors that encourage continuity and stability. It is also important to differentiate between radical restructuring and development that does not necessarily have to be identified with the creation of new production capacity. If a moderate, mainly technological modernization takes place in a traditional industrial region, ensuring greater efficiency of production structures without replacing them, then it is not a case of restructuring but of development.

The emergence of the former is usually the result of development of a new kind of productive activity which leads to structural changes between industries and the emergence of new industries. But even this is an oversimplified approach. In Poland's case, it is also necessary to ensure that changes take place in the system of management and working practices, i.e. in the mode of functioning of both the public and private sectors, which can either stimulate or hamper restructuring in the broader sense of the word.

These are by no means the only barriers which have to be overcome in the process of restructuring the economy. The impediments listed above raise questions as to the role of spatial planning, based on a broad geographical and economic analysis, in meeting the needs of the country, its regions and local communities.

Selected Regional Problems of Restructuring

Collective examination of Poland's regions suggests that numerous factors need to be taken into consideration when it comes to making decisions about local restructuring. They stem mainly from regional limitations to economic growth, which have a varying impact on the geographical environment. This chapter will discuss only four of the more important factors, but it is recognised that in most cases there are a whole range of features that have to be taken into account in any strategy for the development of a given region.

The next section deals with labour shortages, the physical depreciation of fixed assets, the concentration of unfinished investment projects in the public sector of the economy, and ecological degradation. These factors are discussed only in relation to those regions which ranked the highest in this respect. The problem of spatial differences which are characteristic of the typical features of Poland's regions are ignored; attention is focused instead on the problems in selected areas (Figure 17.1).

Regions with labour shortages

The regions with the biggest labour shortages, exceeding 2.5 per cent, of total employment, were identified on the basis of an indicator which relates the number of registered vacancies to total employment in the public sector of the local economy in 1984. The problem regions include both the biggest agglomerations such as Gdańsk, Katowice and Warsaw, industrial regions such as Bielsko-Biała, Jelenia Góra,

Legnica, Tarnów or Wałbrzych and isolated less developed regions such as Biała Podlaska or Chełm (Figure 17.1). The shortage of labour in the biggest agglomerations and industrial centres was so large the problem spilled over into neighbouring regions creating shortages there, for example, in Skierniewice voivodship (Ciechocinska, 1985).

It should be noted that labour shortages occurred even in periods when demographic analysis showed steady growth in the population of working age and a much smaller growth of employment in the public sector in the aftermath of the economic crisis of the early 1980s.

From the point of view of restructuring, this presents a fundamental problem in terms of the management of labour resources, especially in the regions which have traditionally been the target of migration and at the same time have the largest number of vacancies. These problems are completely unknown in countries in which there is unemployment (Kornai, 1980).

It ought to be emphasised that labour shortages were essentially restricted to the public sector of the economy and was its most striking feature throughout the period when the command and quota system was in force in the economy. This was a situation in which the bureaucratic administrative structures managing the economy kept wages and incomes at a low level. On the other hand, the same system encouraged the severance of any relationship between productivity and efficiency and pay, which in combination with the absence of any mechanisms for negotiating pay increases, was responsible for the fact that workers were disinterested in the introduction of new technology and the modernization of production processes.

The bureaucracies were not interested in such changes either, because their introduction could to some extent upset the existing alignments and structures. Moreover, the introduction of new technological and organizational structures could lead to a failure to meet production targets, problems with the availability of components, etc., because the plan was rigid and did not envisage major flexibility. In these circumstances, room for new technology and technological progress was limited to investment and modernization plans which defined the scope of change in great detail. In this way, reviews of the state of technology and its introduction into Polish industry as permanent features in the widest development of production were eliminated as a matter of principle. Thus, the promotion of technological progress, which was to be served by rationalization and innovation on the shopfloor, was strictly subordinated to bureaucratic administrative structures which neutralized such efforts, despite special institutional reforms.

In these circumstances, it was no accident that whenever a possibility emerged

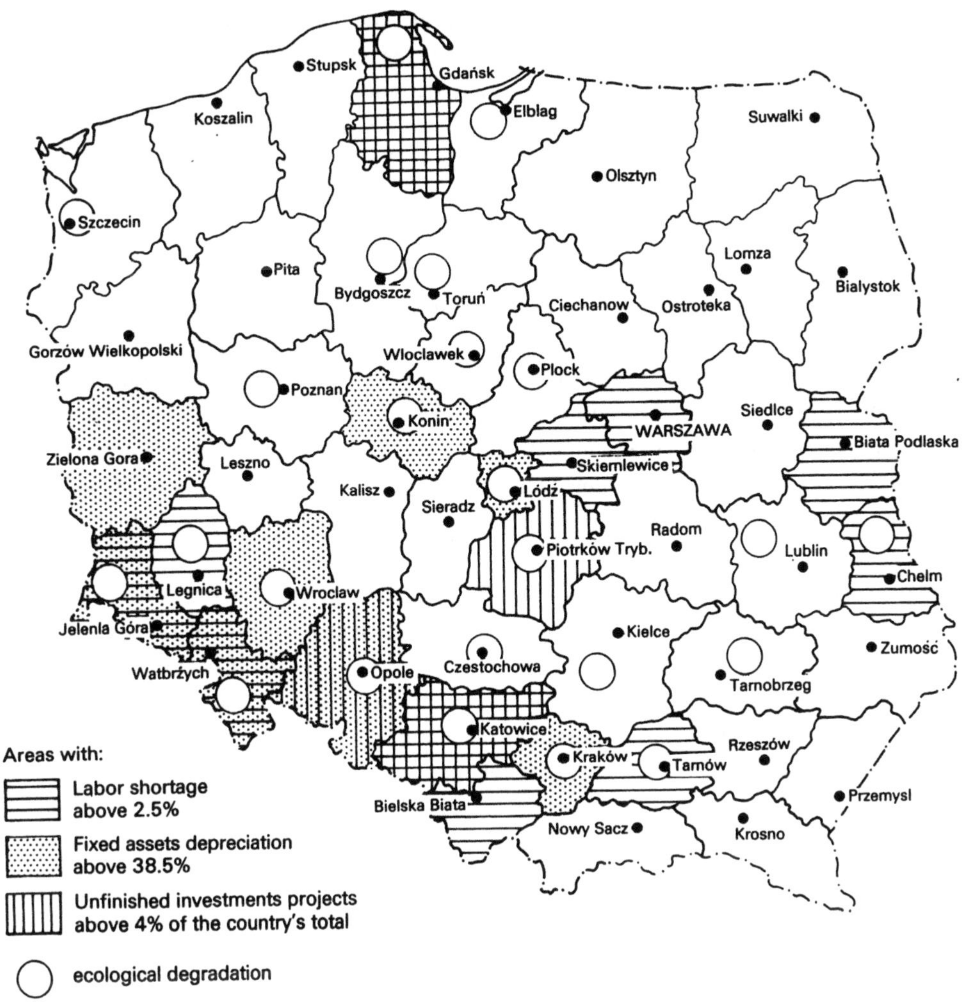

Regions (Voivodships) in Poland according to the territorial organization of 1975.

Source: Own elaboration based on State Planning Commission (1986). The assumptions of spatial development of the country until 1995, Warsaw.

Figure 17.1: Impact of selected factors upon the national strategy of Spatial Planning in Poland in 1986.

of finding employment outside the public sector on the local labour market, workers eagerly capitalized on it. However, it should also be remembered that for decades in many regions such possibilities were very limited, and it was only in the 1980s that the options of working abroad and in the private sector became widely accessible. Employment in foreign firms operating in Poland, which will be referred to later, became the best examples of this movement of labour.

Regions with the highest degree of physical depreciation of fixed assets

The group of regions with the highest degree of physical depreciation of fixed assets in the public sector, exceeding 38.5 per cent, i.e. higher by a half or more than the standard deviation from the national average in 1983, includes old industrial regions such as Jelenia Góra, Opole and Wałbrzych, some agglomerations such as Cracow and Wrocław, and other regions such as Konin or Zielona Góra.

The situation is not the same in all the regions named above. When mineable resources become depleted, such as the lignite deposits in Konin or coking coal deposits in Wałbrzych, restructuring took the form of the introduction of new functions. When it is a case of outdated processes in outdated plants, the decisions are more concerned with the continuation of change in the geographic patterns and modes of production, as in the case of Jelenia Góra, Opole, or Zielona Góra. The economic troubles of the early 1980s are responsible for the fact that depreciation in the broad sense is becoming one of the fundamental problems in the Polish economy. In some cases, it has advanced so far as to make it necessary to close down individual plants as uneconomic or unsafe. It can be expected that the number of closures will increase in the depreciated plants for economic reasons, but also because they have an especially adverse effect on the natural environment. In this way advanced physical depreciation becomes a factor contributing to the restructuring of the economy. It should be added that until recently plant closures and bankruptcies were unknown in the Polish economy where enterprises could exist regardless of their economic performance. It was only after the recent introduction of economic reforms that the closing down of loss-making enterprises became possible.

So long as the concept of closure remained a loose proposal, this view had many supporters. However, when the Lenin Shipyard in Gdańsk was closed down on the grounds of negative profitability and a list was drawn up of other enterprises that should be closed, owing to aged technology, huge environmental damage and a real threat to human life and health, more people began to support an alternative view which emphasised the necessity of modernizing such plants through programmes which included the introduction of new technology.

The present problems of economic development in Poland illustrate the difficulty that can arise from fixing a hierarchy of social, political and economic considerations; where the last traditionally received the lowest priority. Poland's unfavourable economic and industrial situation in general is compounded by the fact that most plants in individual industries date back to roughly the same period of time and are hit by physical depreciation more or less simultaneously; this concerns power plants, steelworks, etc..

It is also significant that the aforementioned process of concentration of production led to a situation in which almost half of the employment in the public sector of industry (47.3 per cent in 1982), which remains the biggest employer, was concentrated in just 8.8 per cent of the plants. Therefore it is not going to be easy to change both the organizational pattern of Poland's industry, and its production and employment profiles, even though it is commonly recognized that from the point of view of prevalent world trade both are very outdated. Without a massive injection of new technology, no way out of the economic and political impasse is in sight.

The high degree of physical depreciation of fixed assets is the result of the adopted strategy of overcoming the economic crisis, which consisted of protecting the level of consumption at the expense of the economy (Graham and Ciechocinska, 1987) This made it possible to reduce considerably social tension in the early 1980s, but the cost of it will have to be borne over many years into the future.

Studies in economic geography of the regional distribution of the greatest concentration of completely depreciated assets in the public sector of the economy in 1983 (based on the criterion of the existence of more than four per cent of fully depreciated assets across the country), revealed that the worst situations were to be found in the five biggest agglomerations which are also the biggest industrial centres; namely in Katowice, Cracow, Lódź, Warsaw and Wrocław. Therefore the restitution of the aged non-productive assets in the agglomerations appears to be a foregone conclusion, whereas the reconstruction of productive assets would create better prospects for restructuring the economy. However, the availability of funds limits the possibility of replacing aged productive fixed assets.

In the years 1986-90 investment spending in the public sector of the economy is expected to total some 10,000 billion Zloty, of which two-thirds will be spent on the productive sector and the rest on public consumption. In constant prices, the average yearly expenditure in the productive sector (1986-1990) will be 18 per cent higher than in 1985. It is estimated that 5,300 billion Zloty must be spent on arresting the depreciation of productive fixed assets in this period, which leaves 1,350 billion Zloty for the growth of such assets.

In the light of these estimates, physical depreciation only appears to facilitate restructuring moves which in reality are very much constrained. It can be regarded as paradoxical that while the need for restructuring is hardly questioned by anyone, economic realities seem to force the authorities to accept the petrification of the existing regional structures.

The concentration of fully depreciated assets in the five regions named above and their importance to the national economy are responsible for the fact that more and more people are beginning to subscribe to the view that embarkation on broad restructuring in the years 1986-90 is unlikely to be feasible. The process can only take off after 1990, when depreciation will no longer be the main concern of virtually every enterprise as it is at the present time. Of course, this view depends on the way the vision of future development of the economy and the preferred strategy are formulated. However, this is not to say that the present economic-spatial structure of Poland can be extrapolated into the future without any limitations.

The regions with the largest concentration of investment

One of the reasons for the aforementioned economic crisis in Poland was the embarking on many particularly capital-intensive investment projects in the 1970s, of which only a few could be completed owing to the country's adverse balance of payments. The bulk of the projects are of very large scale and their economic usefulness is often questioned owing to the scale of the undertaking, the technology to be used, etc.. Nonetheless they remain as realities which tie up funds and thereby limit the restructuring potential.

Among the regions with the largest concentration of unfinished investment projects, each of which accounts for more than four per cent of the overall money available for investment if the projects are to be completed, one can identify three of the biggest agglomerations, i.e. Gdańsk, Katowice and Warsaw, and two industrial regions, Opole and Piotrków Trybunalski. The last named region found itself in this group on account of the lignite strip mine under construction in Bełchatów and the construction of a huge power plant.

The fact that Katowice and Warsaw belong to the regions with both the biggest concentration of such investments and the highest degree of physical depreciation of fixed assets is not tantamount to the solving of the problem of restructuring. This is due to the fact that the funds already spent on the implementation of a given project cannot be moved elsewhere. Once a project is started, it must either be continued or dropped if funds are not available or are needed for other urgent goals. In the 1980s this has been a typical situation facing decision-makers in

Poland.

Regions with the highest degree of ecological degradation

Ecological degradation affects in particular the areas in which there are the biggest and most dangerous sources of dust emission, gas, sewage and waste. In Poland there are 27 areas in which natural balances have been upset as a result of anthropopressure. This means the degradation of biological systems which have lost their resistance and in which there exist threats to health, and the risk of environmentally conditioned diseases. These areas are estimated to occupy a total of 35,200 square kilometers, ie. 11.3 per cent of the land area of Poland which is inhabited by 13 million people, or 35.1 per cent of Poland's population.

Varying degrees of ecological degradation occur in almost all agglomerations and major industrial regions. This is illustrated in Figure 17.1, without specifying the kinds of pollution present in each region. The pollution of Poland's rivers has also been ignored, despite the fact that half the length of them is outside any class of water purity. Similarly, the regions and smaller areas where water is scarce, due to poor management of their resources, have been ignored.

Ecological degradation affects both people and industry as it becomes a barrier to economic development (Carter, 1986). Apart from environmental engineering projects, such as the building of sewage treatment plants or the installation of more efficient dust precipitators, structural changes are also necessary. These changes are necessary in order to limit the use of dirty technology and to ensure its replacement with less polluting processes. These are difficult decisions, which sometimes must be taken in order to comply with international conventions regarding environmental protection. The present pollution levels do not facilitate the introduction of new technology into industry more generally and entail considerable investment. These are important aspects of the restructuring of the economy.

Poland's contribution to the pollution of the environment is, however, not simply one of domestic concern. Pollutants know no boundaries and travel from the country in which they are released to do damage in another. With the availability of satellite photographs the situation in Poland, and in other East European countries for that matter, is being closely monitored abroad, especially in the Scandinavian countries, which become importers of pollution. The accumulation of air pollution coming from Poland, Czechoslovakia and East Germany is being monitored in Holland, for example. As a result Sweden has declared its intention to participate in the funding of pollution control projects in Poland. The thesis about the lack of domestic funds for the introduction of new, cleaner technology ought to be revised in the light of such moves.

In view of Poland's problems with servicing its huge foreign debt, the Swedish government has proposed to convert its slice of it into funding for new technology aimed at improvement in the quality of the environment in Poland. The Swedes are especially interested in a reduction of the pollution and toxicity of the Vistula, which flows into the Baltic and thus causes damage along the shores of Scandinavia. In 1989, a proposal for the establishment of a European Environmental Protection Foundation was submitted to the Geneva-based UN Economic Commission for Europe. Its signatories, led by Sweden's Riksdag, insist that the Foundation finance some environmental projects in Poland and in other countries bordering on Scandinavia. This could be one way of starting to build installations for the protection of the environment. At the same time it would become possible to repay foreign debts by investing Zloty funds in new technology for Poland. This external action, enforcing the modernization and tidying up of the Polish economy, shows how in the contemporary world new technology may be imposed through political influence.

The Search for the Ways of Modernizing and Invigorating the Polish Economy

The efforts launched in the late 1980s to consolidate the nation around a programme of political and economic reforms create, potentially, more favourable conditions for new economic initiatives and the development of new technology in a generally very difficult economic situation. These moves appear to be extremely valuable if very much overdue. Nevertheless, from the point of view of the invigoration and modernization of the economy, it is worth examining three additional actions which are associated with the introduction of new technology in a variety of ways.

The first of these actions is the development of state-owned high technology plants. Such plants already exist - although they account for only 2.3 per cent of total employment - and are promising from the point of view of the economy as a whole. It should be emphasized that some of them, especially in the electronics industry are the product of the implementation of socio-economic plans in the 1970s. The CEMI Semi-Conductors Research and Production Centre in Warsaw is one example of such projects, falling into the class of the so-called 'key state-run industries'.

On the other hand, it was the activity of private firms and individuals that has been responsible for Poland's rapid computerization and which actually surpasses the current needs of the economy. It is hoped that, although the possibilities it offers are not fully used at present, it is a good harbinger of modernization and restructuring.

The third group consists of projects dating back to the same period. They are, however, the result of grass roots economic initiatives launched as a result of advancing decentralization and a gradual growth of the autonomy of enterprises. Many managers took the modernization of their plants into their own hands, taking advantage of opportunities as they presented themselves. There were a variety of reasons for such moves. For example, in Łódź, traditionally the hub of Poland's textile industry, the electronics industry has been developed through a series of small steps with a view to achieving a more balanced labour market in the longer term. Depending on future trends, this could thoroughly change the face of the Lodz Industrial District.

One result of the first kind of activity mentioned above is the emergence of the Toruń Industrial Centre in the Lower Vistula region. High technology requires ties between industry and science. The Toruń Industrial Centre has as its base the Copernicus University, the modern electronics industry built in the 1970s, and the existing infrastructure provided by the town and other industries. In this respect, there are some similarities with Łódź, which has both a university and a technical college.

The third kind of activity is related to the operation of the private sector of economy, which includes foreign-owned enterprises (Manikowska et al, 1985). While relatively small, they are quite expansive and dynamic. However, difficult experiences of the past are responsible for the fact that, despite assurances coming from the authorities, they are reluctant to invest heavily in the local economy, fearing that the authorities might change the rules of the economic game again. Their contribution to the transfer of new technology is rather limited. In the main, they capitalize on the relatively high level of professional qualifications, cheap labour and the availability of many natural resources which, in a processed form, are sought after on the Western markets. Given a proper economic policy, such enterprises will most likely be able to employ many of the latest achievements of Polish science, such as the marked advances in the research into superconductivity, etc..

It is hard to predict the extent to which the assurance on the part of the authorities and progress in the decentralization of management and of the operation of the state sector of the economy, combined with the provision of advantageous conditions for the private sector, can stimulate the transfer of new technology through this channel in the future (Karpiński, 1986). It is both a question of mutual trust and the volume of dividends the Polish authorities will be inclined to guarantee to foreign businessmen.

It is self-evident that, in the case of Poland, any effort to extrapolate into the future from the present organizational structure of industry rules out any possibility of overcoming the difficult political and economic situation, while the introduction of new technology and new arrangements may be one of the key factors that can effectively contribute to such change.

Note

1. Voivodship = administration unit at the regional level.

References

Carter, F.W. (1986) Post-war pollution problems in Poland, paper presented at the Institute of British Geographers/Institute of Geography of the Polish Academy of Sciences, British-Polish Seminar, 7th-9th July 1986, London. Geographia Polonica forthcoming.

Ciechocinska, M. (1980) Wskazniki i zasady lokalizacji przemyslu stosowame w polskiej praktyce planowaoania regionalnego w latach 1944-1979. Gospodarka Planowa, 4, 224-230.

Ciechocinska, M. (1983) Charakterystyka struktur zatrudnienia na obszarze stol. woj. warszawskiego. Proba typologil zroznicowan przestrzennych. Zespół Koordynacyjny PM Podstawy Przestrzennego Zagospodarowania Kraju. Biuletyn Informacyjny, 43, 79-118.

Ciechocinska, M. (1985) Charakterystyka struktury zatrudnienia na obszarze funkcjonalnego makroregionu Warszawy. Zespół Koordynacyjny PM Podstawy Przestrzennego Zagospodarowania Krajn, Biuletyn Informacyjny, 48, 25-56.

Ciechocinska, M. (1986) Niveau descolarisation en Pologne et relation ville-campagne, Les Cahiers de Fontenay, 41-43, 151-168.

Dawson, A.H. (1988) The Economic Geography of Poland in 2025, Geography, 347-351.

Graham, L.S. and Ciechocinska, M. (eds.) (1987) The Polish Dilemma. Views from Within, Westview Press, Boulder, London.

Karpinski, A. (1986) Restrukturyzacja gospodarki w Polsce ie na świecie, PWE, Warszawa.

Kassenberg, A. and Rolewicz, C. (1985) Przestrzenna Diagnoza Ochrony Srodowiska w Polsce, PWN, Warszawa.

Komisja Planowania przy Radzie Ministrów (1986) Program przedsięwzięć stymulujacych zmiany strukturalne. Załacznik nr 2 do Narodowego Planu Społeczno-Gospodarczego na lata 1986/90, Warszawa.

Komisja Planowania przy Radzie Ministrow (1986) Założenia planu przestrzennego zagospodarowania kraju do 1995, Warszawa.

Kornai, J. (1980) Economics of Shortage, North Holland, Amsterdam.

Kozłowski, S. (1985) Procedura badań przyrodniczych na etapie projektowania inwestycyjnego. Biuletyn KPZK PAN, 125, 174-205.

Kruczała, J. (ed.) (1986) Spatial Economy and Regional Development During the Economic Reform. Akademia Ekonomiczna, Krakow.

Manikowska, B., Matykowski, R. and Stryjakiewicz T. (1985) Tendencje lokalizacyjne i struktura przestrzenna przedsiębiorstw polonijno-zagranicznych w Polsce, Czasopismo Geograficzne, 3-4, 305-377.

Thwaites, A. and Alderman, N. (1988) The Location of Research and Development: Retrospect and Prospect, paper presented at the Regional Science Association, European Summer Institute, Arco, Italy,

18. POLICIES TO SUPPORT INNOVATION IN EMILIA-ROMAGNA: EXPERIENCES, PROSPECTS AND THEORETICAL ASPECTS

Claudio Tolomelli

Introduction

Regional policies for technological development are now a consolidated and growing reality in numerous European countries and in the USA. Despite the increasing range of experiences and of empirical research, theoretical reflection appears, nevertheless, to be insufficiently developed, both with regard to consolidated models of regional action, and with regard to the relationship between regional industrial policies and those of other institutional levels (National Governments, EEC, Provinces, Municipalities etc.). Theoretical studies should also be extended to cover the complexity of factors which interact more and more with regional policies, such as economic and financial processes at the international level and social, territorial and institutional dynamics. The achievement of this complex goal requires intermediate steps aimed at building links between empirical studies and theoretical research.

The objective of this contribution is to examine some aspects of change reflected in regional policies promoting technological innovation, with a special emphasis on the relationships between innovative processes, industrial and territorial policies and the institutional role at the regional level. This examination will draw on past and present experiences in Emilia-Romagna; experiences that can be considered of specific interest and significance for more than one reason.

In the first place, this is due to the character of the productive structure of this region. It shows the distinctive trait of a well developed and varied economic system, characterized by a productive organization in which a strong presence of small and medium size firms (SMEs) is wedded to advanced technological and to a complex web of inter and intra organizational relations.

A second reason is that, for over ten years now, the regional government has developed policies of intervention to support technological change, even in the absence of specific legislative powers, and to provide services to enterprises and the transfer of technological innovations. Also of interest, and in many ways unique, is the growing and complex interrelation between the innovative processes inside of these firms and the action of the regional institutions, which have led to the creation of a new relationship between public and private operators.

In conclusion we must mention the keen interest that European and American researchers have shown in this productive and institutional reality (Brusco and Sabel 1981; Putnam et al, 1983, 1985), as well as the level of cultural development of very high order that assisted the development of the regional industrial policy, both at the analysis stage of the economic and territorial issues, and in the planning and realization of the projects. Consequently, the study of Emilia-Romagnan policies draws upon a decade of research useful for the understanding, not only of the evolution of the regional economic structure, but also of the related cultural and institutional dynamics (Bagnasco and Messori, 1975; Brusco, 1982; Nanetti, 1988).

On the basis of such experiences and of the current transformations, it is possible to bring new material to bear upon the analysis.

The following topics will be discussed in the course of this exposition:

- the possibility of intervention by the regional institutions in the industrial system, with some reference to the international debate on the role of the regional and local governments in this process;

- the policies implemented over the past few years and the analysis on which these policies are based;

- the analysis of some relevant transformations still in progress, focusing on the central importance of the relationships between firms;

- the changes in the role of political institutions and in the interrelation between industrial and territorial policies at the regional and local level; and

- the problems arising from the perspective of a more advanced policy of technological change and of the elaboration of a more comprehensive theoretical basis for regional industrial policies.

The Regional Power of Intervention

First of all, it may be useful to consider the power of intervention already existing at the regional level. The Italian case is, indeed, indicative of the increasing importance on the part of the sub-national level of government and of their right to intervene in the promotion of technological innovation and economic development. It is a process in progress both in Europe and in the USA, and is characterized by a wide range of experiences: it is possible to identify situations in which the regional level of intervention has been consolidated for a long time, and others which are still characterized by strong centralization; in other countries the process of decentralization is only beginning.

On these themes the debate has been very lively in the last decade, though it has been primarily concerned with juridic and administrative aspects (Rokkan and Urwin 1982, Hueglin, 1986). The debate, specifically oriented to the decentralization of industrial policies, has been very limited for a long time, and only in recent years has it been possible to see an increase in the number of significant contributions (Fosler, 1988; Schlekner, 1987; Antonelli et al, 1988).

In the Italian case, in the face of strong juridical limitations (Zariski, 1985), there has been a considerable increase in interventions (Lassini, 1985). Italian national legislation does not provide many possibilities for intervention by the Regions in the industrial sphere. Regions have full competence to intervene only in respect to crafts firms where Italian laws define crafts firms as those enterprises which employ less than 20 workers and where the entrepreneur himself works in the 'shop'. The Italian Constitution dictates the national responsibilities regarding industry, allowing very few possibilities for regional intervention in the provision of industrial land or financial guarantees for firms. Some limited opportunities for regional intervention derive from the delegation to regions of the application of some national laws, such as in the case of laws concerning energy saving. This situation has not prevented numerous regions, among them Emilia-Romagna, from developing policies of intervention through the actions of regional companies in which the Regional Government holds the majority of the shares.

This fact has allowed several regions to promote, in the last decade, important and complex actions, specifically in the fields of services to the enterprises, technological transfer, the promotion of innovative processes, and export activities (Regione Emilia-Romagna, 1984, 1986; Regione Lombardia, 1985). For these reasons regions appear to become more and more frequently an important subject of industrial policy.

This fact is also recognized by the European Community, which is increasingly looking to regions for the realization of its programmes which are designed to promote economic development and technological innovation (for example, Integrated Mediterranean Programmes).

Consolidated Policies

Emilia-Romagna has an area of 22,000 square kilometres and a population of four million inhabitants. Its economic structure is made up from more than 300,000 enterprises, 140,000 of which are artisan ones. It is the third industrial region in Italy and it is among the highest in terms of per-capita income. Over the last ten years the policies designed to support innovation processes have been based on a vision of the productive sector as one composed mainly of small, medium and craftsman firms, organized in industrial districts, and often of a mono-sectoral character.

Here we are concerned with a production system where the small and medium firms have been capable of achieving high levels of specialization, while maintaining a high level of productive and commercial flexibility and creating a complex network of relations which link them to other fields of industrial activity, as well as to the service sector. We have, therefore, a structure endowed with an effective adaptability and a capability for quick adjustments to technological changes and market fluctuations.

Regional intervention has been based on an analysis which refers to the themes of the 'peripheral economy', of the crucial role and the diffusion of SMEs, of a productive organization based on the industrial district, and of social and institutional complexity (Piore and Sabel, 1984; Sabel and Zeitlin, 1982). If, on one hand, some of these themes are peculiar to the Italian situation, on the other, similar elements of the economic structure and dynamics can be found in several European and American areas.

The model of the 'peripheral economy' goes beyond the traditional dualistic interpretation of the Italian economy, founded on a dichtomy between the industrialized North, centred on the triangle Milan-Turin-Genoa, and the generically underdeveloped South to encompass the 'Third Italy', characterized by SMEs. The model of the 'Third Italy' (Bagnasco, 1977; Bagnasco et al, 1978; Garofoli, 1981; Fuà and Zacchia, 1983) has emphasized the peculiarity of the development factors in the north-eastern area, from which emerge the following:

- the specific capacity of structuration linked to a high social integration;
- the diffusion of small property in agriculture and industry (Brusco, 1982, 1983);
- the strong labour values and culture;
- the wide consent given to one of the two major policy parties.

In a situation characterized by the 'urbanized country' (Innocenti, 1985), the spread of SMEs has taken place within the industrial districts (Arrighetti, 1982; Becattini, 1987). These districts are endowed with considerable agglomeration economies and with complex, although often informal, relations among firms (Gresse, 1988; Jurgen, 1988). In particular, the analyses of the development process in Emilia-Romagna in the 1970s, have pointed to the fact that the considerable growth of manufacturing activities has not been followed by the adequate development of advanced service sectors.

In the same way, even though the presence of many urban areas of small and medium dimension seemed to be at one time a positive factor for the increase in 'diffuse' industrialization, in the early 1980s such a structure did not appear to be the most suitable territorial configuration to allow the spontaneous development of advanced services, development that usually corresponds with upper level urban functions. From that point of view, the Emilia-Romagna reality confirmed the relationship among metropolitan areas, services, and the importance of these for economic and industrial development (Cappellin, 1988). The Region has therefore considered producer services to be a crucial element on which to concentrate industrial policy intervention.

Based on the vision referred to above and faced by the legislative impossibility of intervening by means of financial incentives, the regional government decided on a line of intervention that aimed to provide firms with a complete range of specialised services, which the firms themselves could only have created with great difficulty within their own organizations. Consequently, the policy has given birth to a series of specialized centres, mostly located within the industrial districts which deal with such different sectors as knitwear, ceramic tiles, agricultural machinery, bulldozers, construction, footwear, the aerospace component industry, and metal works. The services offered by these centres cover information on technologies, fashion trends, foreign markets and technological specifications. Quality control and technical and professional training are also within the scope of the centres.

A peculiarity of this policy is that the centres have been implemented through a partnership of the public and private sectors, with the creation of boards of directors of unique character, where political institutions co-operate with private economic organizations. Over the last two years the action of the Regional Government, apart from improving the quality of the existing centres and creating new ones, met the growing demand for information transfer on intersectoral technologies with a new instrument: a regional agency for technological development.

That agency operates on two lines. It acts as an interface between the firms and the main technological data-banks in the world and fulfils the task of organizing the demand for technological information. It also works to encourage and to assist the firms in the realization of important intersectoral projects in such fields as microelectronics and flexible automation. The latter field of activity is particularly important, taking into account how fundamental intersectoral technologies are for the development of the industrial sectors in the region.

Completing the picture of interventions designed to promote innovation are the incentives for craftsmen and for environmental technologies, as well as for energy saving programmes, oleodynamics and aerospace industry components, and finally high-level training in the field of polymers.

We can see how, over the years, a wide range of intervention was stimulated by the need to satisfy the demands of a highly differentiated productive activity, with a variety of local productive systems, different sizes of firms, kinds of product and different problems to be faced. The Emilia-Romagna model of intervention has been classified by regional scientists as producer services and innovation promotion.

The Evaluation of the Outcomes

The evaluation of the efficacy and efficiency of the interventions may be dealt with only within an advanced planning policy. Only the framework of a clear definition of objectives, resources and instruments of intervention, in fact, allows one to verify the outcomes. Such an approach applies both to regional institutions and to the service centres.

After the substantial failure of the attempts in economic planning at the national level in the 1960s, the Emilia-Romagna region has tried to affirm the strategy of 'possible planning'. That is, essentially, the evolution of planning from the system of objectives of macro-economic character, towards planning related to the crucial points of local development; in this scheme, planning refers to the action of the administration above all.

The adoption of a closer link between planning and budgeting and of a project-based action has proven significantly innovative in terms of the processes of regional action, and has strongly increased the possibility of evaluating these interventions. As regards the policy of promoting innovation, the problem has been dealt within two distinct areas: the evaluation of the interventions realised through laws, and the evaluation of the success of the activities of the service centres and the outcome of the integrated projects at the territorial level. In both cases the problems are complex.

The evaluation of the efficacy of laws is a more frequently studied theme at the theoretical level too. The analysis of regional laws, aimed at promoting technological change in the crafts firms through financial incentives has shown that this instrument is losing the capacity to impact on the regional economic tissue, especially in cases in which the financing procedure has led to the scattering of resources amongst a large number of firms, each of which has received limited sums. More generally, the sole financial incentive has less and less efficacy, because of the growing complexity of the determinant factors of innovative investment choices. As regards the service centres and the territorial projects, the Region has promoted a study that aims to build more formalized systems of evaluation.

In the meantime, an empirical approach has been suggested: the rate of self-financing. This index measures the suitability of supplied services to be sold in the market with respect to the needs of the firms. But, in the case of important projects which are promoted and paid for by the Region, the resultant growth in public financing leads to a lower index of self-financing, even though the centre is carrying out its precise duties. It is therefore necessary to develop a better system of evaluation. The problem is that the objectives of the centres are twofold: one is the supply of tried and tested services (in that case it is possible to use the market as index of suitability), the other is the realization of projects which have to face complex and difficult problems and which have a high risk of failure (in this case the rate of self-financing is not adequate and it is necessary to estimate the degree of fulfilment of the project and to compare the costs against the outcomes).

The solution seems to lie in a comparison of outcomes with respect to the objectives defined by the board of directors, objectives which often are a mix of different scopes and projects, with different risks of failure, financial sources, and time-scales for realization (Tolomelli, 1988). In addition to this kind of evaluation, which is undertaken first of all by the board of directors of the centres, the region verifies the number of enterprises that are involved, their distribution by type and their territorial location, as well as the quantity and quality of the supplied services. The presence of entrepreneurial organizations on the board of directors allows further control of the suitability of the supplied services to meet the needs of the firms.

The eight centres set up so far are associated with several hundred enterprises and supply services to thousands of firms, including co-operative and crafts workshops. The overall evaluation of the outcomes shows widespread involvement of firms operating in the districts and particularly of those that are able to utilize the services. There would also appear to be efficient use of resources provided by the centres. If on one hand the expected outcomes are generally achieved, on the other hand some problems arise with regard to access to the services for the smaller firms and for the ones located far from the centres. This is one of the most important problems the region is facing with regard to this type of intervention.

The More Recent Years

The technological revolution of the last ten years has brought about two specific demands of the regional productive system. Firstly, there is the demand for support in the process of technological change from the most advanced firms that are looking at national and international markets and are operating on the cutting edge of technology. For them, the crucial question is how to find the best operative relation to link scientific and applied research to the production process; in this field the problem is to ensure the highest quality of services and technological transfer by means of national and international linkages.

Secondly, there is the demand that comes from the majority of Emilia-Romagna firms, including the lagging ones, for access to the support for technological change offered by the regional government and various research institutions; in this case the problem is how to ensure the greatest diffusion of services in territorial terms.

To answer the first demand for support in the process of technological change, a project has been set up aimed at creating a scientific and technological pole of national importance in the town of Bologna, along with other specialized structures. The Scientific Pole of Bologna will be the core of a regional system for research, innovation and services to firms. To promote the pole, the Region has issued a specific law that provides financial support for technology transfer projects promoted by private and public organizations.

As for the second demand, several interface structures are now under experiment at the local level. The problem has no easy solution, because it is not sufficient to give information on the activities of the service centres, rather it is necessary to understand the real needs of the firms and to extend the entrepreneurial culture towards the use of advanced services.

The necessity of dealing with the financial aspects of the promotion of technological change is a further challenge. For this purpose a line of intervention has been concerned with the improvement of the financial instruments to support the innovative process. Accordingly, a financial agency is being formed to guarantee investments in innovative projects. The agency will examine and select projects and furnish financial guarantees to the credit institutions.

In anticipation of the European Single Market the Region has planned appropriate programmes, both of information and of intervention. The main one is a project aimed at diffusing the capability of operating with guaranteed quality, both of products and services, because the guarantee of quality will be an increasingly necessary feature of the Single Market. For these reasons, the Region is promoting a specific project together with ENEA, the national Agency for nuclear and alternative energy (see Appendix 1 for a description of relevant agencies and organizations). ENEA, thanks to its field of activity, has the necessary know-how to help firms to adopt a more advanced quality-oriented approach. The first sectors in which this project will operate are mechanics, metallurgy, buildings and electronics.

Finally, we may mention the complex theme of the relationship between development and the environment. This theme is being addressed on the productive side through a project about environmental technologies. The origins of this project are similar to those of regional planning, aiming to link promotion of competitive capacity with environmental safegards, a more and more crucial problem for an industrialized area like Emilia-Romagna. The cultural roots of such a policy are found in important and fruitful debates on the relationship between economics and ecology (Commoner, 1973; Gerelli, 1974; Cervia Ambiente, 1985), which were developed in the early 1980s. In the context of the different themes concerning growth and environment, the Region chose to originate an experimental action to promote the use of clean technologies. This choice does not mean an undervaluation of the political and institutional aspects of the problem, but reflects an awareness of the importance to start and to provide some concrete answers on the technological side. The project is intended to support research into, and the use of, productive technologies with a lower consumption of raw material, with less environmental impact, and with reduced production of toxic and polluting waste. In that field the Region and ENI (National Oil Company) have signed an agreement, to set up in Ravenna a centre designed to promote innovation in environmental technologies.

Another project has been established in the sector of ceramic tiles. This project, a joint venture between national and regional technological agencies (ENEA and ASTER), is aimed at promoting the adoption of more advanced technology and organizational systems. One of the objectives is the reduction of the environmental impact of the tile production process through the recovery of toxic components deriving from production muds.

From this it can be seen that Emilia-Romagna Region has implemented or is setting up a wide range of policy interventions from amongst the ones suggested by academic studies (Garofoli 1988), such as science parks, service centres, technological agencies, local development agencies, consortia of firms, laws promoting job and enterprise creation, financial instruments, information networks, etc.. This is a significant result that cannot be underestimated and which has required more than ten years of effort. The real challenge now for the Region is to govern in an integrated way such a complex system of instruments. It is also important to verify whether any changes in the instruments or organizations are required in the light of the transformations taking place in recent years.

Some Recent Transformations

While the regional government is carrying out the above mentioned projects, the important transformations that have taken place in the economic structure compel us to consider if and how the industrial policies pursued so far will be able to cope with the problems of the 1990s. In the 1980s technological aspects proved to be central to ensuring the competitive edge of firms; the 1990s scenario seems to point to the centrality of internal and external organizational arrangements of firms, with a special emphasis on inter-firm relations and between firms and those entities that have primary roles in the innovative process at the national and international level.

The changes that in different ways substantiate this overall transformation are the following:

i) The regained competitiveness and the strategic role of large firms, which have changed their function towards the aggregation of smaller units. This recovery has been possible mainly through the introduction of automation and the adoption of more flexible internal set-ups and relationships with suppliers. The new characteristics of technological competition in the marketplace has created a situation in which large size seems more appropriate for ensuring effective operation in world markets and in maintaining stable contacts with the main technological and financial flows at the international level.

The enhanced strategic role of the big firms has introduced new problematical relations in a productive structure composed mainly of small and medium firms; more than cutting market shares or bringing about crises and difficulties, it has, for the small firms, provoked changes in the balance of forces, in the respective roles in the market and in the acquisition of quotas of added value, as well as in the capacity to affect the nature of strategic choices regarding products and technologies.

ii) Closely related to the recovery of the leading role by the big firms, the phenomenon of share-holding in, and acquisition of, small innovative firms by big Italian and foreign corporations is increasingly occurring. This process, widely reflected at the national and international level, may induce deep modifications in the regional productive map. In detail we see that the traditional organization of the industrial districts will be overshadowed and replaced by an organization based on groups of enterprises, with a variety of interconnections and tied to Italian and foreign corporations. It is important to observe how, by means of such operations, groups of enterprises not only are taking over individual firms, but through them can enter the web of relations between firms within the districts and integrate them into their corporate network.

These processes can be judged in different ways. On the one hand, they are indicative of the attractive power of the regional productive system, both for the quality of the individual firms and the overall quality of its economic structure. We can judge this phenomenon to be positive, for in this process SMEs are able to link themselves to important trusts, with considerable opportunities for drawing resources from a wider technological and financial pool and for exploiting wider opportunities in the market.

On the other hand, these processes generate many questions about the role that the small firms of Emilia-Romagna can play within bigger groups: in particular it needs to be considered what the implications are in terms of the diminished autonomy of individual firms, as well as the more important role that the regional system will play on the national and international stage.

What, in any case, appears to be undeniable is that these processes show a tendency to affect competitive relations and to erase gradually the borderlines between local and regional productive systems.

iii) The above mentioned processes tend to increase vertical integration, within specific sectors and product lines, between industry and services. After the disintegration of the productive cycles in the 1970s in the 'diffuse-economy' systems, we see nowadays increasing signals of a reversed process of functional integration of the activities preceding and following production.

iv) It is also apparent that, in order to govern such a complex integration of functions, it becomes more and more necessary on the part of the firms to approach the innovative process as a constant and simultaneous intervention in the adjustment of the different functions of technological innovation, finance, marketing and organization. For an economic system based on small firms this is a change of vision that cannot be underestimated. The small firms tend, in fact, to organize their internal functions around production and centering their attention on the product. The adoption of an approach in which the innovative process permeates and informs all the different functions of the firm requires an entrepreneurial and organizational culture that is far from established.

In the same way as for the culture of the firms, the climate and quality of the relations between trade unions and entrepreneurs appears more important in order to ensure the optimal introduction of advanced technology in the products and in the productive process. It is a fact that more and more frequently innovative processes call for the greater involvement of employees and more advanced forms of industrial agreement on the introduction of new technologies. This is true both for the effects on the organization of labour and professional skills, and for the quality and safety of labour.

v) In the light of such transformations, we increasingly see the inadequacy of approaches and comparisons that look at regional economies as separate entities. In order to judge correctly the characteristics and development possibilities of the regional productive system, it is necessary to look with greater attention at the existing networks of active relations, or those that can be put into action, as well as to pay attention to the quality of the culture, competence and organization within such systems.

This change of vision suggests a need to reflect on what changes are needed in industrial regional policies, at a time when the important decision-making capability is tending to be transferred to the strategic boards of big corporations, often external to the region. We shall attempt to sketch some answers to these questions in the final part of this chapter.

Firms, Sectors, and Territorial Systems

In line with the processes described above, other aspects emerge in relation to the increasing differentials in competitive capability among firms, sectors, and territorial areas.

i) Regarding the firms, it can be observed that, beyond the recovery of a strategic role by the big firms, competitive differentials cannot be traced directly back to the size of the firms, especially if size is simply measured in terms of the number of employees. There is, in fact, a relatively broad range of small innovative firms that, aiming at an ideal integration of market positions, selection of technological opportunities and adequate organization, are able to maintain a competitive advantage. But along with an integrated approach to the innovative process, the competitive advantages of SMEs seem to depend on the extension and quality of the network of relations that they can establish with the centres of research and innovation.

The recent transformations, therefore, do not necessarily diminish the role of the SMEs: such a role, is going to increasingly depend on a capability to develop complex relations among firms, on their positions in the networks, and on the culture and structures adopted. Needless to say, in the light of these transformations, it is now necessary to reconsider quite a few established conceptions regarding the features and role of the SMEs in local economic development.

ii) New and significant facts are also emerging with regard to conditions in different sectors. We can talk less and less about mature sectors and it becomes increasingly apparent how, even in these so-called mature sectors, ample scope is left for considerable product and process, innovation. This phenomenon can be easily observed in Emilia-Romagna, where many of the sectors, such as knitwear, footwear, ceramic tiles, food processing and mechanical industries, are usually classified among the mature sectors. It must, in any case, be added that in such situations considerable differences emerge between the firms within sectors that are able to innovate successfully and those that are lagging. Regarding these sectors the emerging importance of the intersectoral nature of technology cannot be overlooked, particularly concerning automation and the utilization of new materials, as well as the more sophisticated trading and marketing techniques and structures, including a certain capacity to customize products.

iii) New configurations are also taking shape in the competitive relations between territorial systems. The recent transformations show a tendency to re-design the development process, bringing forth those local systems that are more able to enter regional and national networks, together with the ones with the highest quality of service and the greatest functional complexity. Specifically, among the determinants of the competitive capability of territorial systems, are the emergence of a new and different role played by the firms' environment, a closer interrelation of the environment with firm culture, and a greater capacity for them to establish relations with advanced areas at the international level. The policy of supplying services springs from an appreciation of the importance of the firms' environment in the determination of their performance and in the development of the industrial districts.

During the 1970s, the firms' environment has been based on services of a general character and on informal relations. The environment acted as a non-selective filter for innovative processes and played a mediating role in information derived, in the first place, from the physical proximity to the information user. Progress in information technology has diminished the advantages deriving from physical proximity, while there is an increase in the importance of the global competitiveness of the services offered. In other words, what matters more than the physical proximity of the services suppliers, is the adequacy of the services offered, their global quality, and the presence of an entrepreneurial culture able to make the best use of the available services, both traditional and specialized.

Complex problems arise in regions with a policentric structure, where the presence of many cities of medium size implies that the attraction and localization of highly specialized services (both technological, financial, commercial and formative) does not occur automatically.

Whilst the relationship between the supply of services and metropolitan areas has been widely studied, the same problems in policentric systems are less well understood. While the metropolitan areas seem to succeed in creating services of a higher character because of the spatial concentration of a demand that is also more aware of its needs, the policentric territorial systems, such as Emilia-Romagna, find themselves facing the two essential problems of quality and integration of the existing services and multiplying advanced services.

The first problem derives from the increasing needs of the firms for high quality and international standard services; the second derives from the necessity of realizing a single urban system through the integration of initiatives of a complementary, but often fragmented and redundant nature. Eventually, the problem lies with the growth of entrepreneurial awareness towards the importance of specialized services.

The Interaction Between Innovative, Territorial and Institutional Policies

The setting for this section on innovative policies is complex and represents the close interrelation between technological and productive questions and territorial and institutional themes. From this point of view, the first demand is for a definition of a set of objectives that is consistent with the transformations in progress and is feasible on the basis of both the institutional and normative framework and the available financial and organizational resources. A second group of questions pertains to the policy instruments that may be adopted, taking account of the significant progress made so far in the Region through a variety of instruments (laws, service centres, development projects, etc.), the usefulness of which seems on the whole to be confirmed, but requiring mainly to be widened in scope, improved in quality and integrated.

Lastly, we have to consider the more global role of innovative policies in the context of regional planning; a particular reference is made to the necessity of pursuing, among others, objectives concerned with the quality of the natural and urban environment, and quality of jobs. It is a theme that, of course, represents a forefront of policies for technological change, requiring a complex approach that largely remains to be tested. It is impossible here to enter into the details of the various programmes of intervention originating from this approach, rather the purpose is to describe and debate some possible guidelines for the evolution of regional industrial policies.

On the basis of the arguments developed above, one of the primary objectives of the Region's intervention is to create a more adequate environment for firms in terms of the availability of globally competitive specialized services. It is, from this point of view, a matter of creating structures of regional and national importance and avoiding the needless proliferation of services in different areas within the regions. The policies of the service centres in Emilia-Romagna have moved in this direction. Now it is a question, on the one hand, of looking for the highest quality and adequacy of services, while on the other, of interfacing those structures with the local productive systems, and easing the approach for those firms that, because of their size and organizational limits, still find it difficult to look to wider horizons.

Moreover, if it is true that the relations network is of increasing importance to firms, public intervention must aim at helping them to improve the quality of their relations with other Italian or foreign firms, as well as with the poles of research and of technological innovation. In this respect there are many operational possibilities. It is a matter of the best utilization of the opportunities offered by the presence of university structures and of centres and laboratories of national research agencies, and also to improve the connection between industry and research. This problem must be faced not only at the local level, but also by establishing relations with research networks at the national and international level.

Similarly, we face the problem of the relationship between local industries and the big Italian and foreign technologically advanced firms. In this respect interesting projects are under way in different fields, including environmental technologies and components supply for the aerospace industry. New instruments intended to link together groups of small firms working on shared technological projects in the fields of CAD and of the design of VLSI electronic circuitry are also under experimentation. One more example regarding the assistance given to small firms is the help given them to find European partners to accede to EEC programmes for technological innovation. There are many other projects and intervention proposals currently under way. What is important to stress here is that the kind of problems and competitive horizons at the international level facing the economic structure of Emilia-Romagna, together with the results achieved by the policies adopted so far, call for a deeper reflection on the steps that need to be taken in the promotion of technological change in the future.

It is immediately apparent how the instruments of intervention based on financial incentives, or on the supply of convenient locations for industry, are increasingly inadequate. This kind of instrumentation has been partially utilized in Emilia-Romagna and is still used in Italy, both at the national and regional level. Without entering into the detail of the matter, it is sufficient to state here that this kind of intervention, at the regional level, seems ineffective as a utilization of public resources, and risks provoking a distortion effect. Only the instrument of offering convenient locations seems to work in certain conditions, particularly if it is integrated with other interventions.

Our experience and research has shown that the first demand of firms operating in the region is for a quality environment, for specialized services of a reliable and adequate nature, for a growing and articulated system of productive and technological relations interfacing with other activities both international and external to the region. And even if we have already taken a step in this direction with the Region's policies of services to the enterprise, there is today the necessity of a further evolution, primarily in the following directions:

- the integration of such different functions as services, research and transfer of innovation in a regional system that is accessible from all over the territory and open to the exterior;

- a new attention to the central role and changes in the relations among firms and in the network system;

- a particular consideration of the importance of entrepreneurial and managerial training, together with a industrial relations climate that is favourable to innovation.

Having passed from financial and locational incentives to information and services to the firms, it seems now that the borderline of public policies runs along the relations between the small firms and the scientific, productive and financial activities, both at the national and the international level. These relations need to be made effective and utilized on the basis of a more advanced entrepreneurial and managerial culture.

The entrepreneurial culture is an increasingly determinant factor in the promotion of technological innovation. The situation appears highly differentiated as regards sectors and territorial areas. The essential questions arising appear to be:

- the perception of innovation as an integrated process affecting the entire enterprise function;

- the awareness of the importance of the network of relationships;

- the crucial role of organizational questions both internal and external to the firm;

- an entrepreneurial culture able to manage a more difficult and complex evolutionary stage of firm development.

The policy promoting technological change cannot but involve the theme of entrepreneurial training and technical teaching. It is evident that an action of this kind inevitably links itself on one hand with labour policies and the policies for the training and development of human resources, and on the other with territorial policies.

As for the latter it seems possible to affirm that nowadays adequate territorial policies at the regional level should be planned not only with the object of a balanced economic growth among the different areas (this is often impossible to achieve and, in any case, is less relevant given growing competition among regional systems), but they should also be aimed at defining the role that each individual area at the subregional level can and must play inside the regional system. This means, for instance, that different roles are defined for different types of area. In the mountain areas, the most important objectives are those of economic and demographic growth and of safeguarding environmental values. For the metropolitan areas, which are to be the qualified seats of the innovative processes, it is the relations with the national and international circles of advanced services in the commercial, financial and cultural fields.

In the process of defining the functions of the different territorial circles, the industrial and territorial policies meet and mix. In this regard, recall the relations between the roles of the different cities, the choices concerned with the creation of a structure for supporting technological change, or the consequent decisions concerning location and urban planning. In a different field, think of the effect of choices concerning the transportation network on the orientation of systems of relation and on the demand for the integration of an urban and productive system based on policentric structures. In this regard we could mention that in the very moment of the elaboration of the Regional Territorial Plan, a great deal of attention was given to innovation policies calibrated according to the territorial reality (Regione Emilia-Romagna 1988).

Moreover, the industrial and territorial policies cannot ignore institutional policies and they cannot ignore the web of relations linking the institution at higher and lower levels to the Region. The Provinces and Municipalities, in fact, are not only increasingly endowed with territorial and infrastructural powers, but tend to play a greater protagonist role in the supporting of technological and economic development. Ever more numerous are the local authorities that create local development agencies, innovation centres for the different sectors, and services centres for small-firm districts.

If such a process shows a certain dynamic at the local level and therefore represents something to be followed with keen interest, it risks, if not adequately governed, engendering a duplication and overlap of initiatives. The main risk here lies with the proliferation of similar initiatives, sometimes in direct competition, none of which is sufficiently specialized to achieve significant results. Here, the problem is how to give coherence to such initiatives in relation to the Regional Plan. What matters is that the initiatives which have originated at the local level, promoted by local authorities and also by the private sector, continue to answer the demand of the local economic structures, while pursuing the objective of hooking the local systems to the regional and national networks.

In this matter there are two kind of problems that need to be analysed. The first is the consequence, of the new and stronger subjectivity of local institutions. The second concerns the role of the Region, which has to ensure co-ordination of local activities and to plan integrated projects of regional scope involving both Provinces and Municipalities. The solution presently adopted in Emilia-Romagna is a co-operative and non-hierarchical relationship among different institutional levels.

In the same way the relation with national and the EEC policies can cause problems. With regard to national policies, the problem is that they are mainly thought out in terms of financial incentives, and are therefore difficult to integrate with regional policy instruments. The relationship between regional and national governments is often difficult. The national government denies the region the power to adopt industrial policies. Regions claim not only the right of policy intervention, but also demand a national legal framework for their actions. In the last few years regions have begun to put pressure on the Italian Parliament to recognise the regional power of intervention. The purpose of this action is not only to have a role in the national instruments of intervention, but to affirm the importance of an innovation policy based on service centres, technological agencies, science parks and so forth.

The EEC policies appear to be of the greatest interest, even if the bulk of the programmes are still directed at the big firms and are hardly utilizable on the part of the small ones. The relationship between the EEC and the regions is generally good; the EEC is considering the role of regions as an important one in the implementation of its projects.

Conclusions

From the foregoing discussion we may draw two kinds of conclusions: one regarding the definition of the desirable features of the regional governments' intervention to promote and adjust technological change, the second regarding some theoretical aspects.

There is, in the first place, a need to upgrade the instruments used so far to cope with the changing technological, organizational and competitive conditions. There are, indeed, fundamental changes occurring in the relations among firms, as well as their growing need to establish better relations between themselves and the scientific, technological and financial activities.

Secondly we must face the need to govern a complex range of instruments that, in different cases, can assume the form of service centres, structures for research and technological transfer, financial incentives, regional laws, the enforcement of national laws, and integrated projects of intervention at the sectoral and territorial levels. It is extremely important, along with the internal consistency of such a complex range of policy interventions, to retain a consistency with territorial choices, in order to coordinate the development of every individual area with that of the whole region.

In addition, such an attitude calls for the presence of a framework of relations among the different institutional levels and between them and other institutions, both public and private, such as universities, research agencies, firms and entrepreneurial associations. Regional industrial policy now has to go beyond the mere supply of services; the leading objective is the best integration of the regional economic structure with the most advanced areas in Europe and in the world. It is necessary to pass from the endogenous development point of view to an integration-oriented approach. This approach requires the establishment of new instruments aimed at building relationships and common projects with institutions, agencies and firms of other regions in Europe and in other continents. The regional level of industrial policy becomes a tessera in a more global picture of relationships between private and public actors.

Finally, it is possible to draw some theoretical indications for further research. This chapter has surveyed the nature of a regional industrial policy that, over the years, has become wider and more complex; a policy which has been derived from analyses of different disciplines and from general scientific debate. Experiences like this are in progress in several European countries and in the USA. There are a great number of studies and investigations of these experiences, but a complete theoretical framework for this important aspect of industrial policy is still to be achieved. The time is ripe to pass from a wide and rich empiricism to a comprehensive theoretical framework. Several elements are sufficiently developed to contribute to this work:

i) An interdisciplinary approach, concerning economics, technology, firm culture, territorial and environmental sciences, and institutional themes.

ii) A systemic approach: the local economies have to be examined within a system of a more and more international character. In the same way, regional policy has to aim at a better integration of the region within larger systems and with other areas, both advanced and similar in their productive structure. The emphasis passes from local resources and self-centred development to integration with the more advanced Italian and European areas; this integration will be a crucial factor in an ever more global economy. The network of relations and the capacity for integration will depend on a growing number of public and private actors and, more generally, on the social climate and entrepreneurial and labour cultures.

iii) The importance of cultural and social behaviour to aid understanding and interpretation of economic changes.

iv) The multiplicity of the actors involved in the transformation process and, consequently, a social complexity greater than the social model usually considered by economics.

v) The central importance, at the local level, of the 'real' economic structure, in respect of the monetary and financial aspects that is prominent at the national level.

vi) The increasing importance of the systems of relations among firms and among territorial systems; particularly the growing integration of local areas within wider scale systems.

vii) The relations between micro- and macro-economics.

viii) The overcoming of the traditional and simplistic division between the public and private sectors.

In this approach the role of the institutions emerges as a focal point, with new dynamics and new questions.

An increasing subjectivity and will for intervention by the regional and local institutions induces the emergence and the strengthening of industrial policy which in many respects is innovative in comparison with the consolidated national policies. The regional policies appear increasingly able to affect productive systems that are mainly composed of SMEs and are strongly rooted to the territorial reality.

A more complex public sector, composed of many different actors, each one with its own logic and objectives, can be seen to be emerging, together with greater institutional complexity, with relations based on hierarchy, competition and co-operation. The relations between institutions and society are also changing, with stronger interrelations, but at the same time, a demand for deregulation and a higher capacity to govern complex systems.

The adoption of planning instruments able to integrate the provision of action and a negotiated regulation of the relations among private and public actors requires a flexible and efficacy-oriented planning system and a closer relationship between economic and territorial planning. In this context, the role of institutions and the concept of 'policy' form a new agenda for research into regional policies for technological change.

References

Arrighetti A. (ed.) (1982) Piccola impresa e politica industriale, Franco Angeli, Milano.

Antonelli C., Cappellin R. and Garofoli G. (1988) Le politiche di sviluppo locale, Franco Angeli, Milano.

Bagnasco A. (1977) Tre Italie, Il Mulino, Bologna.

Bagnasco A. and Messori M. (1975) Tendenze dell'economia periferica, Editoriale Valentino, Torino.

Bagnasco A., Messori M. and Trigilia C. (1978) Le problematiche dello sviluppo italiano, Feltrinelli, Milano.

Becattini G. (1987) Mercato e forze locali: il distretto industriale, Il Mulino, Bologna.

Brusco S. (1982) The Emilian model: productive decentralization and social integration, Cambridge Journal of Economics, 6 (2) 167-184.

Brusco S. (1983) Flessibilità e solidità del sistema: l'esperienza emiliana, in Fuà G. and Zacchia C. (eds.), Industrializzazione senza fratture, Il Mulino, Bologna.

Brusco S. and Sabel C. (1981) Artisan production and economic growth in Wilkinson F. (ed.), The dynamics of labour market segmentation, Academic Press, London.

Cappellin R. (1988) I servizi alla produzione e lo sviluppo locale, in Antonelli C., Cappellin R and Garofoli G, (eds), Le politiche di sviluppo locale, Franco Angeli, Milano.

Cervia Ambiente (1985) Chiudere il cerchio, Maggioli, Rimini.

Commoner B.(1973) Il cerchio da chiudere, Garzanti, Milano.

Fosler R.S. (1988) The new economic role of American States, Oxford University Press, New York.

Fuà G. and Zacchia C. (eds.) (1983) Industrializzazione senza fratture, Il Mulino, Bologna.

Garofoli G. (1981) Lo sviluppo delle aree periferiche nell'economia italiana degli anni settanta, L'Industria, II, 3.

Garofoli G. (1988) Le politiche di intervento a sostegno dello sviluppo locale, in Antonelli C., Cappellin R. and Garofoli G.,(eds.) Le politiche di sviluppo locale, Franco Angeli, Milano.

Gerelli E. (1974) Economia e tutela dell'ambiente, Il Mulino, Bologna.

Gresse X. (1988) Decentralizer pour l'emploi: les initiatives locales de developpement, in Le cas de l'Emilia-Romagna, Economica, Paris.

Hueglin T.O. (1986) Regionalism in Western Europe. Conceptual problems of a new political perspective, Comparative Politics, 18.

Innocenti R. (ed.) (1985) Piccola città e piccola impresa, Franco Angeli, Milano.

Jurgen G. (1988) Innovation oriented policies in Regions with high growth dynamics: a comparison of Baden-Wurttemberg, Massachussetts, Emilia-Romagna, 5th International Conference on Innovation and Regional Development, Berlin.

Lassini A. (1985) Gli interventi regionali per i servizi alle imprese, Franco Angeli, Milano.

Nanetti R.Y. (1988) Growth and territorial policies: the Italian Model of social capitalism, Pinter, London.

Piore M.J. and Sabel C. (1984) The second industrial divide, Basic Books, New York.

Putnam R.D. Leonardi R., Nanetti R.Y. and Pavoncello, F. (1983) Explaining institutional success: the case of italian regional Government, American Political Science Review, 77, 55-74.

Putnam R.D., Leonardi R. and Nanetti R.Y. (1985) La pianta e le radici. Il radicamento dell'istituto regionale nel sistema politico italiano, Il Mulino, Bologna.

Regione Emilia-Romagna (1986) Programma regionale di sviluppo 1986-'88, Bologna.

Regione Emilia-Romagna (1988) Piano territoriale regionale, Bologna.

Regione Emilia-Romagna (1984) 2nd Conferenza regionale sulla politica industriale, Bologna.

Regione Lombardia (1985) Gli interventi regionali per i servizi alle imprese, Franco Angeli, Milano.

Rokkan S. and Urwin D.W. (eds.) (1982) The politics of territorial identity, Sage, London.

Sabel C. and Zeitlin J. (1982) Alternative storiche alla produzione di massa, Stato e mercato, 5.

Schlenker L.H. (1987) Local industries strategies. A key to French economic planning in the eighties, Comparative politics, 19, 267-280.

Stilwell F. (1988) Regional economic policy and local enterprises, Economics department, University of Sidney (mimeo).

Tolomelli C. (1988) La valutazione delle politiche industriali e la dimensione territoriale: il livello regionale, Argomenti, Regione Emilia-Romagna, Bologna.

Zariski R: (1985) Approaches to the problem of Local Autonomy: the lessons of Italian regional devolution, West European Politics, 8 (3), 64-81.

19. THE CHANGING ROLE OF MUNICIPALITIES IN LISBON METROPOLITAN REGION AS A RESULT OF TECHNOLOGICAL CHANGE

Paulo V.D. Correia

Introduction

Major changes in the social, economic and political context of Portugal have occurred since the political changes of the 25th of April 1974. However the new scope for development provided by the democratic regime has been seriously constrained by a number of different factors (Correia and Nunes da Silva, 1986):

- The international economic recession triggered by the 1973 oil crisis has hit Portugal harder than most European countries.

- The sudden and large increase in real wages in 1974 and 1975 without the corresponding economic growth brought high inflation rates until 1984, a persistent negative balance of payments and an increasing foreign debt.

- The importance of oil in total imports (approximately 30%) during the seventies, and of food supplies, coupled to the independence of the former Portuguese colonies, increased the dependence of the Portuguese economy on factors influenced by the international scene.

- The structural unemployment in sectors such as shipbuilding and other large scale heavy industries, particularly in Lisbon Metropolitan Region (which accounts for one quarter of a national population of 10 million), caused serious problems in some of the municipalities of this region, to the extent of losing part of their active population.

The changes taking place in the structure of the Portuguese economy during the last decade show that the time scales for the creation of new firms and in the introduction of innovation are shorter than ever. Moreover, the cyclical changes in the world economy are having a growing impact on international flows of trade and on the adjustment of demand profiles, stressing the need for flexibility in production and increased productivity. In Portugal these aspects are now of increasing relevance owing to both the increasing challenges arising from membership of the EEC and the urgent need of some regions and some sectors to innovate and restructure. New technologies can provide part of the solution. The near future will therefore be marked by deep structural changes, especially in the industrial sectors, through innovation and the diffusion of innovation, and by the changes in market demand, assisted by the growing (but still small) national spending on R & D.

The locational choices for activities using new technologies have, so far, been favouring the more developed regions of the country, strengthening the already strong disparities between these developed regions and the other still developing regions. Although the economies of scale which gave origin to the major urban conurbations no longer seem to be the driving factor behind the location of new economic activities, national and regional strategies seem to fail to have much influence on the location of activities using innovative processes or producing new products to reduce regional disparities. In addition, new firms adopting new technologies also tend to locate in the major conurbations, where skilled labour and synergies are easier to obtain.

However, it is difficult to measure and to forecast the impacts of new technologies in the modernization of industry, services and trade, since the pace of development of new technologies is extremely fast, both through increasing investment in R & D itself, and its applications in production. The adoption of new technologies by a growing number of firms and sectors shows an unclear boundary, constantly evolving. Moreover, the adoption of new technologies is determined by many variables such as incentive policies, features of the promoters, the state of the economy, market dynamics, etc.

It is however certain that R & D, innovation diffusion and the location of activities using new technologies will play a key role in development in the next five years, to an extent that is difficult to foresee. This implies that the public administration at all levels must not only ensure and stimulate all the conditions for innovation and diffusion, but also support adequate locational choices for the restructuring of urban and regional systems.

The Structure of Employment

Portugal is now recovering from the deep crisis triggered by the 1974 revolution and the international economic recession. Since 1985 there have been indications of steady recovery with falling inflation and reduced levels of unemployment (see Figure 19.1). However, the period after 1974 has been marked by profound structural changes, especially in Lisbon metropolitan region. In structural terms Portugal remains divided with a large rural and agrarian population outside Lisbon, while the metropolitan area is mainly industrial, as a comparison between the share of the primary sector nationwide and in the metropolitan area shows (see Figure 19.2). Other structural differences are identifiable in the relative importance of scientific and technical professions in the Lisbon region, which is nearly double its level nationwide, and does not vary significantly within the metropolitan area. This concentration of expertise should increase the relative potential for the introduction of new technologies within the metropolitan area, where most of the highly skilled labour is located (see professional groups 0, 1, 2 and 3 in Figure 19.2). This also implies that the opportunities for restructuring employment will be less constrained within the metropolitan area (though varying at the local level) and lead to an increase in the already existing development gap between Lisbon and the rest of the country.

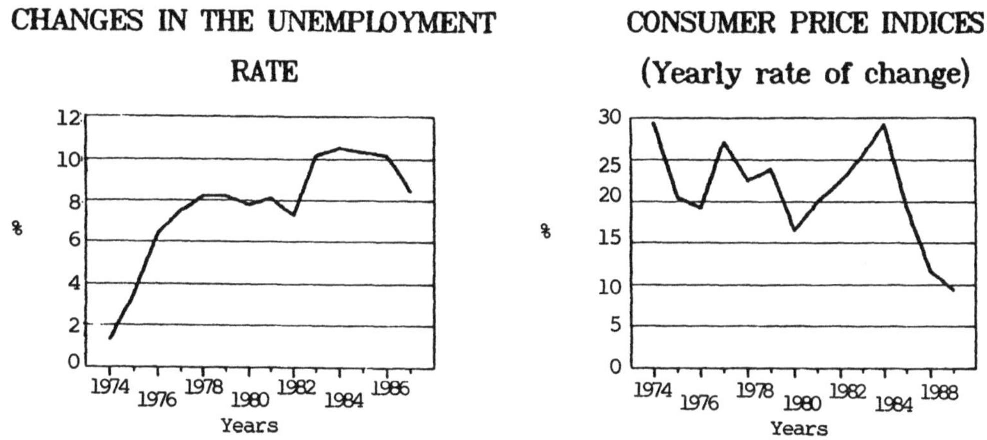

Figure 19.1: Changes in the Unemployment Rate and Consumer Price Indices.

The electronics and electrical equipment sector further illustrates the relative importance within the national economy and national technological development of the coastal regions in general, and of Lisbon in particular, being a sector where new technologies have been adopted first and around which expectations for further development are set. The perceived advantages are, however, not always realised. For example, the computer industry is still in an embryonic stage in Portugal and in 1984 there were only two factories in this sub-sector experiencing unstable growth (see Figure 19.3), due to the policies of the multinationals which control them. Thus, while both these factories are located in the south Lisbon metropolitan and therefore most advantageous areas within the Portuguese economy, their progress is uncertain. Between 1983 and 1985, when the economic crisis reached its peak, the worst affected sectors in terms of job losses, however, were the ones with the highest percentages of jobs and the second highest in gross added value: 'metallic

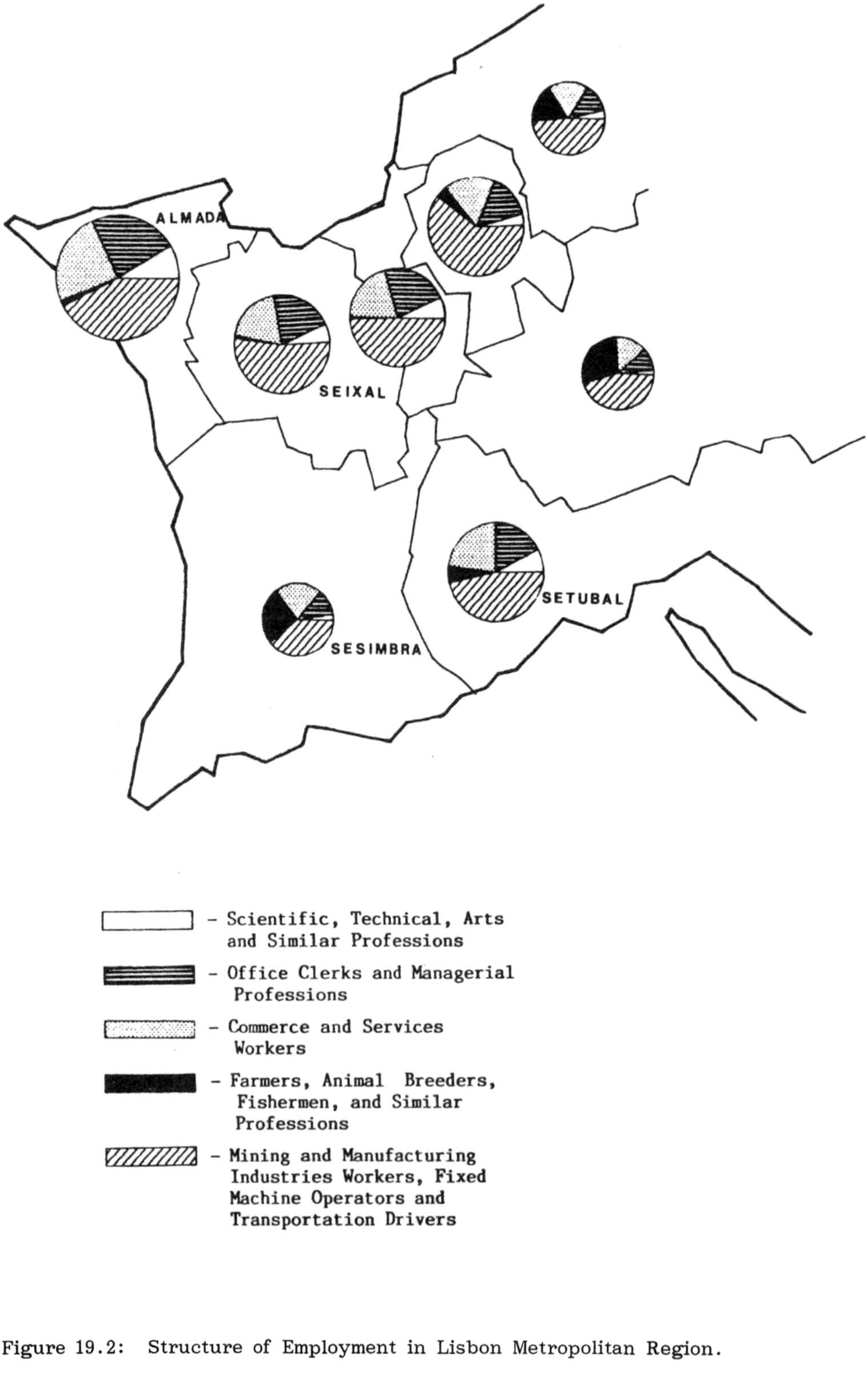

Figure 19.2: Structure of Employment in Lisbon Metropolitan Region.

products and machinery, and transportation equipment' (see Tables 19.1 and 19.2). As a result the highest growth rates in gross added value are still to be found in the traditional sectors, which, by eventually adopting new technologies in the production of traditional products, are expected to play an important role at the regional and national levels in future. These traditional sectors include 'food and drinks', 'oil and chemical industries', and 'metallic products and machinery'.

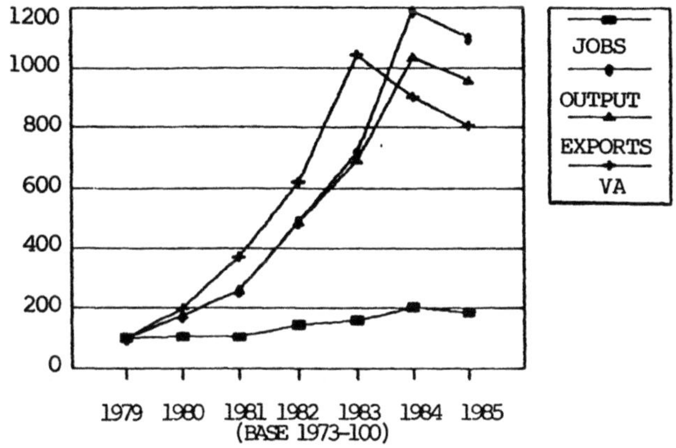

Figure 19.3: Portuguese Computer Industry.

Though the rates of change for this period do not yet show any general or steady trends, it appears that the eventual growth in the number of jobs will not be significant, even in those sectors where the increase in gross added value is very high. Evidence shows that the average ratio between the gross added value and the number of jobs has increased over time, suggesting a general search for higher productivity and competitiveness, rather than for job generation. There is however considerable variation in experience between sectors.

The sub-sectors which had higher positive ratios during this period were: in north Lisbon metropolitan area 'food and drinks' and 'oil and chemical industries', while in south Lisbon metropolitan area it was 'non-metallic mineral product industries' (excluding oil and coal) and 'metallic products and machinery, and transportation equipment'.

The sub-sectors with the worst negative change in the north and south Lisbon metropolitan areas were, respectively, the 'base metallurgical industry' and 'paper, graphic design and publishing'. These were not important sub-sectors at the sub-regional level, except for the considerable drop in gross added value in the latter one.

The data required for a more consistent and reliable analysis of the

Table 19.1

Manufacturing Industries in Lisbon Metropolitan Region

(1983, 1984 and 1985)

DISTRITO (Sub-region)	CODE * CAE	WORKERS % 1983	1984	1985	V.A. * 1983	1984	1985
Lisboa	31	15.4%	16.0%	16.3%	27.9%	28.4%	29.4%
	32	10.0%	9.8%	10.3%	4.4%	4.3%	4.5%
(North Metrop.	33	2.9%	2.9%	2.8%	1.5%	1.2%	0.9%
Region)	34	11.7%	12.1%	12.0%	10.6%	10.3%	10.3%
	35	16.6%	17.3%	17.5%	16.8%	20.1%	21.7%
	36	8.5%	8.5%	8.3%	7.8%	7.2%	6.5%
	37	1.3%	1.0%	1.3%	1.1%	0.9%	0.9%
	38	33.2%	32.1%	31.1%	29.7%	27.4%	25.6%
	39	0.3%	0.3%	0.3%	0.2%	0.3%	0.2%
TOTAL		**100.0%**	**100.0%**	**100.0%**	**100.0%**	**100.0%**	**100.0%**
Setubal	31	10.8%	10.5%	10.9%	9.8%	12.6%	9.0%
	32	5.9%	5.9%	7.0%	31.%	4.3%	5.0%
(South Metrop.	33	...	...	...	...	...	...
Region)	34	3.5%	3.8%	4.0%	9.0%	14.1%	1.1%
	35	14.4%	15.2%	16.6%	24.5%	27.1%	25.3%
	36	5.2%	4.8%	3.8%	6.4%	6.7%	6.2%
	37	10.0%	10.0%	10.7%	14.1%	7.1%	7.5%
	38	42.6%	41.5%	39.9%	28.4%	23.6%	32.1%
	39	...	...	...	...	...	...
TOTAL		**100.0%**	**100.0%**	**100.0%**	**100.0%**	**100.0%**	**100.0%**

For Key to codes see Appendix 19.1.

restructuring of the economy at a regional level, and on how new technologies are affecting this transformation, are not easy to obtain for various reasons.

(a) The latest version of the classification of firms by sectors of economic activity (CAE), is from 1973, and has not been revised since. There are currently therefore many problems in classifying firms producing new products, not to mention the difficulties in making international comparisons.

(b) The Current National Classification of Professions (CNP) was published in 1981. The professions related in some way to new technologies are dispersed within and between many sub-sectors, and it is only possible to detect them through a high disaggregation of the data, which is not always feasible or reliable.

Table 19.2
Growth of the Manufacturing Industries in Lisbon Metropolitan Region

DISTRITO (Sub-region)	CODE * CAE	1983-84	1984-85	83-84	84-85
Lisboa	31	0.52%	0.30%	0.47%	1.02%
	32	-0.19%	0.48%	-0.15%	0.21%
(North Metrop.	33	-0.07%	-0.04%	-0.30%	-0.30%
Region)	34	0.36%	-0.06%	-0.30%	-0.02%
	35	0.68%	0.24%	3.28%	1.62%
	36	0.05%	-0.22%	-0.61%	-0.69%
	37	-0.30%	0.30%	-0.18%	-0.02%
	38	-1.07%	-1.01%	-2.25%	-1.85%
	39	0.01%	0.01%	0.05%	-0.02%
Setubal	31	-0.26%	0.36%	2.72%	-3.56%
	32	0.04%	1.09%	1.27%	0.68%
(South Metrop.	33	...	...	...	...
Region)	34	0.27%	0.18%	5.14%	-13.09%
	35	0.75%	1.44%	2.64%	-1.77%
	36	0.47%	-0.98%	0.24%	-0.53%
	37	0.05%	0.73%	-6.92%	0.41%
	38	-1.14%	-1.58%	-4.80%	8.46%
	39	...	...	...	...

For key to codes see Appendix 19.1.

* Classified economic activities
Classification Internationale Type des Professions)
... Confidencial information
V.A.- Value Added (Gross)

Source: Estatisticas Industriais 83, 84 e 85 - INE Lisbon

(c) The use of new technologies in the production of new products is also difficult to detect owing to both the difficulties in defining the meaning of new technology and operationalizing this concept, and to the wide variety of sub-sectors which may adopt new technologies.

(d) Finally, the available data is normally not spatially disaggregated, making it very difficult to interpret the changes and trends of high technology impacts at the local level without carrying out specific surveys.

Potential Factors for Development

The disparities between the developed regions, particularly the metropolitan areas of Lisbon and Porto, and the rest of the country are likely to increase without policy intervention owing to the existence in those regions of a series of highly interrelated factors, which are normally essential for development (Stohr, 1986):

(a) **A highly skilled labour force** is a key location factor for high-tech industry and services. Its relative importance in the work force of the metropolitan regions is very high and growing.

(b) **The qualities of the natural and built environment** present different and some times conflicting issues. The natural environment, particularly along the sea coast, still has great potential, particularly for housing, tourism and recreation. However the built environment has been partly produced through illegal and unplanned development, particularly in peripheral areas of Lisbon Metropolitan Region, which still lacks basic infrastructure and urban amenities. This limits their attractiveness and is likely to incur the use of valuable resources to bring them up to standard.

(c) **A diversified urban base** exists and is consistently developing in the urban cores of the metropolitan regions, combining administrative, commercial, service and cultural functions with industry. However, the original urban base was the result of the location of the large heavy and traditional industries which are now in crisis. This is most noticeable in the case of the Lisbon region. The later development of a diversified urban base is now more closely linked to employment in other urban centres; either secondary employment centres with strong links with the core of the metropolitan regions, or other regional employment centres experiencing high rates of growth over the past decade and which small scale industries, not necessarily high-tech, prefer to locate due to accessibility factors and lower costs.

(d) **Access to an airport, national motorways and railways, and to tele-communications networks** is a key factor which clearly differentiates the developed Portuguese coastal regions and the developing inland areas.

Those factors with a less obvious link to local development include **R & D activities** such as universities (science and technology departments) and research institutes. The creation of new universities and polytechnics in some regional urban centres, has taken place in recent years. They provide courses linked to the economic sectors relevant to each region. Whilst initial evidence suggests they are producing positive results, it is still too early to assess the extent of factual and potential links between these institutions and new technology industries in neighbouring areas.

The **availability of venture capital** is another important factor for innovation, though not always thought of as crucial at the local rather than the national level. However there have been suggestions that venture capital easily follows the emergence of innovations on an international scale. The growth of foreign investment in Portugal, coupled to a widening of the European markets has already produced new (fóreign and national) investments in parts of the country, outside of the metropolitan centres, containing strong innovative components.

A recent survey of new technologies in Portugal (Fonseca, 1986) tried to assess their effects on production processes, as well as the introduction of new products into industry. Five major groups were considered, taking into account functional and technological similarities: micro-electronics information technologies, production technologies, bio-technology and materials technologies.

These five technological areas were examined in the context of 26 industrial sub-sectors, leading to the conclusions (see also Tables 19.1 and 19.2) that at a global level, information and production technologies have already been playing a fundamental role in most sub-sectors, introducing large increases in competitiveness and flexibility. The traditional sub-sectors or the specialised sub-sectors of the Portuguese economy, namely textiles and clothing, and food and drinks, will have to adapt to the challenge made possible by new technologies if they are to remain competitive and expand existing, or seek out new, markets.

The sector producing electro-mechanical equipment will be the one where the impact of new technologies will be more systematic and widespread, both in the introduction of new products and of new production processes, leading to higher quality and competitiveness. The new material technologies will affect a large number of sectors in particular the non-metallic, metallurgical and electromechanical industries, both in terms of improvements and economies in production and inequality, as well as through the creation of new products. The Portuguese textile and clothing industries were already seen to be making use of new technologies in production, where the increasingly widespread use of information technologies is to be expected, bringing higher flexibility and efficiency to these traditional sectors. On the evidence available, product innovation is expected mainly to affect the electromechanical industry, as well as the chemical (for example, pharmaceutical products), food and drinks industries.

The prospects for positive changes in each sub-sector should not be attributed simply to technology but will also be affected by the financial means available for change deriving from different governmental and EEC incentive programmes. Such is the case of PEDIP - Specific Programme for the Development of the Portuguese Industry (Ministerio da Industria e Energia, 1988). PEDIP is a programmed system of selective financial support for certain projects which contribute to the technological innovation within, and modernization of, Portuguese industrial firms. The Programme has operated since 1988, and includes several qualifying components for financial support: basic and technological infrastructure development; professional training; incentives for investment; financial engineering; consultancy visits aimed at raising productivity; industrial quality and design consultancy visits and information collection, implementation and control.

The first general policy of the incentives programme (SINPEDIP) aims at providing financial support through grants of between 20 per cent and 70 per cent of the estimated cost of each project, according to its features and merit. Other policies of this programme aim at different goals, such as energy conservation, the restructuring or modernization of industry on a sectorial and/or regional basis, and the production of new equipment and improved efficiency in the use of natural resources through the introduction of new technologies. Linked to SIMPEDIP, a separate system was created in order to allocate grants to firms located or locating in less favoured regions: SIBR - Incentive system of regional basis.

Scope for Municipal Intervention

The foregoing sections of the chapter have illustrated the divergent sectoral and regional experience in Portugal with regard to economic and technological development, revealing the more positive nature and prospects of the metropolitan economy of Lisbon. The Lisbon metropolitan economy is, however, made up of a number of sub-regions whose individual capabilities also differ greatly from the regional average.

Among the 17 municipalities in the Lisbon Region, Almada stands out as one of the most important after Lisbon itself and to which it is linked by a road bridge and by ferries. It has the largest unpolluted beaches in the metropolitan region, though large peripheral areas still present serious shortfalls in infrastructure and other urban amenities (see Figure 19.4). With a population of approximately 150,000 inhabitants, and an activity ratio of 0.40, Almada has the largest active population among the municipalities of the south metropolitan area and also the largest share of employment in industry, commerce and services.

The distribution of the active population in the south Lisbon metropolitan region (see Figure 19.5), shows that the largest percentage of people work in the services and trade sector. Almada is the municipality of the south metropolitan region with the largest services sector, both in absolute and relative terms. However, a comparison between the active population living in Almada and the number of jobs within the municipality shows that 45 per cent of the employed workers commute to other municipalities, and that 36 per cent commute to Lisbon alone, mainly to the services and trade sectors operating there.

Since 1974, local authorities have been engaged in promoting the general upgrading of community facilities, and have progressively adopted a planning process to support their actions. However so far not much has been done beyond their traditional areas of intervention, such as the provision of technical and social infrastructures. Local financial resources are not sufficient to rapidly overcome existing shortfalls. The absence of regional authorities and the dependence on central government on many issues still constrains new local initiatives. Moreover, except for the nationalised industries, and those of strategic importance, most job creation is left to the private sector. For example, over 90 per cent of all new housing is promoted by the private sector. What, then, could be the role of a local authority in the promotion of development, and how can it help to face the challenges to local economic development of new technologies?

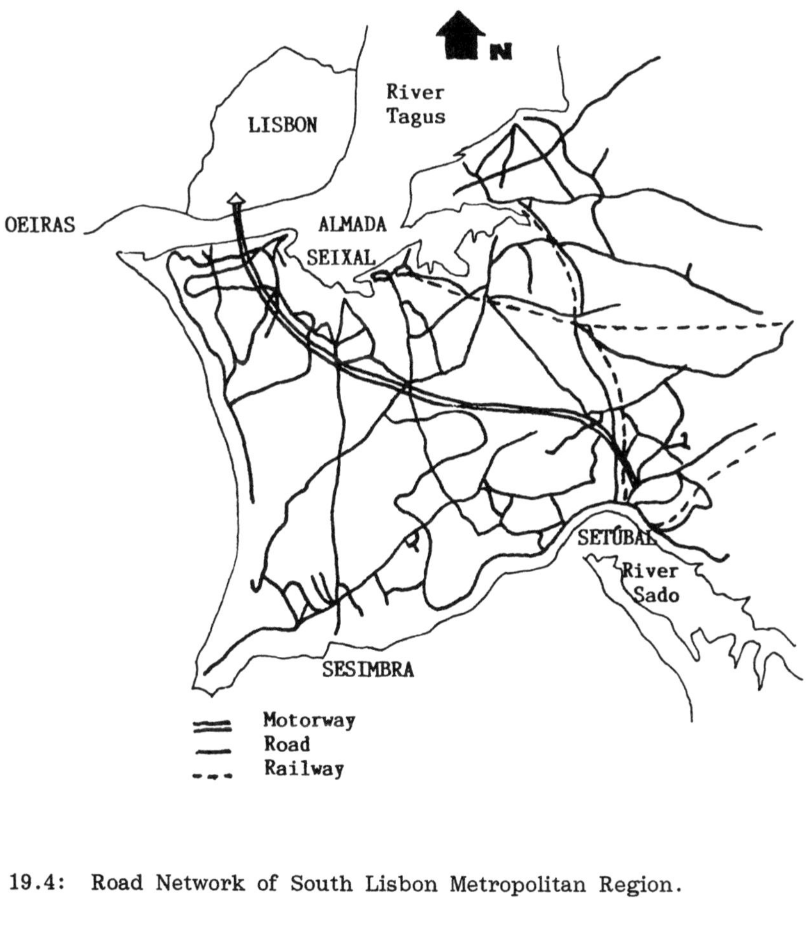

Figure 19.4: Road Network of South Lisbon Metropolitan Region.

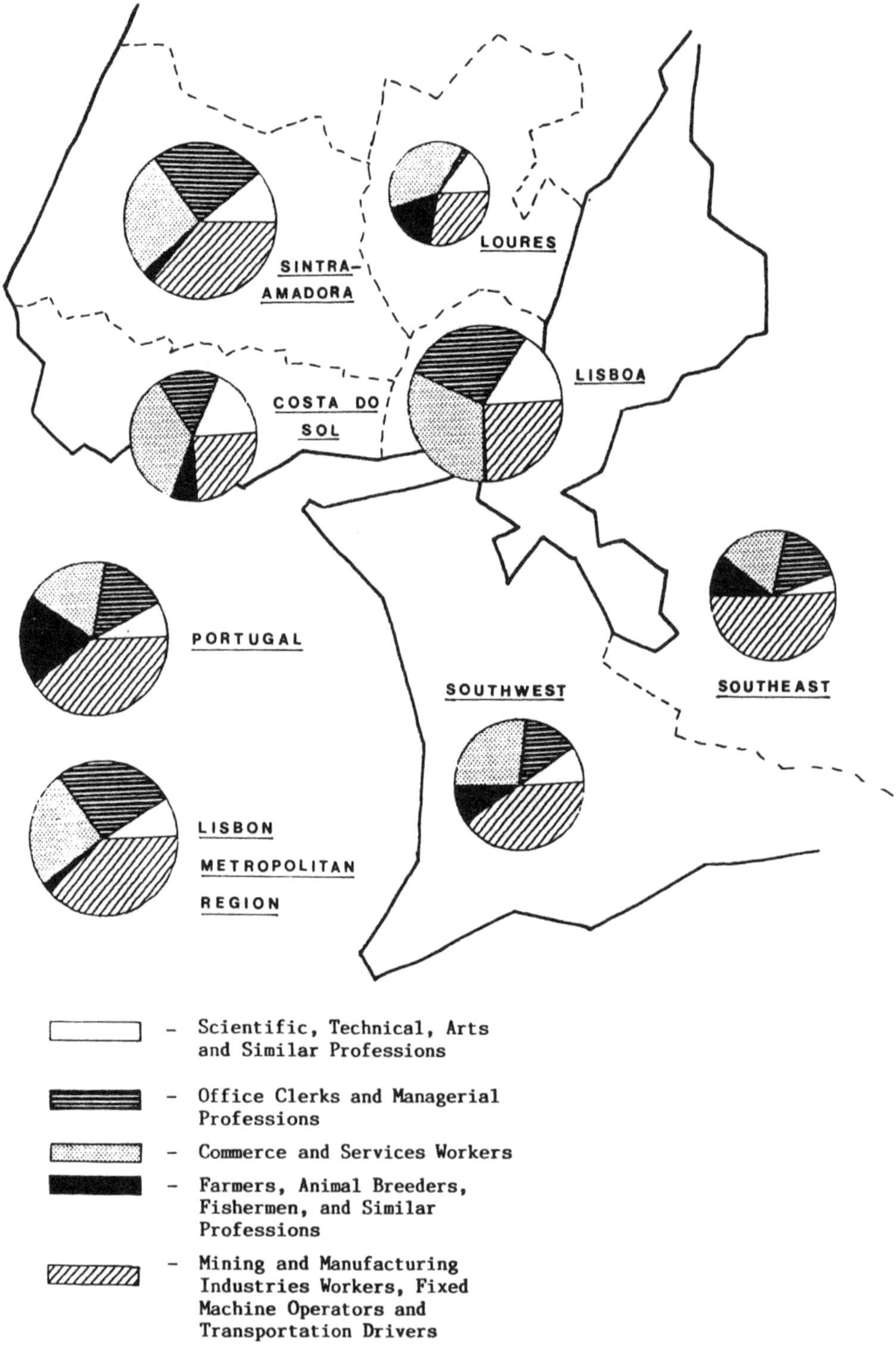

Figure 19.5: Structure of Employment in South Lisbon Metropolitan Region.

In the south metropolitan area of Lisbon, municipalities have formed an association to plan and implement integrated strategies for the sub-region. But this initiative is still at a very early stage, where parochial matters still prevail, and the cost of regional initiatives are often prohibitive if not co-financed by the central government and/or the EEC structural funds. Regions with elected bodies, however, are expected to play a key role in development in the near future (Costa, et al., 1988), when the process leading to a new regional administrative structure is completed.

However the municipality of Almada is changing its approach to local development. A direct and positive role in local development does not necessarily imply financing it, but rather being able to attract the right investors, provide them with adequate infrastructure, and also with adequate information. The municipality is exploring, and may eventually adopt, joint ventures as a means of encouraging local investment, thus also improving its own resources, as well as providing a sounder control over development. Institutional improvement is also accompanied by technological opportunities.

The factors considered to be the most relevant for the introduction of new technologies already exist in Almada, although they are not yet fully exploited.

(a) The work force of Almada has a relatively high proportion of high skills which is growing (see Figure 19.5).

(b) The natural environment, particularly along the sea coast, still has great potential being particularly attractive for housing, tourism and recreation. The municipality has been very active, since 1974, in building and restructuring the sub-standard areas.

(c) The urban core of Almada combines administrative, commercial and cultural functions with industry, providing a diversified urban base, which is still developing steadily. However, the original urban base was the result of the location of the large industrial units now in crisis, such as the ship repair yards where nearly 5000 jobs have been lost over the last 10 years. The later development of the diversified urban base is now more linked to employment in other urban centres, namely in Lisbon itself, (due to accessibility factors) and to small scale industries, not necessarily high-tech.

(d) The access to airport, national motorways and railways, and to tele-communication networks is a key factor not yet fully developed in Almada. Though accessibility by road is very high (see Figure 19.4), both at metropolitan and national levels, with Lisbon International Airport 20km away from Almada, the capacity of the bridge over the Tagus river is still seen as a potential constraint (though its enlargement is planned for the near future) and railways do not yet serve the region adequately (there are no railway bridges over the river).

(e) A new deep sea harbour has been planned for the south bank of the river near its mouth. The postponement of its building because of the international and national contexts, may be reversed with a new role for Lisbon as a port within a transport network viewed on an EEC wide scale and providing a significant increase in jobs at Almada and in the south metropolitan region more generally.

(f) The existence in Almada of the Faculty of Sciences and Technology of the Lisbon New University, and its research centre, has not yet produced significant high technology industry in neighbouring areas. The inter-dependence of this with Lisbon universities and research centres seems to explain this fact. However the future may bring closer links between the role of this knowledge-creating institution and future local industries.

(g) The availability of venture capital in Almada in the near future will depend not only on the policies of the central administration and on R & D funding institutions aimed at the creation of high technology industrial clusters, but also on the capacity of the local authorities to provide adequate facilities at the right time and at the right locations.

The existence in Almada of one high-tech industry (micro-electronics), the recent interest shown by a major scientific research funding foundation and the municipal policy to promote the urban development of a new industrial area, on municipal land, coupled to the university, should set off a new stage of development for the municipality in which synergies resulting from the clustering of R & D institutions and existing firms with new enterprises might become a reality.

However, new technologies are not the only promising sector in the case of Almada. Tourism and recreation and associated activities are an unexplored potential, not only due to the natural features of the sea coast, but also because the municipality owns large estates along it. Although the impact of large scale developments in this sector would affect the whole region, job creation would easily cover the job losses in traditional sectors of the past decade.

The adoption of a new strategy starts by a new attitude towards both market forces and the planning process. Although financial means are usually regarded as the most important, political means (political will for the change, bargaining power, capacity for negotiations and fast but efficient decision making) are crucial. Such decisions however, usually require support from adequate data bases, without which strategic planning and the whole process are pointless.

The local authority may also play a very important role as a catalyst inspiring new investment opportunities within the municipality, which may attract internal and external investors, as well as playing a role in the diffusion of the new job opportunities that may arise.

The interdependence of Almada with other sub-sectors in the metropolitan region in terms of employment and division of labour, together with the international dependence of the Portuguese economy, may interact in a variety of ways to either constrain or develop further local policy-making, the outcome of which is currently uncertain. However the importance of the services sector, where the introduction of new technologies has been very significant, may well be a key factor in the creation of new jobs related to new technology activities within the municipality, thus significantly reducing the dependent on the city of Lisbon.

The Emergence of a New Municipal Development Strategy

Local economic development can no longer rely on growth sustained by simply defined growth poles (Perroux, 1955) or growth centres (Boudeville, 1966). The underlying inter-industrial relations at a local level cannot explain local growth on their own. The same can be said about innovation at local and regional levels (Stohr, 1986). What is obvious is that local development results from a complex interaction of variables specified at a variety of functional and spatial levels.

Development with innovation calls for an approach which considers dynamic regional interaction among the key factors in a network (sectors and locations), essential for sustained regional innovation. There is always the danger of functional-specialization from core to peripheries due to four main bottlenecks which may hamper or promote innovation. Empirical analysis by Ewers and other authors in West Germany (Ewers et al., 1980) identified these as human capital, financing, information on markets and technology and organizational and management potential.

The creation of new activity centres ensuring the spread of the 'developed and innovative network' to less favoured regions can only be achieved through active and consistent policies across all levels which ensure the introduction of new synergies in new locations (Perrin, 1986). Such an approach implies that innovative activities are not only able to overcome the eventual bottlenecks, but also to integrate with local and/or regional traditional activities, in order to reach new, but balanced, diversified productive structures.

Almada is not far from overcoming the bottlenecks that still make it, to some extent, a peripheral sub-region. However a new strategic approach is required in order to implement new local policies consistent with regional and national policies, which may ensure success in local development and modernization.

The main challenge is the municipal ability to link its strategy dealing with the physical environment (natural and built) with an active role in the economic and promotional relationships.

The new planning process which is being set off in Almada has defined a set of strategic goals which are grouped in the five following policy areas (source: 'Gabinete do Plano Director Municipal' - Municipality of Almada):

(1) Re-enforcement and equilibrium of the municipal urban system in a regional setting.

(2) Diversification of the economic base and modernization of production processes.

(3) Improvements to the natural and built environments.

(4) Social development and professional training.

(5) Introduction of a new planning and management practice.

The objectives and policies which relate more directly to modernization and innovation are:

(a) Besides the general improvements in accessibility, including all modes of transportation, the municipality aims at becoming a node in the national and international networks of telecommunications and teledata processing, and not just an extension of Lisbon.

(b) Support for the creation of small and medium size industrial and services firms, especially those adopting new technologies. Encouragement of R & D through existing and future research institutions locally based, in technology parks clustering high-tech industries, and linked to the university. The setting up of a municipal information system on new job opportunities resulting from the new activities, as a support to the new division of labour.

(c) Besides the general upgrading of technical and social infrastructures, the municipality aims to make available urban land adequate for the location of new activities, producing high quality built environments, protecting the values of natural landscape, and attracting national and foreign investment through carefully negotiated development contracts.

(d) Support for training in skilled professions, especially when related to new technologies, in order to provide locally qualified manpower, with concomitant reductions in commuting and unemployment.

(e) Innovative municipal management and planning, through the introduction of adequate information systems and inter-institutional relationships, implying new organizational models.

The policies to be adopted will largely depend on the capacity of local decision-makers to involve themselves in a process of permanent negotiation with other levels of the Public Administration in order to attain multi-level consistent policies and adequate means, as well as on the agreements with investors and promoters, in order to achieve the desired development standards.

The levels of achievement possible through the adoption of such a strategy with such objectives are not easy to measure. Development is achieved through changes in a complex system combining traditional and innovative activities, and many of its factors are non-economic and non-quantitative.

The methodological and technical aspects of policy impact assessment will also require further research (see Folmer and Nijkamp, 1985), though criteria and causal relationships can progressively be identified in order to build a conceptual model.

Conclusion and Future Research

New technologies are causing deep changes in the spatial and sectoral division of labour, with serious consequences for the hierarchy of the urban and regional system, and a growing international dimension to production and market processes. The increasingly international dimension of the economy, stimulated by the new information technologies, is generating a new space, the boundaries of which constantly evolve. The capacity of each region or urban centre to control its own evolution will increasingly depend on its position in that complex and widening space.

Depending on the capacity of the public administration at all levels to exercise different forms of control, as well as of promotion over physical and economic spaces, development will widen or reduce the existing disparities between the developed regions and the disadvantaged ones. Within the Lisbon metropolitan area, the municipalities where the factors which normally determine the location of activities adopting or introducing new technologies occur, will undergo major changes in their structure of employment and reach new levels of development, particularly if they are able to become very dynamic promotion agents. This implies a new concept for a municipal authority, not only as a provider of technical and social infrastructures, but also as an efficient information agency and a promoter/co-ordinator of new initiatives leading to job creation and modernization (reconversion of existing economic structure in terms of skills and processes, etc.).

Such new practices call for adequate databases at the local, regional and national levels, with adequately specified sector and locational variables. The specification of the causal structure and of the external environment of each spatial system is also necessary as well as a careful definition of the different facets of the planning/development process (goals, policies, instruments and means) (Folmer and Nijkamp, 1985). Impact assessment of regional and local policies, though always a difficult task, will have to explore the capabilities of different methods ranging from micro to macro levels in order to bring about any necessary adjustments.

References

Boudeville, J. (1966) Problems of regional economic planning, The University Press, Edinburgh.

Correia, P. and Nunes Da Silva (1986) Liberalism or community participation - Two ways to deal with the housing problem in Portugal, paper presented at the International Housing Conference, Gavle, Sweden.

Costa, J. et al. (1988) Regionalization, pouvoir politique et planification regionale au Portugal, working paper of the Faculty of Economics: University of Porto.

Ewers, H.J., Wettman, R., Kleine, J.H. and Bade, F.J. (1980) Innovationsorientierte regionalpolitik, Schriftenreihe 'Raumordnung' des BM fur Raumordnung, Bausesen und Stadtebau 06042.

Falmer, H. and Nijkamp, P. (1985) Methodological aspects of impact analysis of regional economic policy, Papers of the Regional Science Association, 57, 165-181.

Fonseca, F.S. (1986) Novas tecnologias: avaliacao do impacto na estrutura industrial, paper presented at the first Innovation and Technological Development Symposium.

Ministerio da Industria e Energia (1988) PEDIP - Programa 3 - Incentivo ao investimento produtivo, Special bulletin of the Ministerio da Industria e Energia, Lisbon.

Perrin, J.C. (1986) Technologies nouvelles et synergies locales. Elements de theorie et d'analyse, in Technologies nouvelles et developement regional, actes du Colloque de l'A.S.R.D.L.F., Universite de Paris, Paris.

Perroux, F. (1955) Note sur la notion pole de croissance, Economie Applique, 7, 307-20.

Stohr, W. (1986) Regional innovation complexes, Papers of the Regional Science Association, 59, 29-44.

Appendix 19.1

(Key to Tables)

CODE CAE *	
31	Food and Drinks Industry
32	Textiles, Clothing and Leather
33	Cork and Wood Industry
34	Paper, Graphic Design and Publishing
35	Oil, Coal, Rubber and Plastic products
36	Non-Met, Mineral Prod. excl. Oil and Coal
37	Base Metallurgical Industry
38	Metalic Prod. and Machinery, Transport. Eq.
39	Others
31	Food and Drinks Industry
32	Textiles, Clothing and Leather
33	Cork and Wood Industry
34	Paper, Graphic Design and Publishing
35	Oil, Coal, Rubber and Plastic products
36	Non-Met, Mineral Prod. excl. Oil and Coal
37	Base Metallurgical Industry
38	Metalic Prod. and Machinery, Transport. Eq.
39	Others

LIST OF CONTRIBUTORS

Neil Alderman is a Research Fellow at the Centre for Urban and Regional Development Studies, University of Newcastle upon Tyne, UK.

Bjørn T. Asheim is Associate Professor of human geography at the Department of Geography, University of Oslo, Norway.

Raphael Bar-El is Director of the Settlement Study Centre, Rehovot, Israel and lecturer at Ben Gurion University of the Negev, Beer-Sheba, Israel.

Fiorenza Belussi is a Researcher at the Institute of Economic and Social Research, Venice, Italy.

Michael Bradfield is Associate Professor of Economics at Dalhousie University, Halifax, Nova Scotia, Canada.

Enrico Ciciotti is Associate Professor of Urban and Regional Economics at the University of Venice, Italy.

Maria Ciechocinska is a Professor at the Institute of Geography and Spatial Organization, Polish Academy of Sciences, Warsaw, Poland.

Ulf Christiansen is Senior Research worker at the Danish Building Research Institute, Hørsholm, Denmark.

Paulo V.D. Correia is Assistant Professor in Urban and Regional Planning and researcher at the Centre of Urban and Regional Systems (CESUR), Lisbon Technical University, Portugal.

Olivier Crevoisier is a Research Assistant at the Institute of Regional and Economic Research (IRER), University of Neuchâtel, Switzerland.

Daniel Felsenstein is a Researcher at the Settlement Study Centre, Rehovot and a Teaching Assistant at the Hebrew University of Jerusalem, Israel.

Marina van Geenhuizen is a Research Associate at the Economic Geography Institute, Erasmus University, Rotterdam, Netherlands.

Sunwoong Kim is Assistant Professor of Economics at the University of Wisconsin, Milwaukee, USA.

Peter Klekner is Head of the Regional Research Section at the Institute of Economic Planning, Budapest, Hungary.

Bert van der Knaap is Professor of Economic and Social Geography and Director of the Economic Geography Institute, Erasmus University, Rotterdam, Netherlands.

Tomaso Pompili is a Research Associate in the Department of Economics, Bocconi University of Milan, Italy.

Michel Quevit is Professor of Regional Development and Director of the Interdisciplinary Research Group for Regional Development (RIDER) at the University of Louvain-la-Neuve, Belgium.

Roberta Rabellotti is a Researcher at the Institute of Economics, Bocconi University, Milan, Italy.

Juan Sanchez-Garcia is Lecturer in Regional Economics in the Department of Applied Economics, University of La Laguna, Canary Islands, Spain.

Alfred Thwaites is Deputy Director of the Centre for Urban and Regional Development Studies, University of Newcastle upon Tyne, UK.

Franz Tödtling is Associate Professor of Regional Economics at the Interdisciplinary Institute of Urban and Regional Studies, University of Economics and Business Administration, Vienna, Austria.

Claudio Tolomelli is Planning Director and Project Manager at the Department of Industry, Regione Emilia-Romagna, Bologna, Italy.

MIX
Papier aus verantwortungsvollen Quellen
Paper from responsible sources
FSC® C105338

If you have any concerns about our products,
you can contact us on
ProductSafety@springernature.com

In case Publisher is established outside the EU,
the EU authorized representative is:
Springer Nature Customer Service Center GmbH
Europaplatz 3, 69115 Heidelberg, Germany

Printed by Libri Plureos GmbH
in Hamburg, Germany